Drug Development Assay Approaches, Including Molecular Imprinting and Biomarkers

METHODOLOGICAL SURVEYS IN BIOANALYSIS OF DRUGS

(Former series title: Methodological Surveys in Biochemistry and Analysis)

Series Editor: Eric Reid
Guildford Academic Associates
72 The Chase
Guildford GU2 5UL, United Kingdom

This series was divided into Subseries A: Analysis, and B: Biochemistry up to Volume 22. Enquiries concerning Volumes 1–18 should be sent to the above address.

Recent Titles

Volume 16(A): Bioactive Analytes, Including CNS Drugs, Peptides, and Enantiomers
Edited by Eric Reid, Bryan Scales, and Ian D. Wilson

Volume 17(B): Cells, Membranes, and Disease, Including Renal
Edited by Eric Reid, G.M.W. Cook, and.J.P. Luzio

Volume 18(A): Bioanalysis of Drugs and Metabolites, Especially Anti-inflammatory and Cardiovascular
Edited by Eric Reid, J.D. Robinson, and Ian D. Wilson

Volume 19(B): Biochemical Approaches to Cellular Calcium
Edited by Eric Reid, G.M.W. Cook, and J.P. Luzio

Volume 20(A): Analysis for Drugs and Metabolites, Including Anti-infective Agents
Edited by Eric Reid and Ian D. Wilson

Volume 21(B): Cell Signalling: Experimental Strategies
Edited by Eric Reid, G.M.W. Cook, and J.P. Luzio

Volume 22(A): Bioanalytical Approaches for Drugs, Including Anti-asthmatics and Metabolites
Edited by Eric Reid and Ian D. Wilson

Volume 23: Biofluid and Tissue Analysis for Drugs, Including Hypolipidaemics
Edited by Eric Reid, Howard M. Hill, and Ian D. Wilson

Volume 24: Biofluid Assay for Peptide-related and Other Drugs
Edited by Eric Reid, Howard M. Hill, and Ian D. Wilson

Volume 25: Drug Development Assay Approaches, Including Molecular Imprinting and Biomarkers
Edited by Eric Reid, Howard M. Hill, and Ian D. Wilson

Methodological Surveys in Bioanalysis of Drugs, Volume 25

Drug Development Assay Approaches,

Including Molecular Imprinting and Biomarkers

Edited by

Eric Reid
Guildford Academic Associates, Guildford, United Kingdom

Howard M. Hill
Corning Hazleton (Europe), Harrogate, United Kingdom

Ian D. Wilson
Zeneca Pharmaceuticals, Macclesfield, United Kingdom

THE ROYAL SOCIETY OF CHEMISTRY
Information Services

Based on the proceedings of the 12th International Bioanalytical Forum mounted by the Forum Syndicate of the Chromatographic Society, entitled "Drug Development Assay Approaches, including Molecular Imprinting", held 2–5 September 1997, in Guildford, UK

Special Publication No. 226

ISBN 0-85404-748-4

A catalogue record for this book is available from the British Library.

Published by The Royal Society of Chemistry,
Thomas Graham House, Science Park, Milton Road,
Cambridge CB4 4WF, UK

For further information see our web site at www.rsc.org

Printed by Athenaeum Press Ltd, Gateshead, Tyne and Wear, UK

Senior Editor's Preface *with Abbreviations*

This volume maintains the tradition of strong editing, the aim being a user-friendly shelf book with 'taut' but informative presentations and with compact layout, conducive to minimizing book size as stipulated by the publisher. Compactness results partly from obviating sprawling Tables such as prevail in other publications, especially for validation data (which generally can be housed neatly within the text). The increasing prevalence of sprawling mass spectrograms has to be lived with, subject to any remedial work as often done to put Figures into good shape.

The Bioanalytical Forum series (now under Chromatographic Society auspices), a novelty in 1975, still flourishes although nowadays there are other meetings that likewise tempt bioanalysts (mostly in companies) to share 'know-how'. Plans are in hand for 1999 (31 Aug. to 3 Sept.). The 1997 Forum, co-organized by Derek Stevenson, gave rise to the present book (not mere 'Proceedings') in a series that is gaining recognition as a bibliographic asset. It was gratifying that busy speakers 'found' writing-up time.*

Some features of the book.- There has been the traditional rigorous indexing, with an Analyte Index based on some chemical kinships which may allow extrapolation to analytes not listed. The usual Assay Compendium (based on therapeutic categories) has alertings, commended by a reviewer, to some points of technique. The main text features timely surveys of biomarkers (bearing on drug efficacy) and of drug-specific polymers, a powerful analytical aid. There are also Forum remarks and editorial insertions.

A Vol. 24 reviewer, surmising forceful editing, remarked on consistency amongst the articles - and also welcomed nomenclature guidance, which largely concerned MS terms as now recapitulated.

Abbreviations.- 1995 Forum Abstracts showed 'MS diversity':

LC/MS/MS LC/MS-MS LC-MS/MS LC-MS-MS HPLC-MS/MS HPLC-MS-MS

as reflected in Journals, wherein guidance to authors is scant. Consultations in the MS area have now led to adoption of **hyphens** [not slashes // unless 'options' are signified, e.g. FTICR/MS], and a further policy that saves a string of hyphens.- For **sub-descriptions**, e.g. 'atmospheric pressure (chemical) ionization', parentheses () are used: LC-MS(API) [not LC-API-MS], LC-MS(APcI), sometimes LC-MS(APncI) [n = negative]. The electrospray mode is denoted ESP or ESI (variant: ISP, ionspray). Other examples: LC-MS-MS(TSP), tandem MS with thermospray; GC-MS(EI) if electron impact; MS(FAB), fast atom bombardment. 'Monitoring' (M) terms:

SIM/SRM, selected ion/reaction; MRM, multiple reaction.

Any attempt to achieve full consensus and consistency, rather than amelioration, would have been presumptuous and over-earnest. Nomenclature is evolving, e.g. 'protonated molecule' to replace 'molecular ion'.

---

*Art. D-8, which Tim Olah volunteered to compile, is a tribute to a car-crash victim, John Gilbert. John was a keen participant in 1995 (with Mary) as well as in 1997.

Other abbreviations.- NP/RP, normal/reversed-phase (HPLC); ODS is a disfavoured synonym for C-18. Detection: fluor = fluorimetric - ex[citation]/em[ission] nm maybe stated; EC, electrochemical [**never** ECD, = electron-capture detection as in GC-ECD; cf. FID, flame ionization; NPD, nitrogen-phosphorus (AFID or thermionic)].

Ab, antibody (mAb, monoclonal)
BSA, bovine serum albumin
ELISA. Enzyme-linked immunosorbent assay
IA, immunoassay (RIA, radio-)
i.s., internal standard
i.v., intravenous; s.c., sub-cutanous; i.m., intramuscular
LOD, limit of detection; LLOQ, lower limit of quantitation
MeCN, acetonitrile
C(Z)E, capillary (zone) electrophoresis (variant: MEKC/MECC, micellar electrokinetic capillary chromatography)
C.V., coefficient of variation[*] (relative std. deviation, RSD)
MIP, molecularly imprinted polymer
PK, pharmacokinetic(s)
QC/QA, quality control/assurance
S/N, signal-to-noise ratio
SPE, solid-phase extraction

Other 'special shorthand' is defined where it occurs. N/A so far: the in-vogue instrumental term PQ = Performance Qualification. 'LC' (seldom gravity-flow nowadays) generally connotes HPLC.

Some textual points.- #All temperatures (°) are Celsius.
#A now-prevailing preference is respected: sulfur, NOT sulphur.
#Disfavoured: the term 'hyphenated' - which, rather wildly, may embrace HPLC coupling to a pre-column operation or to detection. It features in a useful sketch [Brinkman (1996) *Analusis* **24**, M12] which also clarifies a vague term: multi-dimensional (cf. '2-D'). Guessed meanings: linkage of two separatory/discriminatory steps.
#A re-iterated plea: nature of packings (jargon!) to be stated.

This Editor (who fostered the term 'bioanalysis') seeks comments:
#The term 'bioassay' does *not* suit cell-free tests, e.g. enzymic.
#A 'biosensor' *may* be enzymic, but the term is deplorably vague (arguably should connote 'bio' *use*); plain 'sensor' may suffice.
#'Microdialysis' ambiguous: *in vivo?* cf. sample-preparation use.
#Unhappy terms: '*in*accuracy' (bias), '*im*precision'; see art. D-6.

Bibliographic points (may affect Libraries besides readers):
#For *J. Chromatog*. (variants now used: J. Chr'y J Chr) specify **A** *or* **B.**
#Present policy: citations confined to widely stocked 'hard' refs.
#Vol. 23, present series: erroneous 'Cumulative Index' phrase on spine (was in Vol. 22!).- *Mask over! (by tape?)*.
#Earlier vols.- Nos. 1-4, 6, 8 and then odd nos. up to 21 were in in a now-ceased sub-series, *Biochemistry;* now solely *Analysis*.
#Series title pre-Vol. 23 [affects shelving or (info. requestable) purchase]. *Methodological Surveys in Biochemistry and Analysis*. The *Biochemistry* sub-series largely concerned pharmacological aspects.

[*] Depending on context, 'C.V.'/'CV' may mean: #coefficient of variation; #cardiovascular; #curriculum vitae. Hence for cyclic voltammetry (art. C-4 and a cognate art. in Vol. 24) authors' term 'CV' altered to 'CyV'.

72 The Chase, Guildford GU2 5UL, U.K.
Tel./Fax: [+44](0)1483-565324
5 May 1998

ERIC REID
Forum Syndicate Secretary,
The Chromatographic Society

CONTENTS

Besides main contributions there are short articles - 'NOTES' (denoted **n** as in #A-7n) - and 'COMMENTS' comprising Forum discussion remarks and Editor's 'annotations'.

'n' as in #B-3n signifies a Note

List of Authors with p. nos.

Primary author

L.I. Andersson: 2-12
Astra Pain Control, Södertälje, Sweden

K.D. Barrell: (i) 86-87, (ii) 90, (iii) 95-96
Simbec Res., Merthyr Tydfil

S.D. Clarke: 91-94, & see Noctor
Covance, Harrogate

J.D.H. Cooper: 64-65
Clinical Innovations, Stareton

A.M. Edgington: 82-85
Pfizer Central Res., Sandwich

P. End: 275-276
Novartis, Basel

J.J. Gardner: 104-111
Astra Charnwood, Loughborough

B.A. Griffin: 258-262
Univ. of Surrey, Guildford

A. Guenzi: 155-158
Hoffmann-La Roche, Basel

C.M. Heard: 37-43
Welsh Natl. Sch. of Pharmacy, Cardiff

P. Heizmann: (i) 196-200, (ii) 201-204
Hoffmann-La Roche, Basel

M.J. Hemsley: 149-154
Covance, Harrogate

H.M. Hill: (i) 223-229, (ii) 268-274, & see Clarke, Hemsley
Covance, Harrogate

H. Lingeman: 55-63
Free Univ. Amsterdam

P. Martin: 21-27
Zeneca Pharm'ls, Alderley Park

A.G. Mayes: 28-36
Inst. of Biotechnology, Cambridge

Co-authors, with relevant name to be consulted in left column

C.J. Allender - Heard
C. Ampasavate - Stobaugh
W.J. Bachman - Stewart (ii)
R.A. Biddlecombe - Pleasance
M.A.E. Bowman - Heard
K.R. Brain - Heard
R.J. Briggs - Stevenson (i)
J. Brown - Starkey
M.A. Bryant - Barrell (ii)
J. Burnett - Hill (ii)
R.G. Carlson - Stobaugh
R.M. Carr - Wring (ii)
E. Clayton - Wilson
D.C. Delinsky - Stewart (ii)
D. Dell - Heizmann (ii)
D. Doppler - Guenzi

H. Eschey - Schmid
I. Esteban - Whitcombe
C. Gardner - Hill (i)
J.D. Gilbert - Olah
R.J. Goody - Venn (i)
D. Gygax - End
P. Hamilton - Perrett
D.T. Hardy - Starkey
J. Hay - Stevenson (i)
R. Hochel - Wring (ii)
A. Howard - Venn (ii)
A.H. Jenkins - Barrell (iii)
W.N. Jenner - Wring (ii)
G.R. Jones - Martin
K. Jones - Martin
Jae-Yeong Ju - Whitcombe

M.F. Katmeh - Stevenson (ii)
A. Khan-Raja - Barrell (i)
K.E. Kilpatrick - Wring (ii)
W. King - Perrett
J-U. Klein - Whitcombe
P. Kristjansson - Stobaugh
P. Kwaskowski - Stevenson (ii)
G.A. Laban - Barrell (ii) & (iii)
J. Lapp - Skrzipczyk
L. Laugher - Noctor
T.N. Liddicoat - Clarke
C.R. Lowe - Mayes
M. Lübke - Whitcombe
D.A. McLoughlin - Olah

Primary author

T.A.G. Noctor: 99-103, & see Clarke, Hemsley, Hill (ii)
Covance, Harrogate

T.V. Olah: 235-245
Merck Res. Labs., West Point, PA

R. Payne: 230-234
Zeneca Pharm'ls, Alderley Park

D. Perrett: 251-257
Sch. of Medicine & Dentistry, Bart's Hosp., London

S. Pleasance: 205-212
Glaxo Wellcome, Ware

P.R. Robinson: (i) 88-89, (ii) 97-98, & see Barrell (i)
Simbec Res., Merthyr Tydfil

J. Schmid: 213-218
Boehringer Ingelheim, Biberach/Ris, Germany

K. Selinger: 219-222
Glaxo Wellcome, Research Triangle Park, NC

H.J. Skrzipczyk: 72-77
Hoechst MR, Frankfurt

B.J. Starkey: 263-267
Roy. Surrey County Hosp., Guildford

D. Stevenson: (i) 49-51; (ii) 52-54
Univ. of Surrey, Guildford

J.T. Stewart: (i) 78-81, (ii) 170-176
Univ. of Georgia, Athens, GA

J.F. Stobaugh: 177-180
Univ. of Kansas, Lawrence, KS

R.F. Venn: (i) 13-20, (ii) 166-169, & see Edgington
Pfizer Central Res., Sandwich

M.J. Whitcombe: 44-48
Inst. of Food Res., Reading

I.D. Wilson: 142-148 (& see Martin, Payne)
Zeneca Pharm'ls, Alderley Park

S.A. Wring: (i) 159-165, (ii) 181-189
as for Selinger

Co-author, with relevant name to be consulted in left column

J. Maddock - Barrell (all 3), Robinson (both)
H. Martin - Heizmann (ii)
L. Martin - Whitcombe
B.K. Matuszewski - Stewart (ii)
H.C. Morgan - Robinson (ii)
F.J. Müller - Skrzipzyk
R. Pemberton - Wring (i)
R. Pittson - Wring (i)
A.S. Prangle - Hemsley
B. Rashid - Stevenson (both)
L.W. Rees - Robinson (both)
W.F. Richter - Heizmann (i)
G. Ringsdorf - Skrzipzyk
M.J. Rose - Stobaugh
D. Russell - Venn (ii)
J.M. Russell - Gardner
J. Sandow - Skrzipzyk
E. Saulnier - Heizmann (i)
T. Sauter - Schmid
P. Schenk - Heizmann (i)
R. Schindler - End
D. Schütze - Schmid
C. Serabjit-Singh - Wring (both)
Chul Soo Shin - Whitcombe
M. Siluveru - Stewart (both)
C. Smith - Payne
L. Takahashi - Wring (i)
J.E. Taylor - Cooper
S. Taylor - Wilson
K.A. Tennant - Hill (i)
F.P. Theil - Heizmann (i)
D. Thomas - Noctor & Hemsley
T. Thompson - Hill (ii)
J. Tolan - Wring (i)
T.G. Venkateshwaran - Stewart (ii)
S. Verity - Noctor
E.N. Vulfson - Whitcombe
A.J.C. Wakefield - Payne
K. Weimer - Skrzipzyk
I. Waterhouse - Wring ii)
A.J. Woodward - Barrel (ii) & (iii)
B. Wright - Wilson
J.W. Wright - Griffin

Section #A

MOLECULAR IMPRINTING
AND OTHER SAMPLE-TREATMENT APPROACHES

The suffix 'n' as in #A-7n distinguishes a 'NOTE' from main articles. COMMENTS (starting on p. 66) comprise discussion remarks and Editor's 'annotations'.

#A-1

MOLECULAR IMPRINTING AS AN AID TO DRUG BIOANALYSIS

Lars I. Andersson[†]

Bioanalytical Chemistry, Pharmaceutical & Analytical R & D, Astra Pain Control, S 151 85 Södertälje, Sweden

Molecular imprinting is becoming increasingly recognized as a technique for the ready preparation of polymeric materials containing recognition sites of predetermined specificity. The technique entails polymerization around an imprint molecule using monomers containing chemical functionalities complementary to those of the template. Subsequent removal of the imprint molecule exposes imprints within the polymer possessing a permanent memory for the imprint species. These imprints enable the polymer (MIP) selectively to re-bind the imprint molecule from a mixture of closely related compounds. In many instances MIP's show binding affinities approaching those demonstrated by antigen-Ab systems. MIP's have been applied in a range of applications requiring selective ligand binding, including chromatography, SPE and 'immunoassay'. The application of molecular imprinting to drug bioanalysis is discussed.*

Molecular imprinting [1-5] is an emerging technique for the preparation of synthetic polymers with high selectivity for a predetermined ligand. The technique entails polymerization of functional monomers in the presence of a templating ligand, or imprint species (Fig. 1). Subsequent removal of the imprint molecules leaves behind 'memory sites', or imprints, in the solid, highly cross-linked polymer network. It is believed that the functional mnonomers become spatially fixed in the polymer, *via* their interaction with the imprint species, during the polymerization reaction. The result is the formation of imprints in the polymer, which are complementary, sterically and chemically, to the templating ligand. This enables the polymer later to selectively re-bind the imprint molecule from a mixture of closely related compounds. In some instances binding affinities and selectivities approaching those demonstrated by antigen-Ab systems have been achieved [6]. MIP's have been applied in a range of applications requiring selective ligand binding, such as chromatography, sensor technology, immunoassay, directed synthesis, and enzyme and receptor mimicry [1-4]. Following the

---

†e-mail: Lars.Andersson@pain.se.astra.com

**Abbreviations*.- Ab, antibody; CE, capillary electrophoresis; CEC, capillary electrochromatography; MIA, molecularly imprinted sorbent assay; MIP, molecularly imprinted polymer.

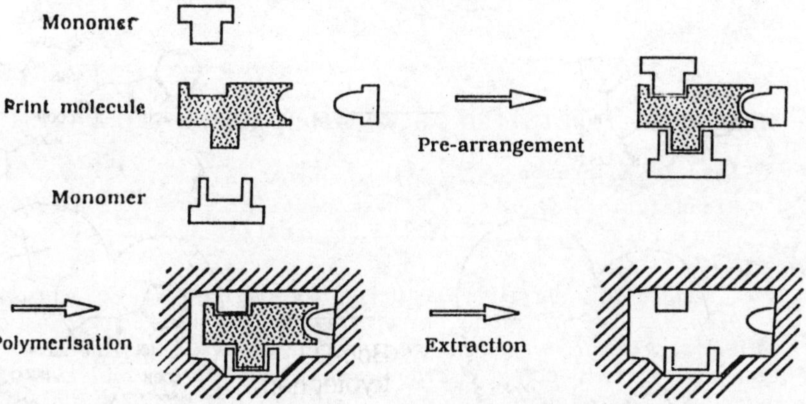

Fig. 1. Schematic depiction of the preparation of molecular imprints. The imprint species is dissolved in an appropriate solvent and mixed with functional monomer(s) and cross-linking monomer. Addition of initiator starts the polymerization, and a hard, insoluble polymer is formed. Solvent extraction removes the imprint molecules, leaving behind recognition sites which possess a 'memory' for the original imprint molecule.

first report in 1993 of the use of MIP's as selective binding species in immunoassay-type analyses [7], not till 1997 did other publications dealing with biological samples appear. Presently, the number of groups working in the field is expanding, and soon there will be many more publications. This review discusses the potential of using MIP's in bioanalysis and highlights some of the problems that need to be addressed.

PREPARATION OF IMPRINTED POLYMERS

As indicated above, the key step in the synthesis of an MIP is the polymerization of monomers in the presence of a templating ligand. The monomers are selected for their ability to interact, by either non-covalent interactions [2,3], e.g. hydrogen bonding, ion-pair interactions, etc., or reversible covalent interactions [1] or metal ion-mediated interactions [5], with the imprint species. Of these approaches the non-covalent strategy is the most widely used [2]. The imprint molecule is simply allowed to pre-arrange with the monomers in solution before initiating polymerization (Fig. 2). Pre-derivatization with functional monomers is not needed; moreover, a notably high number of compounds are amenable to non-covalent imprinting [2]. Several monomers exist that can interact with specific functionalities in the imprint molecule.- **Acidic monomers**: methacrylic acid, trifluoromethylacrylic acid, itaconic acid, acrylic acid. **Basic**: 2- and 4-vinylpyridine, 1-vinylimidazole. **Neutral**: acrylamide, hydroxyethylmethacrylate,

Available **cross-linking monomers**: ethylene glycol dimethacrylate, trimethylolpropane trimethacrylate, divinylbenzene, pentaerythritol triacrylate, pentaerythritol tetraacrylate.

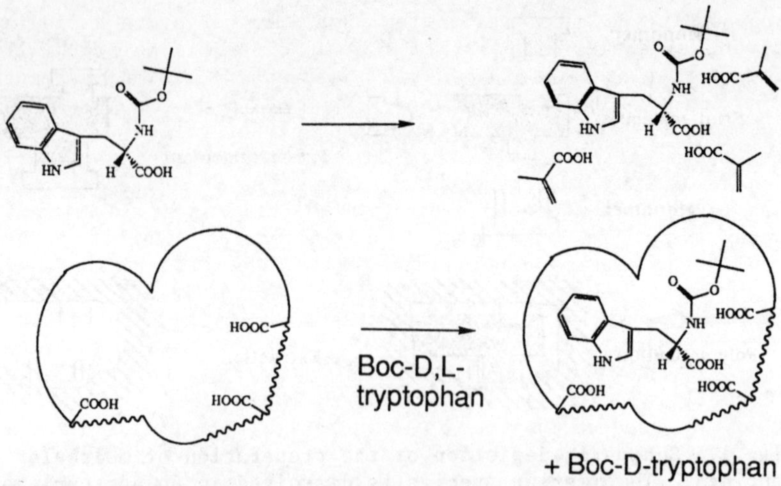

Fig. 2. Preparation of enantio-selective imprints by the molecular imprinting of the L-form of Boc-tryptophan [Boc is N-tert-Butoxycarbonyl]. Through H-bonding between complementary chemical functionalities the methacrylic residues become arranged around the imprint molecule. Upon incubation with a mixture of the optical isomers the enantiomer used as imprint molecule becomes preferentially bound. Accordingly, chiral resolutions on MIP-based stationary phases will occur with predictable elution order of the enantiomers.

Several monomers contain a carboxylic acid functionality, which binds through ionic interactions to basic functionalities, e.g. amines, and H-bonds to amide, alcohol, carboxylic acid, carbamate etc. functionalities of the imprint molecule [8]. Of these, methacrylic acid in particular has proved to be versatile in most situations. Basic monomers, e.g. vinylpyridines, can be employed as complementary functionalities to carboxylic acids of the imprint molecule [9]. Combinations of methacrylic acid and vinylpyridine sometimes work better [10]. Amongst neutral monomers, recently introduced, methylacrylamide offers most promise due to its strong H-bonding abilities [11]. To preserve the integrity of the imprints in the resultant MIP a high cross-linking of the polymer network is required [1]. Amongst cross-linking monomers, as listed, the first two are the most popular so far, due to their mechanical and thermal stability, ease of removal of the imprint molecule and the high selectivities that have been observed [1].

Some operational points

The polymerization mixture is prepared by dissolving the imprint compound in a suitable organic solvent and adding the cross-linking monomer and one or more functional monomers. For maximal efficiency of imprint formation, at least for MIP's of non-covalent type, polymerization should be performed employing a solvent as apolar as possible without compromising solubility of the imprint species [2]. This is to ensure maximal strength of the non-covalent interactions employed, e.g. hydrogen and ionic bonds, which depend strongly on solvent polarity. After adding azobis-nitrile initiators of conventional type polymerization is effected by temperature elevation or UV-irradiation. It has been shown repeatedly that MIP's prepared at lower temperatures (-20° to 0°) using photo-initiation exhibit higher molecular recognition capabilities [1]. This is attributed to the complexes of monomers and imprint molecules in the pre-polymerization mixture being more stable at low temperatures due to a more favourable entropy term, leading to better-defined imprints in the resultant polymer. To date, bulk polymerization followed by grinding and particle sizing has generally been the most often used technique for imprinted polymer preparation [1,3]. Uniformly shaped beads have recently been made available by the development of a novel suspension polymerization technique, compatible with the non-covalent imprinting approach [12].

The removal of the imprint species from the resultant polymer is critically important, since a more thorough extraction yields a superior material where more of the high-avidity sites are free. Complete extraction requires extensive washing using solvents with strong elution power, such as aqueous ethanol containing acid or base. Intermittent acid and base washings may be beneficial. For chromatographic-type applications the extraction can be done on-line. The prepared material can be stored in the dry state at ambient temperatures before use.

Molecular imprinting is a simple and cheap procedure that can be performed in any reasonably equipped laboratory environment. An MIP can be made reproducibly, each batch being near-identical in properties with the previous one. In the author's experience, the batch-to-batch variation of parameters such as selectivity, binding-site density and avidity has usually been negligible. In the few cases where results have been variable, the cause has been variable quality of the imprint species.

CHROMATOGRAPHIC AND CAPILLARY TECHNIQUES

The polymer systems used are characteristically of high mechanical, thermal and solvent stability, which permits packing of MIP's into columns and their use as specific chromatographic sorbents. Many MIP studies have focused on chiral separation problems, where the specificity of a given MIP is pre-determined by the enantiomer of the ligand that was present during its preparation (Fig. 2). Many different racemic compounds have been successfully resolved, including drugs [9, 13, 14], amino acid derivatives [1, 2] and sugars [1, 2]. Often good separations can be achieved, and in one case the separation factor attained was 17.8 for the resolution of the enantiomers of the dipeptide N-acetyl-Trp-Phe-OMe [15]. In one study, MIP's were coated onto glass plates and shown to effect TLC separation of the imprint molecule and its optical antipode [16]. Some recent reviews [1, 2, 17] consider more fully the use of MIP's in LC.

CE and CEC provide a high degree of separation efficiency and short separation times, which leads to high throughput and the ability to analyze samples available only in small amounts. Molecular imprinting provides a means for preparing stationary phases of pre-determined selectivity. Recently a simple and quick method for the *in situ* preparation of monolithic, enantio-selective MIP's inside fused silica capillaries has been described [18]. The MIP's could be rendered superporous by using 1-25% iso-octane as a porogenic agent [19, 20] or by careful timing of the polymerization reaction [18]. The introduction of superpores permits rapid replacement of the solvent of polymerization by electrolyte, by hydrodynamic pumping at low pressures. This exchange is important, since the use of apolar solvents, which are incompatible with electrophoresis, is highly conducive to imprinting efficiency. Molecular imprinting of the β-adrenergic antagonists metoprolol and propranolol [18, 19] and the local anaesthetic ropivacaine [20] has been done using a photo-initiated polymerization process. In the CEC mode, the within-capillary MIP's gave baseline separations of the enantiomers of propranolol and metoprolol in <3 min; several other β-blockers were resolvable by the propranolol column [18], and structurally similar local anaesthetics by the ropivacaine column [20].

Other CE approaches involve immobilization of MIP particles, as used for LC, inside capillaries using a polyacrylamide gel [21], or use of MIP particles as a chiral mobile phase additive [22] or of MIP's prepared by a dispersion polymerization process [23]. An interesting strategy is illustrated by the molecular imprinting of phenylalanine anilide and, with the resultant capillary columns, separation of the phenylalanine enantiomers [21, 24]. Being insoluble in the pre-polymerization mixture, amino acids are not normally amenable to non-covalent imprinting. Instead a structurally similar compound, one that gives rise to imprints that can bind phenylalanine and is soluble in the polymerization mixture, was used as the imprint molecule.

Although the peak performance is better than that obtained in LC, separation efficiencies characteristic of CEC have yet to be achieved. For MIP-based stationary phases to become a complement to the selectors presently employed in CEC, further improvements in peak shape are required, calling for further research into both the methodology for preparation of MIP-based columns and optimization of the conditions for MIP-based CEC.

SAMPLE PRE-CONCENTRATION

Molecularly imprinted polymers have been used as specific binding matrices for SPE and clean-up of biological samples. This application relies on the ability of an MIP to selectively bind an analyte, or a group of structurally similar compounds, from a mixture. The biological sample, or a solvent extract of the sample, is treated with an MIP. This can be done batchwise by adding it directly to the sample. More appealing is the use of a flow system where the MIP is packed in an SPE cartridge. After washing the MIP, the analyte is eluted and the eluate, which can be evaporated and the residue redissolved in a smaller volume, is analyzed by HPLC or GC. Analyses have been performed on propranolol [25], sameridine - an analgesic/local anaesthetic [26] - and the triazine-type herbicides atrazine and simazine [27, 28]. The sample matrices have been biofluids, e.g. dog plasma, human urine and rat bile [25], and solvent extracts of tissues and biofluids, e.g. chloroform extracts of beef liver [27] and heptane extracts of human plasma [26].

A requisite for the use of MIP's in SPE applications is that remaining traces of the imprint species do not leak from the polymer and contaminate the analyte, entailing uncertainty in the response. This risk can be avoided completely by using for the molecular imprinting a close structural analogue of the analyte(s) of interest. Provided that the imprint species and the analyte(s) can be separated by the subsequent HPLC or GC - an assumption that is usually valid - the leakage appears as a separate peak and presents no problem. This alternate-imprint species approach was demonstrated for sameridine (Fig. 3) [26]. The SPE sorbent was prepared with a derivative close enough to sameridine in structure to allow the requisite binding. With sameridine as a model compound, the versatility of this sorbent for SPE was shown for batchwise pre-concentration prior to GC; the specificity of the extraction led to distinctly cleaner chromatographic traces and the ability to improve sensitivity by using larger sample volumes (Fig. 3). Moreover, in analytical performance the MIP-SPE was as good as or surpassed the standard method entailing liquid-liquid extraction for clean-up.

RADIOASSAY

The high affinities and selectivities often achievable make imprinted polymers attractive as alternative binding entities in protocols of immunoassay (IA) type. Advantages of the polymers over biological Ab's are their high stability, easy preparation and ability to be used for assay with both aqueous and solvent sample media. An MIP-based radioassay is analogous in design to the conceptually identical RIA approach [29]. The analyte and a radiolabelled analogue are presented to a restricted amount of imprint sites. As the two ligands compete for binding to the same sites, the amount of radio-ligand bound to the imprints, as well as the amount free in solution, is quantitatively related to the amount of analyte present in the incubation medium.

This approach ('MIA') was pioneered with human serum using molecular imprints against theophylline and diazepam [7]. Both could be determined at clinically significant levels (Table 1) with an accuracy comparable to that obtained using a traditional

IA technique. Comparison with a commercial IA technique showed good correlation for theophylline determination on patient samples [7]. The theophylline-selective MIP's showed excellent selectivity for theophylline (1,3-dimethylxanthine) in the presence of the structurally related caffeine (1,3,7-trimethylxanthine), which showed <1% cross-reactivity although it differs by only one methyl group. Prior to the actual assay, wherein organic solvents comprised the incubation medium, the analyte was extracted from the serum using liquid-liquid extraction. Briefly, an organic solvent, e.g. heptane, diethyl ether or methylene chloride, is added to the aqueous sample (serum in this instance). The aqueous sample is adjusted to acidic, basic or neutral pH depending on which pH is most favourable for quantitative extraction of the particular analyte into the organic phase. Since this first demonstration, several further studies have been presented (Table 1), focused on improving the assay sensitivity and extending the assay to aqueous conditions.

Organic solvents *vs.* aqueous conditions

The use of organic solvents can be beneficial in situations where the analyte is poorly soluble in aqueous buffers, as for cyclosporin [30] - a potent immunosuppressant used in transplant surgery to prevent organ rejection. Its therapeutic range being narrow, from adequate immunosuppression to toxicity, therapy requires periodical monitoring of the whole-blood concentration of the drug. In many IA methods an organic solvent such as methanol is added to the whole blood sample to release the analyte from the red cells and keep the hydrophobic analyte soluble in the incubation mixture. A model MIP-based assay that can be directly performed in organic solvents after solubilization and extraction has been developed, enabling the total concentration of cyclosporin A and its metabolites to be determined in human whole blood. The assay might serve to measure this total concentration, and coupled with a specific assay such as Syva's EMIT the total concentration of metabolites might be measurable.

In most studies the sample was in an organic solvent at the outset or through a solvent extraction stage. Some of the more recently developed analyte-MIP systems work equally well with an aqueous buffer or an organic solvent for incubating the polymer particles, radio-ligand and analyte [31-33]. This applies to a sensitive assay system developed for *S*-propranolol; optimization of the assay conditions led to 100- to 1000-fold improvement (to ~6 nM) in the lower limit of quantification (LLOQ) and 100- to 1000-fold reductions in the amount of imprinted polymer used (only 50 μg needed; previously 1-12 mg), compared with previous studies of this type of assay. The work was pursued to development of a model assay for direct determination of *S*-propranolol in biofluids [32]. Thus, in human plasma and urine it could be measured in the 20 to 1000 nM range with good accuracy and precision, demonstrating the feasibility of MIP-based assays on biological samples without prior sample clean-up.

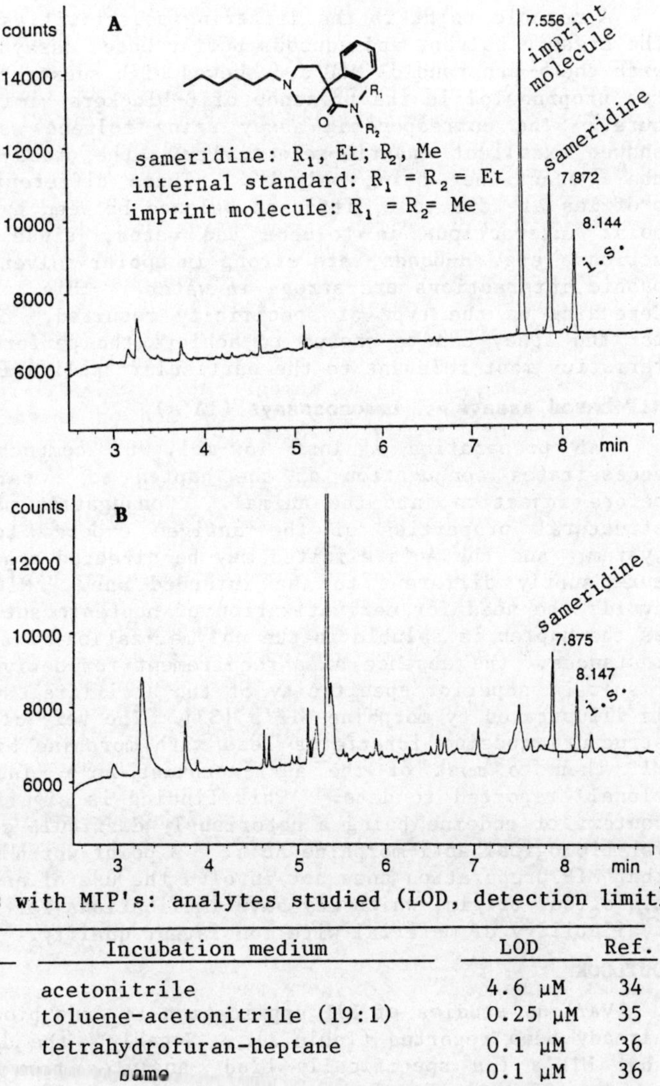

Fig. 3. Representative GC traces of spiked human plasma samples subjected to liquid-liquid extraction with (**A**) or without (**B**) subsequent MIP-based SPE.

Table 1.
IA-type assays with MIP's: analytes studied (LOD, detection limit).

Analyte	Incubation medium	LOD	Ref.
Atrazine	acetonitrile	4.6 μM	34
	toluene-acetonitrile (19:1)	0.25 μM	35
Corticosterone	tetrahydrofuran-heptane	0.1 μM	36
Cortisol	same	0.1 μM	36
Cyclosporin	diisopropyl ether	4 μM	30
Diazepam	toluene-heptane (3:1)	0.2 μM	7
Morphine	citrate buffer pH 6	0.22-0.79 μM	33
	toluene-acetic acid (49:1)	0.15 μM	33
Leu-enkephalin	acetonitrile-acetic acid (19:1)		33
S-Propranolol	human plasma + phosphate buffer pH 7.4 (6:4)	20 nM	32
	citrate buffer pH 6	6 nM	31
	toluene-acetic acid (49:1)	5.5 nM	31
Theophylline	acetonitrile-acetic acid (99:1)	3.5 μM	7

A notable point is the differing selectivities obtained for the organic solvent and aqueous buffer based assays: the latter, with the *S*-propranolol MIP's, showed high substrate-selectivity for propranolol in the presence of β-blockers similar in structure. The corresponding assay using toluene as the solvent showed excellent enantio-selectivity, the cross-reactivity of the *R*-enantiomer being only 1%. These different selectivity profiles are due to a different balance between hydrophobic and polar interactions in toluene and water, since polar interactions, e.g. H-bonds, are strong in apolar solvents and hydrophobic interactions are strong in water. This indicates that, depending on the type of specificity required, the conditions for the assay can be chosen to achieve the performance characteristics most relevant to the particular application.

MIP-based assays *vs.* immunoassays (IA's)

Ab preparation against low-mol. wt. compounds (haptens) necessitates conjugation of the hapten to a carrier protein before injection into the animal. Conjugation may change the structural properties of the antigen exposed to the immune system, and the Ab's elicited may be directed against a structure subtly different to the intended one. MIP preparation avoids the need for derivatization of haptenic antigens as long as the hapten is soluble in the polymerization mixture. In some instances, the absence of a requirement for derivatization may result in superior specificity of the artificial system. This is illustrated by morphine MIP's [33]. The very closely related structure codeine interferes less with morphine binding to the MIP than to most of the anti-morphine Ab's (including monoclonal) reported to date. This finding is significant in the context of codeine being a notoriously difficult cross-reactant for biological anti-morphine Ab's. A point worth mentioning is that MIP preparation does not involve the use of any material of biological origin, which may have implications for the long-term availability of material with consistent quality.

OUTLOOK

Various studies of MIP application to drug bioanalysis have already been reported (Table 2). Notably, the demonstrations that MIP's can specifically bind analytes from a biological sample [25, 32] further the development of selective imprint-derived systems for direct use in combination with such samples.

The drug-bioanalysis MIP application closest to practical realization is probably SPE. This application is less prone to drawbacks generally associated with MIP's, e.g. peak broadening and leakage of imprint molecules. The on-off chromatography type employed is less sensitive to poor chromatographic performance of the sorbent than LC and CEC. A strategy to circumvent problems with imprint molecule leakage during elution, leading

Table 2. Applications of MIP's in analysis of biosamples.

Analyte	Sample	Application	Ref.
Atrazine	chloroform extr. of beef liver homogenate	SPE	27
Chloram-phenicol	acetonitrile after extraction of bovine serum	competitive dis-placement sensor	37
Cyclosporin	diisopropyl ether extr. of human whole blood	MIA	30
Diazepam	toluene-heptane (3:1) after extraction of human serum	MIA	7
Glucose	porcine plasma	'sensor'	38
Propranolol	dog plasma, rat bile, human urine	SPE	25
S- "	human plasma and urine	MIA	32
Sameridine	heptane-ethanol (9:1) after extraction of human plasma	SPE	26
Theophylline	acetonitrile-acetic acid (99:1) after extraction of human serum	MIA	7

to uncertainties in the concentration measurement, is the alternate imprint molecule approach. Non-specific adsorption to the polymer can be reduced by the use of small amounts of MIP and proper washing schemes prior to elution. Because of the strong avidity of the MIP for the analyte, a potential problem may be the harsh conditions required for quantitative analyte elution.

The use of MIP's as substitutes for Ab's in IA is a second area with great interest. MIP's provide a combination of polymer mechanical and chemical robustness with highly selective molecular recognition comparable to biological systems. The ability to be used in both organic solvents and aqueous buffers adds flexibility to assay development. Further investigations into the direct assay of biosamples is, however, warranted and will extend the applicability of MIP-based assays. The lack of sensitivity, compared with traditional IA's, due to leakage of imprint molecules, needs attention. Appropriate washing protocols in the MIP preparation are imperative. MIP preparation being simple and rapid, with no need to derivatize haptenic antigens, the use of MIP's may be considered a viable alternative to biological Ab's in assay development.

References [Chromatograph.. abbreviated to **Chr; J Chr'y** denotes **J. Chromatography]**

1. Wulff, G. (1995) **Angew. Chem. Int. Ed. Engl. 34,** 1812-1832.
2. Mosbach, K.& Ramström, O. (1996) **Bio/Technology 14,** 163-170.
3. Ansell, R.J., Ramström, O. & Mosbach, K. (1996) **Clin. Chem. 42,** 1506-1512.
4. Steinke, J., Sherrington, D.C. & Dunkin, I.R. (1995) **Adv. Polym. Sci. 123,** 81-125.
5. Mallik, S., Plunkett, S.D., Dhal, P.K., Johnson, R.D., Pack, D., Schnek, D. & Arnold, F.H. (1994) **New J. Chem. 18,** 299-304.

[See head of list for the **Chr** abbreviation

6. Andersson, L.I. & Mosbach, K. (1997) in **Principles and Practice of Immunoassay,** 2nd edn. (Price, C.P. & Newman, D.J., eds.), Macmillan Ref., London, pp. 139-152.
7. Vlatakis, G., Andersson, L.I., Müller, R. & Mosbach, K. (1993) **Nature 361,** 645-647.
8. Sellergren, B., Lepistö, M. & Mosbach, K. (1988) **J. Am. Chem. Soc. 110,** 5853-5860.
9. Kempe, M. & Mosbach, K. (1994) **J. Chr'y A 664,** 276-279.
10. Ramström, O., Andersson, L.I. & Mosbach, K. (1993) **J. Org. Chem. 58,** 7562-7564.
11. Yu, C. & Mosbach, K. (1997) **J. Org. Chem. 62,** 4057-4064.
12. Mayes, A.G. & Mosbach, K. (1996) **Anal. Chem. 68,** 3769-3774. [Cf. art. #A-3.- Ed.]
13. Fischer, L., Müller, R., Ekberg, B. & Mosbach, K. (1991) **J. Am. Chem. Soc. 113,** 9358-9360.
14. Matsui, J., Nicholls, I.A. & Takeuchi, T. (1996) **Tetrahedron: Asymmetry 7,** 1357-1361.
15. Ramström, O., Nicholls, I.A. & Mosbach, K. (1994) **Tetrahedron: Asymmetry 5,** 649-[656.
16. Kriz, D., Berggren-Kriz, C., Andersson, L.I. & Mosbach, K. (1994) **Anal. Chem. 66,** 2636-2639.
17. Nicholls, I.A., Andersson, L.I., Mosbach, K. & Ekberg, B. (1995) **Trends Biotechnol. 13,** 47-51.
18. Schweitz, L., Andersson, L.I. & Nilsson, S. (1997) **Anal. Chem. 69,** 1179-1183.
19. Nilsson, S., Schweitz, L., Petersson, M. (1997) **Electrophoresis 18,** 884-890.
20. Schweitz, L., Andersson, L.I. & Nilsson, S. (1998) **J. Chr'y B,** in press.
21. Lin, J.M., Nakagama, T., Uchiyama, K. & Hobo, T. (1996) **Chr'ia 43,** 585-591.
22. Walshe, M., Garcia, E., Howarth, J., Smyth, M.R. & Kelly, M.T. (1997) **Anal. Comm. 34,** 119-122.
23. Nilsson, K., Lindell, J., Norrlöw, O. & Sellergren, B. (1994) **J. Chr'y A 680,** 57-61.
24. Lin, J.M., Nakagama, T., Wu, X-Z., Uchiyama, K. & Hobo, T. (1997) **Fresenius J. Anal. Chem. 357,** 130-132.
25. Martin, P., Wilson, I.D., Morgan, D.E., Jones, G.R. & Jones, K. (1997) **Anal. Comm. 34,** 45-47. [Cf. art. #A-5, this vol.- Ed.]
26. Andersson, L.I., Paprica, A. & Arvidsson, T. (1997) **Chr'ia 46,** 57-62.
27. Muldoon, M.T. & Stanker, L.H. (1997) **Anal. Chem. 69,** 803-808.
28. Matsui, J. Okada, M., Tsuruoka, M. & Takeuchi, T. (1997) **Anal. Comm. 34,** 85-87.
29. Price, C.P. & Newman, D.J., eds. (1997) - see Ref. 6.
30. Senholdt, M., Siemann, M., Mosbach, K. & Andersson, L.I. (1997) **Anal. Lett. 30,** 1809-1821.
31. Andersson, L.I. (1996) **Anal. Chem. 68,** 111-117.
32. Bengtsson, H., Roos, U. & Andersson, L.I. (1997) **Anal. Comm. 34,** 233-235.
33. Andersson, L.I., Müller, R., Vlatakis, G. & Mosbach, K. (1995) **Proc. Natl. Acad. Sci. USA 92,** 4788-4792.
34. Muldoon, M.T. & Stanker, L.H. (1995) **J. Agric. Food Chem. 43,** 1424-1427.
35. Siemann, M., Andersson, L.I. & Mosbach, K. (1996) **J. Agric. Food Chem. 44,** 141-[145.
36. Ramström, O., Ye, L. & Mosbach, K. (1996) **Chem. Biol. 3,** 471-477.
37. Levi, R., McNiven, S., Piletsky, S.A., Cheong, S-H., Yano, K. & Karube, I. **Anal. Chem. 69,** 2017-2021.
38. Chen, G., Guan, Z., Chen, C-T., Fu, L., Sundaresan, V. & Arnold, F.H. (1997) **Nature/Biotechnology 15,** 354-357.

#A-2

SYNTHESIS AND PROPERTIES OF MOLECULAR IMPRINTS OF DARIFENACIN - DOES MOLECULAR IMPRINTING HAVE A FUTURE IN ULTRA-TRACE BIOANALYSIS?

Richard F. Venn† & Robin J. Goody
Department of Drug Metabolism,
Pfizer Central Research, Sandwich, Kent CT13 9NJ, U.K.

An MIP was developed and was subsequently shown to be able to selectively retain darifenacin (UK 88,525 S) from aqueous MeCN when used a stationary phase in HPLC columns and as an SPE packing. The MIP's could distinguish between separate sub-structures of the drug molecule as well as between the R- and S-enantiomers. A ring-opened metabolite was distinguishable from the parent, whereas the mono-hydroxylated metabolite was not. The capacity for the drug was >100 μg per 100 mg polymer. The MIP could extract the drug directly, with 100% recovery, from human plasma deproteinized by MeCN (1 vol.). Other applications of the MIP included repurification of ^{14}C-labelled darifenacin. The drawbacks of MIP's for ultra-trace analysis are considered; they centre on the difficulty of removing all the template from the polymer and the consequent effects of template bleed on assay precision and accuracy when the MIP is used as an SPE cartridge. Possible solutions are discussed.*

Molecular imprinting is a fast-growing area of research with especial relevance to analytical problems [1] but also of interest in connection with modeling of catalysis [2, 3], of molecular recognition in Ab's [1, 4] and of receptors [5-7]. Our interests in this technique are based on the ability of MIP's* to provide chiral separation in column [8-13] or planar [13] chromatography, to provide specific extraction media for clean-up of bio-matrices or other compromised samples, and to generate Ab-like [1, 4] recognition sites to facilitate development of specific competitive 'immuno'-assays for target molecules quickly and at will. We are also interested in the implications of such imprints for other uses, e.g. high-throughput screening. In general, aiming to improve the selectivity and sensitivity of the bioanalytical methods carried out in our laboratories, we investigated the potential of molecular imprinting to provide fast access to specific affinity extraction media without recourse to immuno-affinity methods.

---

†addressee for any correspondence
**Abbreviations include:* Ab, antibody; MIP, molecularly imprinted polymer; MeCN, acetonitrile; TFA, trifluoroacetic acid.

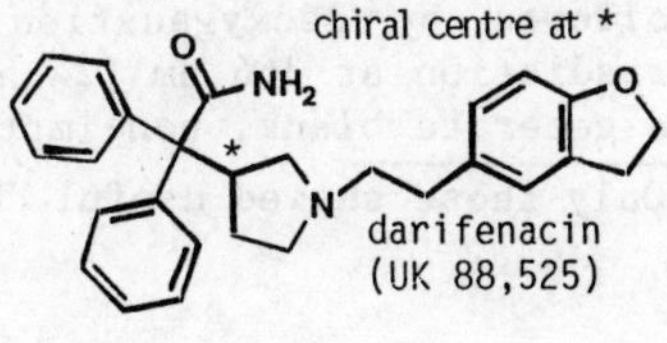

darifenacin
(UK 88,525)

Our model compound was darifenacin, *(S-)*2-{1-[2-(2,3-dihydrobenzofuran-5-yl)ethyl]-3-pyrrolidinyl}-2,2-diphenylacetamide. It is an M3-selective muscarinic receptor antagonist [14], under clinical development for urinary incontinence [15, 16]. An MS technique for its assay has previously been described [17], using SPE in a 96-well block format. The compound has a chiral centre, and on administration to human volunteers and patients gives rise to a number of metabolites.

We describe below the synthesis of specific and selective molecular imprints of darifenacin by polymerizing methacrylate and ethylene glycol dimethacrylate around the template molecule. The MIP was usable as an HPLC or SPE packing. Its ability to extract darifenacin from buffer, plasma and MeCN-precipitated plasma is outlined. Its selectivity with regard to analogs, drug sub-structures and metabolites is discussed, as is the potential for such imprints to be used as analytical tools for ultra-trace analysis. One application is purification of radiolabelled drug.

EXPERIMENTAL (general)

Materials and analytical HPLC

Solvents (methanol '205 Super Purity' and MeCN '195 HPLC') came from Romil (Loughborough, UK) and water from a Waters Milli-Q-plus cartridge system fed with deionized house water. Other reagents came from Fluka (Gillingham, Dorset, UK); they were the best analytical grade available and used without further purification. Pfizer compounds were furnished by colleagues in-house.

HPLC was carried out using a Varian 9012 ternary gradient pump, a Merck-Hitachi AS4000 autosampler, and a Merck-Hitachi F1080 fluorescence detector (nm settings: ex, 285; em 325) or else a Shimadzu LC10 UV detector (at 230 nm). The column was a 250 × 4.6 mm S5 CN (Hichrom, Reading, UK), using as mobile phase MeCN/water/TFA (50:50:0.1 v/v; 1.0 ml/min; room temp.).

Preparation of darifenacin MIP's

Preparation and polymerization was done in 50 ml polypropylene beakers. For 5 MIP batches the mmol composition of the functional monomers 4-vinylpyridine (4-VP) and methacrylate (MAA) was as follows (but this art. concerns only #6 and #9*).-

#5, 3:2 (320 and 172 µl) — #8, 1:4 (106.7 and 344 µl)
#6, 4-VP only: 5 (533.3 µl) — #9, MAA only: 5 (430 µl)
#7, 2:3 (213.3 and 258 µl).

The monomers were mixed with ethylene glycol dimethacrylate (60 mmol, 11.32 ml), α-azo-isobutyronitrile (150 mg) and ethyl acetate-extracted darifenacin (0.5 mmol, 213 mg). This mixture was dissolved in tetrahydrofuran (9 ml) and ethyl acetate (9 ml) followed by deoxygenation for 15 min (N_2 stream) and then by irradiation at 366 nm (24 h, 4°). Identical conditions were used to generate blank, non-imprinted polymers (darifenacin omitted).

*Only these showed useful HPLC or SPE interaction with darifenacin.

Solid polymer masses were physically broken up before milling with a Centrifugal mill (Glen Creston Ultra) fitted with a 200 µm screen. Sieving was carried out using a sieving machine fitted with a 75 µm sieve (Glen Creston). The collected <75 µm particles were washed with MeCN containing 0.1 TFA (50 ml × 6; thorough vortexing each time); each washing step was accelerated using a centrifuge (Jouan; 3500 rpm, 15 min) to sediment particles. Each wash supernatant was saved for HPLC analysis to allow calculation of template recovery. A 20 µl aliquot from each wash sample was analyzed for darifenacin using calibration standards in the range 10-600 µg/ml. Autosampler vials containing wash solution were centrifuged (10,000 rpm, 15 min) to obviate polymer particles being injected onto the column.

Particles were then slurry-packed (in MeCN) into an HPLC column, 150 × 4.6 mm, using 0.15-1.0 ml/min flow rates to ensure uniformity of this stationary phase through the column. Alternatively, 100 mg aliquots of polymer were dry-packed into empty 1 ml SPE cartridges with a 200 µm frit above and below the polymer. These LC columns and SPE cartridges were then characterized as detailed below.

RESULTS (and further experimental points)

HPLC using imprinted Polymer #6

Fig. 1 shows the HPLC of darifenacin enantiomers, injected sequentially, on a column packed with finely ground and sieved Polymer #6, for which the template was *S*-darifenacin. Evidently the latter, but not *R*-, was retained. Further work on the HPLC was hampered by the relative fragility of the MIP (necessitating a low flow-rate) and the regularity with which the column outlet frit blocked. It was apparent, however, that the two enantiomers were recognized differently by this column. Amongst the MIP's, #6 gave the clearest indication of chiral separation by LC.

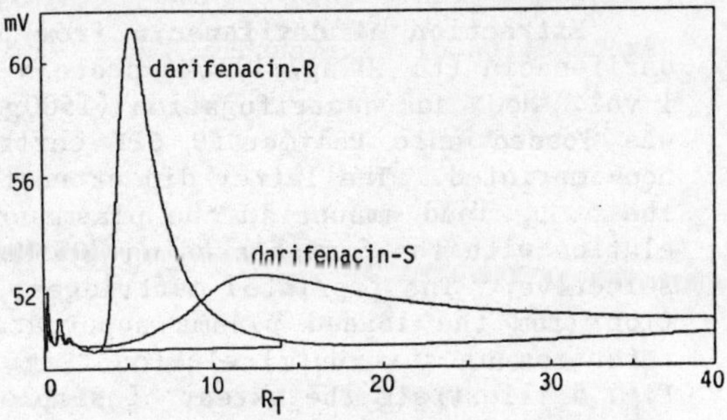

Fig. 1. HPLC of *R*- and (used as the template) *S*-darifenacin on a 150 mm column packed with MIP #6 (40-80 µm particles); equilibrated in mobile phase, 80% aq. MeCN. Flow rate 0.15 ml/min (room temp.); detection at 230 nm.

SPE using MIP #9

Empty 1 ml syringes were fitted with porous polyethylene frits, loaded with 100 mg dry Polymer #9, then washed with 50% aqueous MeCN, 0.1% TFA in MeCN and finally 50% MeCN. Samples (1 ml) in 50% MeCN were then loaded onto the cartridges and allowed to pass through under gravity or light vacuum. After washing with 50% then 70% MeCN, MeCN containing 0.1% TFA was used as eluant. All run-offs were collected and analyzed by HPLC.

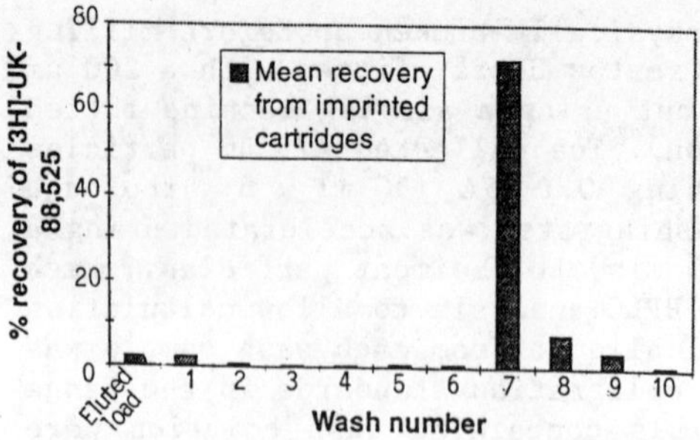

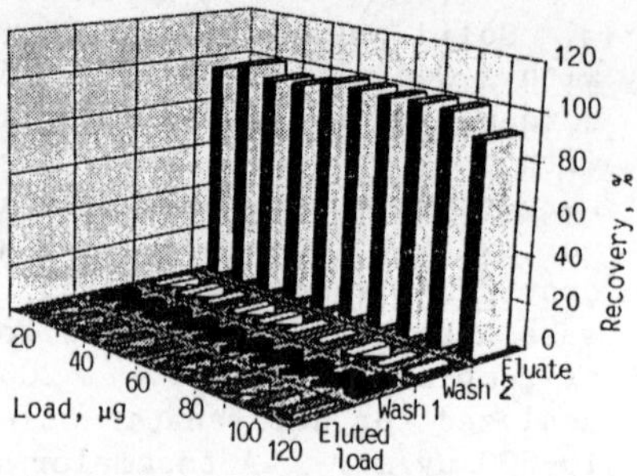

Fig. 2. Recovery, from a cartridge (MIP #9), of ^{3}H-darifenacin applied (1 ml, 0.1 µCi) as a dilution in 50% MeCN. Fractions collected: initial load effluent and subsequent 1 ml washes: 50% MeCN (1-3), 70% MeCN (4-6), and 0.1% TFA in MeCN (as eluant; 7-10). Aliquots were counted in a liquid scintillation counter.

Fig. 3. Loading capacity of SPE cartridges made from 100 mg dry MIP #9. Increasing amounts of darifenacin (the template) were applied to the cartridges and the recovery in the eluates was determined by HPLC.

Extraction of ^{3}H-darifenacin from 50% MeCN.- The recovery data for cartridges in Fig. 2 show that the MIP had a very high affinity for darifenacin, being completely retained and needing strongly acidic MeCN for elution. Loading increasing amounts of darifenacin (Fig. 3) showed very high capacity, each cartridge being able to adsorb, retain and then release 100 µg quantitatively. Only above this amount did recovery begin to fall, e.g. to 95% when 200 µg was loaded at a correspondingly increased concentration. The theoretical maximum capacity for 100 mg polymer is 180 µg darifenacin if an imprinted binding site is produced by every template molecule and all are available subsequently for binding interactions.

Extraction of darifenacin from plasma.- Plasma spiked with darifenacin (to 50 µg/ml) was protein-precipitated by addition of 1 vol. MeCN and centrifugation (1500g, 10 min). The supernatant was loaded onto Polymer #9 SPE cartridges, both imprinted and non-imprinted. The latter did extract a very small percentage of the 50 µg load amount in the plasma supernatant; but subsequent elution with the 6 washes using 70% MeCN showed that this was not selective. The imprinted cartridges, however, gave >90% extraction from the loaded plasma supernatant, and good on-cartridge retention of this until elution (Fig. 4). The chromatograms in Fig. 5 illustrate the extent of sample 'clean-up' obtained by the MIP cartridge wash procedure. No darifenacin was detectable in the load run-off, i.e. there was 100% extraction from the 50% MeCN plasma supernatant. The eluate (0.1% TFA in MeCN) showed a very clear baseline with no plasma-derived interferences.

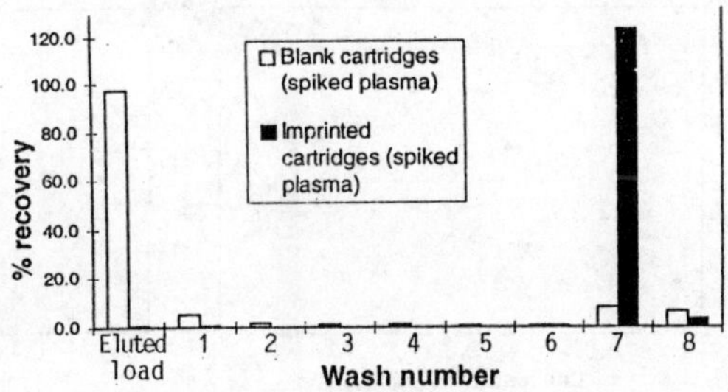

Fig. 4. Recovery (%) by SPE (MIP #9, or non-imprinted polymer; 100 mg) of 50 µg darifenacin from 1 ml plasma. See text for deproteinization and for the cartridge wash steps and HPLC analysis.

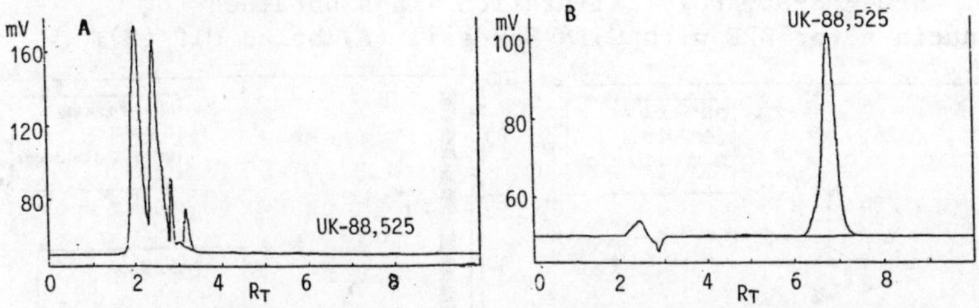

Fig. 5. HPLC analysis of darifenacin (50 µg/ml) in 1 ml plasma, deproteinized and then extracted by a cartridge (100 mg MIP #9), with washing (see text). Eluates (injected directly): **A**, load run-off; **B**, first application of 0.1% TFA in MeCN.

Comparison with our published method, to investigate the potential of the MIP's for ultra-trace analysis.- The method [17] is based on 96-well SPE (C-18) with MS-MS detection. It measures down to 25 pg/ml (LLOQ) in 1 ml plasma, spiked (1 ng/ml) with ^{2}H internal standard (i.s.), viz. darifenacin-d_5, as was likewise added in the present investigation along with standard amounts of darifenacin, 25-1600 pg/ml. As can be seen from the standard curves in Fig. 6, the MIP-SPE calibration line is unusable, with a large intercept and scatter. However, for the i.s. in 7 trials the recovery and precision results (area ratios) were acceptable:

C-18: 199257, 211272, 162440, 189673, 175340, 203542, 190798; mean 190332 ±8.8 (C.V.).
MIP: 119916, 110136, 98238, 103109, 106380, 130924, 107891; " 110942 ±10.0 (").

Selectivity of the MIP's

A number of structurally akin compunds were tested so as to assess the specificity of the imprint sites. They were dissolved in 50% MeCN and loaded onto 100 mg Polymer #9 cartridges.

UK-73,689 and UK-88,862 are essentially the two halves of the UK-88,525 (darifenacin) molecule and hence together provide all the functional groups of the imprint species. Both passed rapidly through the polymer bed with no retention (Fig. 7), suggesting that specificity had been generated by imprint-site shape as well as by interactions such as hydrogen bonds.

UK-148,993, a hydroxy metabolite of darifenacin, is differentially retained compared with darifenacin when loaded in 50% MeCN. Although it is almost completely extracted from the load

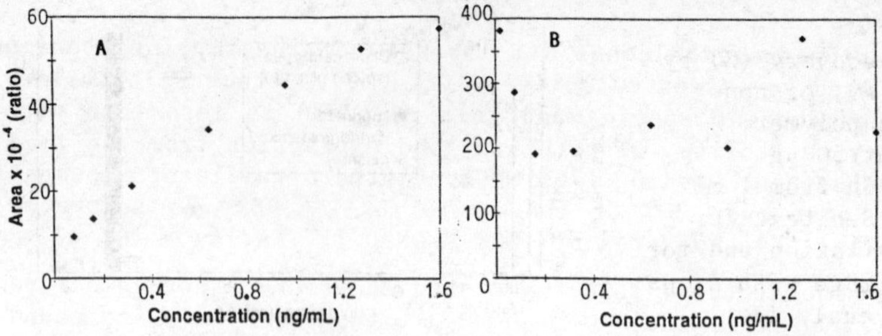

Fig. 6. HPLC-MS-MS(APCI) calibration lines obtained for darifenacin after SPE with C-18 Bondesil (**A**) or an MIP (**B**).

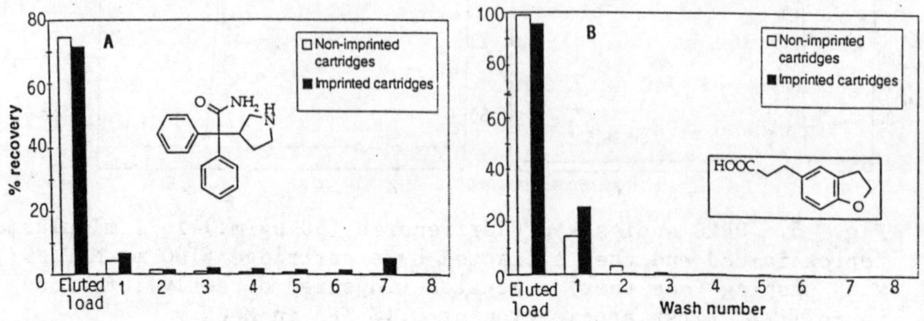

Fig. 7. Recovery of UK-88,862 (**A**) and UK-73,689 (**B**) from 50% MeCN using darifenacin-imprinted SPE cartridges.

Fig. 8. Recovery of UK-148,993 from 50% MeCN using darifenacin-imprinted SPE cartridges.

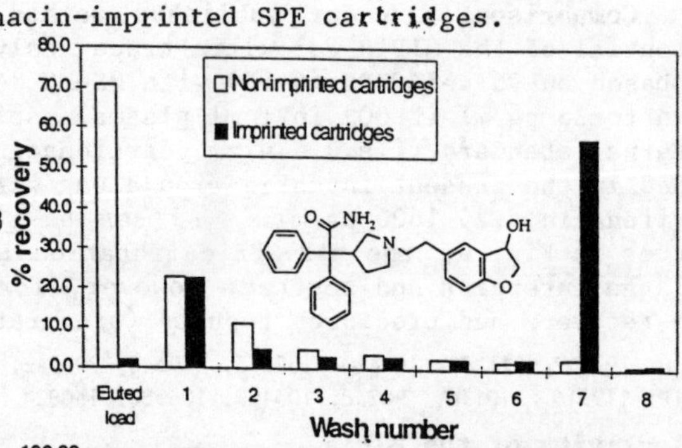

Fig. 9. Recovery of UK-156,981 from 50% MeCN using darifenacin-imprinted SPE cartridges.

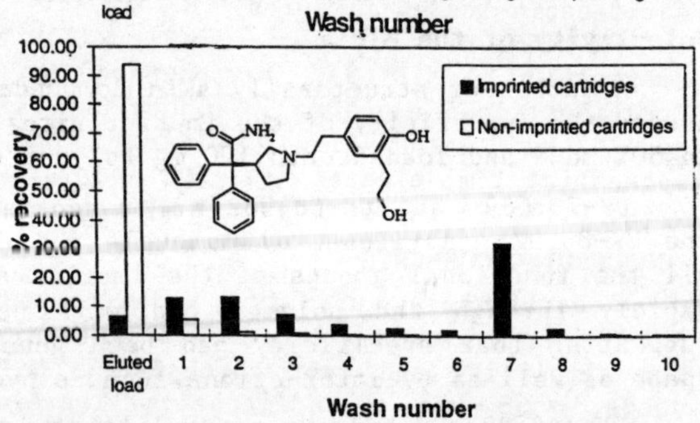

medium, some material is then lost in the first 70% MeCN wash and loss continues slowly until switching of the eluant at wash 7, where the bulk of the load is removed from the polymer rapidly.

The ring-opened alcohol (UK-156,981) of the imprint species was also used to assess selectivity (Fig. 9). Structural similarity between imprint species and this compound is less than with the hydroxy metabolite. Despite its good extraction from the load medium, subsequent washing (70% MeCN) eluted significant amounts. Thus only just over 30% of the loaded UK-156,981 was recovered with the first elution step at 100% MeCN/0.1% TFA.

With none of the 4 structures studied here was any selective retention seen with the non-imprinted cartridge, each compound being either not extracted at all or removed very rapidly with 70% MeCN. The presence of imprint sites within the polymer conferred selectivity, dramatically affecting the extraction of UK-148,993 and UK-156,981 but not of UK-88,862 or UK-73,689.

Purification of ^{14}C-darifenacin

Imprinted SPE cartridges were used to free [^{14}C]-UK-88,525 from associated radioactive impurities. TLC analysis of stock ^{14}C-darifenacin showed at least 3 bands and a fall in purity to ~70%. Loading, washing and elution were done as above. For the fractions Fig. 10 shows recovery of counts (liquid scintillation) and Fig. 11 shows TLC autoradiographs, indicative of >95% purity. For such re-purifications MIP-SPE is simpler than normal methods.

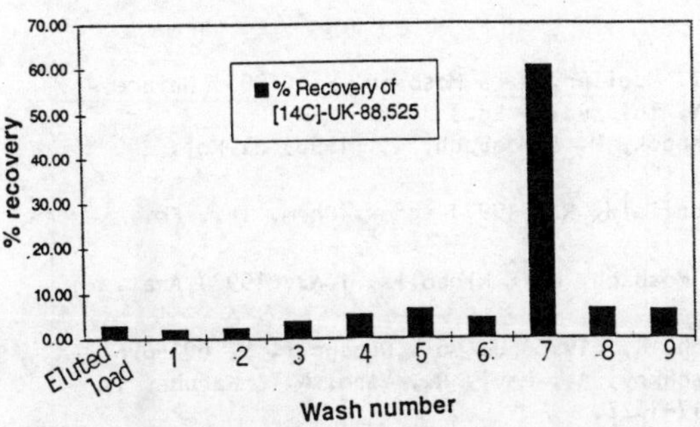

Fig. 10. Recovery of ^{14}C-darifenacin loaded (impure) in 50% MeCN, with SPE cartridges (100 mg MIP #9). Fractions: load run-off; 50% and 70% MeCN washes (1-6); sequential 1 ml eluates (7-9; 0.1% TFA in MeCN).

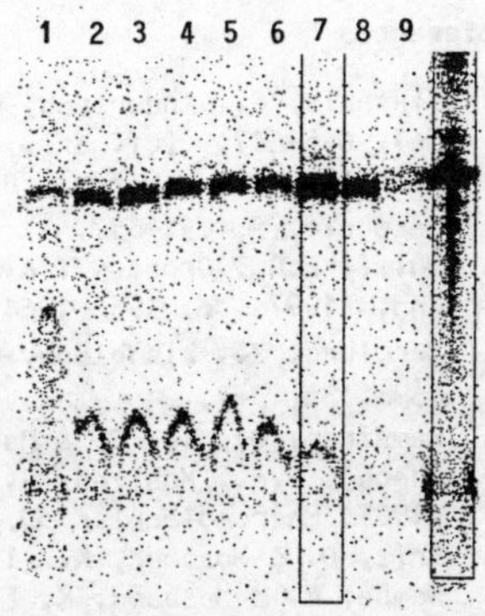

Fig. 11. TLC scan of fractions prepared as in Fig. 10. Far-right lane: unpurified starting material; lane 1, load run-off; 2-6, washes at 50% and 70% MeCN; 7-9, sequential 1 ml eluates at 0.1% TFA in MeCN.

CONCLUDING COMMENTS

Darifenacin, used as a model compound, was extractable by MIP-SPE from plasma after MeCN deproteinization. Advantageously for simplicity and fast throughput, no step in this approach needs drying-down of an extract. There was recognition of some but not other cognate molecules, and (shown by HPLC) chiral selectivity also. Evidently selectivity arises from not only non-covalent interactions (particularly H bonding) between template and functional monomer but also the imprint site's 3-D configuration.

The attempts to use the MIP's for ultra-trace analysis of darifenacin (aiming at LLOQ 25 pg/ml) were, however, thwarted because of the continued leaching of low amounts of template. Since we used 200 mg darifenacin to produce 10 g MIP, each 50 mg cartridge would contain 1 mg template if unwashed, necessitating reduction to 0.000001% to achieve <10 pg potential contamination. Washing was extensive, and extra washing with solvents chosen to swell the polymer gave no remedy (data not shown). As recently shown, much smaller amounts (by a factor of 10-50) can be used for imprinting; but this is no answer. Yet the ^{2}H-darifenacin results show that precision and accuracy can be good. Thus the analytical way forward appears to be through the use of close analogs of the target molecule as imprints, as could here apply to the hydroxyl metabolite of darifenacin. Possible alternative strategies include the use of surface-imprint technologies or different chemistries. However, effective methods for ultra-trace bioanalysis seem assured through imprinting with a close analogue, distinguishable from the target molecule by HPLC or MS-MS such that leaching is not a problem.

Acknowledgements.- We thank Dr B. Kaye for advice and encouragement, Mr Paul Macrae for expert analytical assistance with the MS assay and Mr Alan Mollatt for expertly running the TLC-autoradiography.

References

1. Vlatakis, G., Andersson, L.I., Mueller, R. & Mosbach, K. (1993) Nature **361**, 645-647. [Cf. art. #A-1, this vol.- Ed.]
2. Nicholls, I.A., Matsui, J., Krook, M. & Mosbach, K. (1996) J. Mol. Recognit. **9**, 652-657.
3. Wulff, F.G., Gross, T. & Schonfield, R. (1997) Agnew. Chem. Int. Ed. Eng. (1997) **36**, 1962-1964.
4. Berglund, J., Lindbladh, C., Mosbach, K. & Nicholls, I.A. (1998) Anal. Comm. **35**, 3-7.
5. Ramstrom, O., Yu, C. & Mosbach, K. (1996) J. Mol. Recognit. **9**, 691-696.
6. Cheong, S.H., McNiven, S., Rachkov, A., Levi, R., Yano, K. & Karube, I. (1997) Macromolecules **30**, 1317-1322.
7. Kriz, D. & Mosbach, K. (1995) Anal. Chim. Acta **300**, 1-3.
8. Kempe, M. & Mosbach, K. (1995) J. Chromatog. **A 694**, 3-13.
9. Kempe, M. & Mosbach, K. (1994) J. Chromatog. **A 664**, 276-279.
10. Nilsson, K., Lindell, J., Norrlow, O. & Sellergren, B. (1994) J. Chromatog. **A 680**, 57-61.
11. Svec, F. & Frechet, J.M.J. (1992) Anal. Chem. **64**, 820-822.
12. Kempe, M. & Mosbach, K. (1995) J. Chromatog. **A 691**, 317-323.
13. Kriz, D., Kriz, C.B., Andersson, L.I. & Mosbach, K. (1994) Anal. Chem. **66**, 2636-2639.
14. Alabaster, V.A. (1997) Life Sci. **60**, 1053-1060.
15. Graul, A. & Castaner, J. (1996) Drugs Future **21**, 1105-1108.
16. Beaumont, K.C., Cussans, N.J., Nichols, D.J. & Smith, D.A. (1998) Xenobiotica **28**, 63-75.
17. Kaye, B., Herron, W.J., Macrae, P.V., Robinson, S., Stopher, D.A., Venn, R.F. & Wild, W. (1996) Anal. Chem. **68**, 1658-1660.

#A-3

SELECTIVE SPE's USING MOLECULAR IMPRINTED POLYMERS

[1]P. Martin, [1]I.D. Wilson†, [2]G.R. Jones & [3]K. Jones

[1]Dept. of Safety of Medicines, Zeneca Pharmaceuticals, Mereside, Alderley Park, Macclesfield, Cheshire SK10 4TG, U.K.

[2]Dept. of Chemistry, University of Keele, Keele, ST5 5BG, U.K.

[3]Affinity Chromatography Ltd., Ballasalla, Isle of Man, U.K.

MIP's were prepared using propranolol as template, MAA as monomer and EGDMA as cross-linker. Their extraction characteristics were assessed using radiolabelled compounds structurally related to propranolol. Extraction experiments revealed that appropriate selection of solvent and acidic or basic modifier was essential for good recovery and selectivity. Thus eluents which contained TEA selectively recovered propranolol from the polymer whilst those containing TFA gave excellent recoveries but were not selective. Incomplete removal of the template led to the slow leaching of propranolol from the MIP. Experiments with radiolabelled template revealed that ~80% could be recovered relatively easily whilst the remaining 20% leached slowly from the polymer. Residual propranolol in the polymer precluded its use for trace analyses on biofluids, as confirmed in pilot studies. The polymer was therefore used to establish an analytical method for a close structural analog of propranolol. Other approaches to developing assays using MIP's are discussed, and the selectivity of these MIPs for other propranolol analogs described.*

The technique of molecular imprinting has been widely used to prepare MIP's for use as chromatographic stationary phases, particularly for the separation of enantiomers. Areas of application of MIP's have been reviewed in accompanying articles and elsewhere [e.g. 1]. The potential for such materials to be used to develop highly specific SPE media has, however, been less well explored, with so far only 5 applications in the literature [2-6]. The major potential advantage of the MIP approach in sample preparation is the possibility of producing exquisitely selective extractions. In some respects the approach can be reckoned akin to the use of immunoaffinity-based extractions, but without the need to go through the often lengthy, tedious and by no means certain process of raising antisera in animals. The chemistry involved is relatively trivial, requiring little specialized knowledge or equipment and based on well established procedures.

†addressee for any correspondence

**Abbreviations*.- Ab, antibody; EGDMA, ethyleneglycol dimethacrylate; MAA, methacrylic acid; MIP, molecular(ly) imprinted polymer. TEA, triethylamine; TFA, trifluoroacetic acid; MeCN, acetonitrile.

Therefore where the analyte is available in reasonable quantity (100 mg would suffice to prepare 1-2 g of polymer) the preparation of an MIP for use in SPE becomes feasible.

Here we present results obtained with polymers prepared to the drug propranolol, used as a model, that illustrate some of the potential of the technique besides highlighting pitfalls that may inhibit the use of MIP's in bioanalysis.

METHODS

Materials.- Methanol, MeCN, TEA and TFA were bought from Fisons, and MAA, EGDMA and unlabelled propranolol from Sigma. Our in-house Isotope Chemistry Unit furnished ^{14}C versions of all compounds shown in Fig. 1.

Polymer preparation.- Two MIP's were prepared by the method of Andersson ([7], & art. #A-1, this vol.) using either unlabelled or ^{14}C-labelled propranolol as the template. A blank polymer was also prepared without template. MAA was used as the functional monomer, and EGDMA as the cross-linker. The polymer was ground using a pestle and mortar and then sieved through a copper sieve with 45 µm apertures. It was then thrice sedimented in ethanol to remove fines. The product was dried in a vacuum oven and stored at room temperature. For SPE, 30 mg portions were packed into 1 ml SPE reservoirs (supplier: IST, Hengoed, UK).

Extractions, in experiments to assess the characteristics of the MIP's produced.- The generic SPE method involved solvation with 1 ml methanol, 1 ml water and 0.5 ml 50 mM phosphate buffer pH 7.0. The sample for extraction was propranolol in the buffer (2.5 µg/ml; 0.5 ml). After sample application the MIP was washed with 1 ml water. Analyte was recovered using solvents containing increasing proportions (0 to 100%, in 10% increments) of organic solvent. The combinations used included, in varying proportions, methanol/water, MeCN/water, methanol/water/TFA, methanol/water/TEA, MeCN/water/TFA and MeCN/water/TEA. Where included, TEA or TFA accounted for 1% of the solvent volume. Deviations from this generic protocol are detailed in the Results & Discussion section where applicable, as when alternative elution strategies were used in experiments with polymer prepared using ^{14}C-propranolol. Here the polymer was analyzed (BELOW) for ^{14}C content before use.

HPLC was performed using 250 × 4.6 mm HiRPB column (Hichrom) and, as mobile phase (1 ml/min flow), methanol/water/TFA (600:400:1 by vol.) containing 0.77% (w/v) ammonium acetate. Detection was at 290 nm or by fluorescence (ex 290, em 340) depending upon the experiment. Data collection and analysis were performed by Beckman Peakpro. An LDC Analytical Constametric 3200 pump was used and a Perkin Elmer ISS 200 autosampler. The UV detector was a Shimadzu SPD-10A and the fluorescence detector a Jasco FP920.

Scintillation counting.- SPE eluates were collected directly into 20 ml glass scintillation vials; 10 ml Ultima Gold cocktail was added. A Packard TRI-CARB 1900 CA scintillation counter was

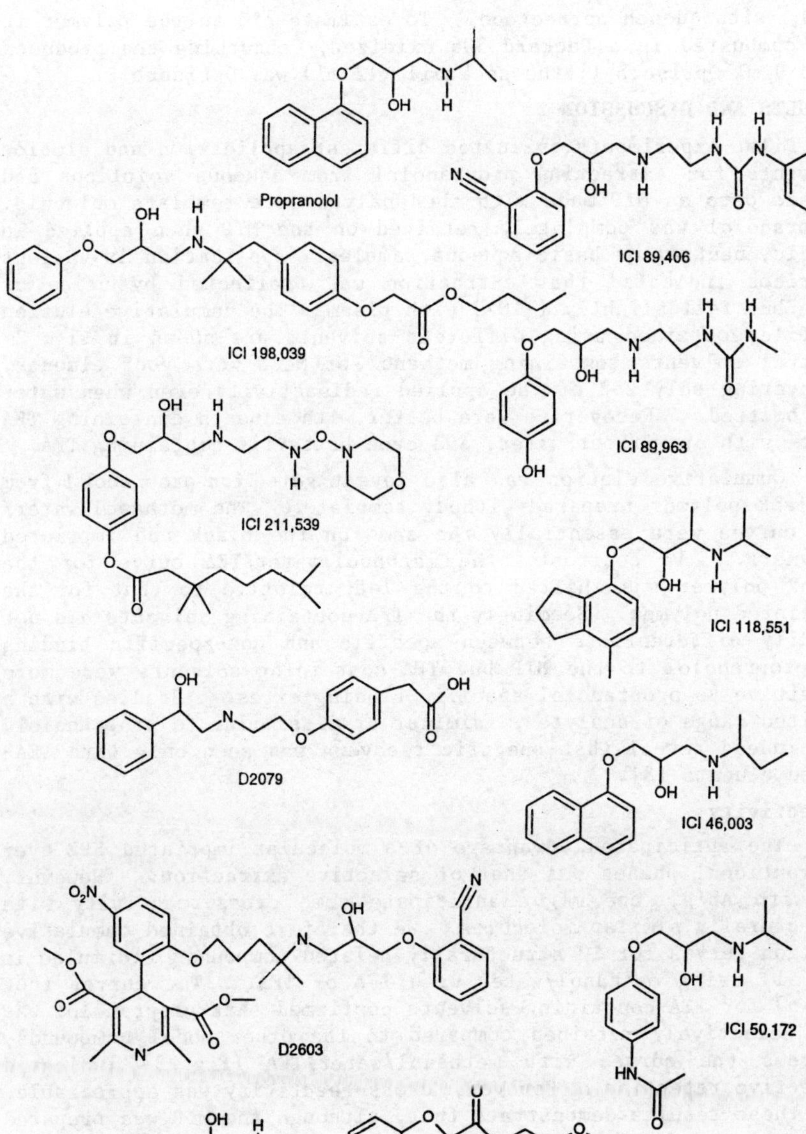

Fig. 1. Structures of propranolol and structurally related compounds used to assess the selectivity of the propranolol MIP. In Fig. 3 below, most of the curves have a 6-figure ref. not preceded, as above, by 'ICI'.

used, with quench correction. To estimate ^{14}C in the polymer it was combusted in a Packard 306 oxidizer, absorbing the products into 9 ml Optisorb 1; the cocktail (12 ml) was Optisorb S.

RESULTS AND DISCUSSION

Pilot experiments evaluated different application and elution solvents for extracting propranolol from aqueous solutions and plasma onto an MIP made with the analyte as a template molecule. Propranolol was completely retained on the MIP when applied in acidic, neutral or basic aqueous samples. Application in various matrices indicated that extraction was unaffected by urine or bile but fell slightly (~10%) with plasma. The cumulative elution profiles obtained using different solvents are shown in Fig. 2. Neutral solvents containing methanol or MeCN were poor eluents, recovering only 25% of the applied radioactivity even when water was omitted. Recoveries were better with eluents containing TEA also, with or without water, and even better if containing TFA.

Cumulative elution was also investigated for propranolol from a blank polymer prepared without template. The methanol/water/TFA curves were essentially the same on the blank and imprinted polymers. In contrast, the methanol/water/TEA curve for the blank polymer was shifted to the left relative to that for the imprinted polymer. Seemingly the TFA-containing solvents did not readily differentiate between specific and non-specific binding of propranolol to the MIP but TEA-containing solvents were more sensitive to propranolol-specific binding sites. Studies with a limited range of analytes, similar or dissimilar to propranolol, did indeed reveal that specific recovery was seen only with TEA-based eluents [8].

Selectivity

The anticipated advantage of a molecular imprinted SPE over conventional phases was that of selective extraction. However, as with Ab's, one might anticipate some cross-reactivity with structurally similar molecules. We therefore obtained cumulative elution curves for 10 structurally related compounds (formulae in Fig. 1) using methanol/water with TFA or TEA. The curves (not shown) for TFA-containing solvents confirmed that propranolol was not selectively retained compared to the other model compounds, whereas the curves with methanol/water/TEA (Fig. 3) indicated selective retention. However, cross-reactivity was appreciable, and these results demonstrate that, although the MIP was prepared to propranolol, it could perhaps be used to develop specific extractions for related compounds. Interestingly, studies on several different propranolol-derived MIP's (prepared in different laboratories or by slightly different routes; unpublished work) showed very similar selectivities, suggesting that batch-to-batch variation may be less of a problem than one might have anticipated.

Methodological problems for trace analysis with MIP's

Clearly the main aim in preparing the propranolol-selective MIP was to use it for sample preparation. Accordingly, having

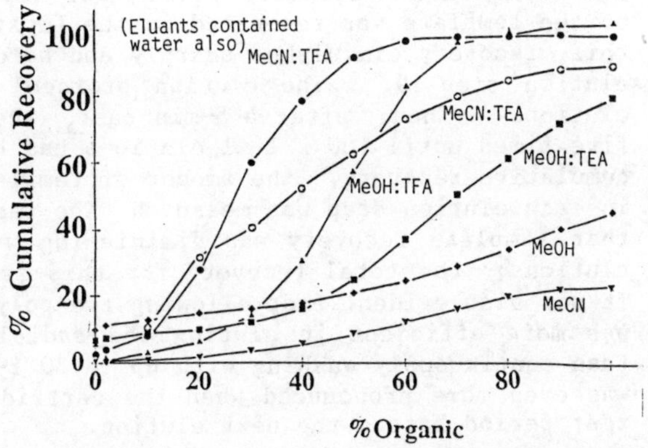

Fig. 2. Cumulative elution profiles for propranolol from a propranolol MIP using different organic components (water too).

Below: **Fig. 3.** Profiles as in Fig. 2 but always methanol/water/TEA: comparison of different compounds (see Fig. 1), viz. (in order of recovery with 100% methanol):

◇ propranolol
o ICI 118551
Δ ICI 198039
∇ ICI 89963
■ ICI 89406
▼ D 2603
◆ ICI 50172
□ D 7114
▲ D 2079
● ICI 211539

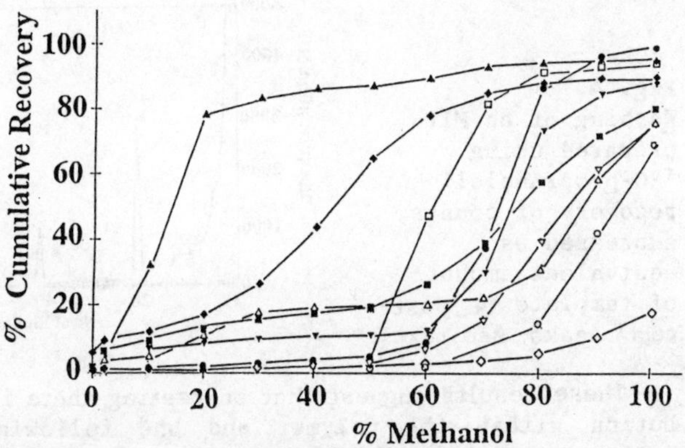

established that selective extraction was feasible, the application of the MIP's to trace analysis of propranolol in biological fluids using conventional chromatographic assay was evaluated. HPLC with fluorescence detection was the chosen basis. However, during development of the assay it became evident that there was still leakage of the propranolol used as the template for the MIP. This was too great to allow quantification at the trace levels (~10 ng) required for bioanalytical work.

In order to study the retention of the template in the MIP, and ensuing leakage, a polymer was prepared using ^{14}C-propranolol for imprinting. The radiolabelled polymer was ground, sieved and freed from fines in the usual way but was not extensively washed. It was then packed into an SPE cartridge and eluted with numerous successive washes using 99:1 methanol/TFA. Finally the balance of the original imprinting template was shown to be present in the MIP in the cartridge, by combustion and liquid scintillation counting.

The cumulative % recovery curve (not depicted) showed that ~80% of the template was recovered in the first two elutions. Beyond this, recovery diminished sharply and hardly any was recovered by elution step 19. The elution protocol involved 20 successive elutions. Then, after a 5-min rest, the procedure was repeated five times until 100 × 1 ml elutions had been performed. Besides cumulative recovery, the amount of template radiolabel recovered in each elution step was measured. The results (Fig. 4) confirmed that template recovery was diminishing within each series of 20 elutions; the total recovery for this series of washes was 86%. It was also evident that allowing the polymer to 'rest' for 5 min was more efficient in eluting the radiolabel in a small volume than continuously washing with up to 20 1-ml washes. This trend was even more pronounced when the cartridge was allowed a 1-week rest period before the next elution.

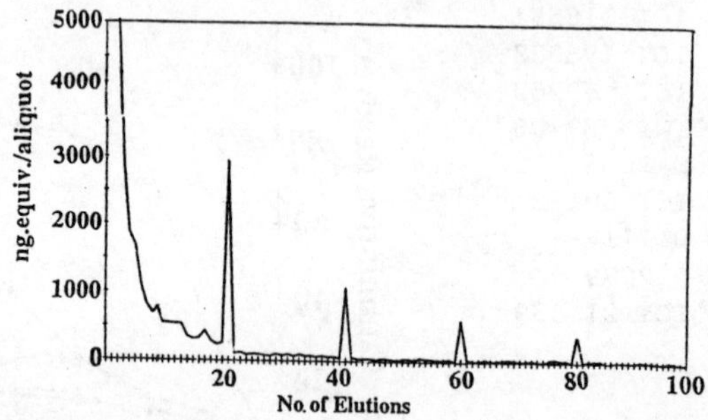

Fig. 4. Washing of an MIP prepared using ^{14}C-propranolol: recovery of counts, expressed as equivalent amount of template (4 'post-rest' peaks; see text).

These results suggest that on resting there is template redistribution within the polymer and the following elution achieves better removal. This type of problem, shown to be a general property of MIP's, could seriously limit their use if effective removal of the template molecule cannot be achieved. However, some possible solutions to this apparent difficulty are apparent.

Firstly, as shown by Fig. 3, it appears that although this MIP showed the highest selectivity for propranolol it could also be used to selectively extract some close structural analogs. Hence it seems feasible to prepare MIP's with suitable selectivity using compounds similar to the analyte in structure but separable by HPLC. An example is ICI 46,003 - a very close structural analog of propranolol - for which Fig. 5 shows an HPLC trace. Careful selection of the imprint molecule and HPLC conditions should enable this type of approach to be readily implemented.

It may also be possible, if MS is the end-step in the analytical procedure, to prepare the MIP using stable isotope-enriched analyte, essentially 'invisible' to the detector. Another option might be applicable where the analyte is thermally or photochemically unstable: the polymer might be treated in such a way as to destroy or alter the unrecovered template molecule so that it does not interfere in subsequent analyses.

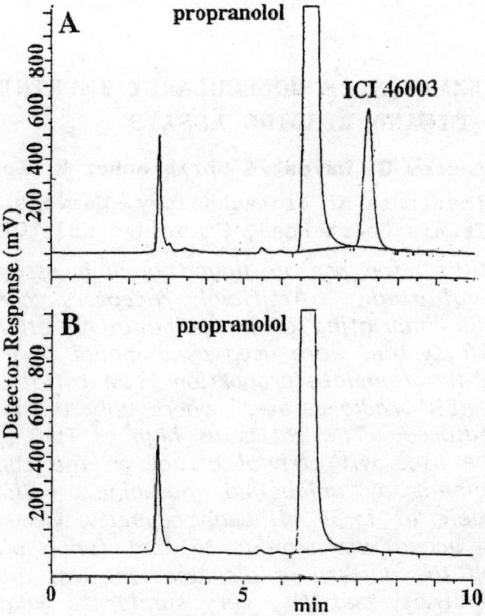

Fig. 5. Chromatogram showing the extraction of the propranolol analog ICI 46,003 (Fig. 1) from rat plasma using an MIP prepared with propranolol as the template. (Unspiked blank also shown.)

CONCLUDING COMMENTS

The potential for the use of MIP's in trace bioanalysis is currently highly speculative. Clearly such polymers can be prepared with affinities apt for analytes such as propranolol, but potential problems such as leakage of the template molecule must be addressed. It seems likely, however, that the attractiveness of the approach, offering as it does Ab-like selectivity without the long lag times associated with Ab production, does offer an advantage.

References

1. Wulff, G. (1995) Angew. Chem. Int. Ed. **34,** 1812-1832.
2. Sellergren, B. (1994) Anal. Chem. **66,** 1578-1582.
3. Matsui, J., Okada,M., Tsuruoka, M. & Takeuchi, T. (1997) Anal. Comm. **34,** 85-87.
4. Muldoon, M.T. & Stanker, L.H. (1997) Anal. Chem. **69,** 803-808.
5. Andersson, L.I., Paprica, A. & Arvidsson, T. (1997) Chromatographia **46,** 57-62.
6. Rashid, B., Briggs, R., Hay, J. & Stevenson, D. (1997) Anal. Comm. **34,** 303-305.
7. Andersson, L.I. (1996) Anal. Chem. **68**, 111-117.
8. Martin, P., Wilson, I.D., Morgan, E.D., Jones, G.R., & Jones, K. (1997) Anal. Comm. **34,** 45-48.

Acknowledgement

We thank Mr J. Clews (Keele University) for technical assistance.

#A-4

OPTIMIZATION OF MOLECULARLY IMPRINTED POLYMERS FOR RADIO-LIGAND BINDING ASSAYS

Andrew G. Mayes† & Christopher R. Lowe

Institute of Biotechnology, University of Cambridge, Tennis Court Road, Cambridge CB2 1QT, U.K.

MIP's for use in ligand binding assays have been optimized and evaluated. Artificial receptors created by non-covalent molecular imprinting of morphine in an ethyleneglycol dimethacrylate/MAA system were used as a model to investigate the effect of varying the template proportion (T:M ratio) during imprinting. For use in 'RLB' radio-assays, where only the highest affinity receptors are measured, T:M ratios as high as 1:50 (1:4 being more typical) could be used with trivial effect on the capacity and the IC_{50} for displacement by unlabelled morphine. Binding curves were still measurable at 1:150 although capacity was reduced. At 1:500 the amount bound was similar to that for a blank reference polymer, although the pattern of displacement was different. The 1:50 MIP had selectivity and IC_{50} very similar to what had been reported for morphine imprinting with a 1:4 ratio. With 1:50 the constraint imposed by the solubility limit of morphine was removed; hence it could be imprinted in a range of solvents that were previously inaccessible. As expected, IC_{50} decreased as the polarity of the porogenic solvent was changed from MeCN to chloroform or xylene. The MIP synthesized in cyclohexane, however, had a very different morphology and showed no specific morphine binding. Thus it seems that, once the solubility constraint of the template is removed, selection of polymerization solvent can be governed instead by the goal of desired polymer characteristics such as high porosity.*

Molecular imprinting is a 'growth area' which is surveyed in accompanying articles (also in [1]). As depicted in Fig. 1 with morphine as the template, the monomers are 'frozen' around it in the presence of a substantial excess of cross-linker; its removal leaves cavities that are complementary in shape and functionality and can re-bind it with high affinity and selectivity. Following the demonstration that MIP's can replace Ab's in ligand binding assays [2], the affinity and structural selectivity of MIP's has been probed with various model systems [3-6]. Thus, re-binding of propranolol to polymers imprinted with propranolol in toluene can be achieved in both organic solvents and aqueous buffers with only small effect on affinity or selectivity [6]. Both analytical applications [3-5] and basic studies are being widely pursued.

---

†correspondent; e-mail: a.mayes@biotech.cam.ac.uk

**Abbreviations*.- MAA, methacrylic acid; MeCN, acetonitrile; MIP, molecularly imprinted polymer; RLB(A), radioligand binding (assay) [variant of 'RIA' - Ed.]; T:M, template/functional monomer (ratio).

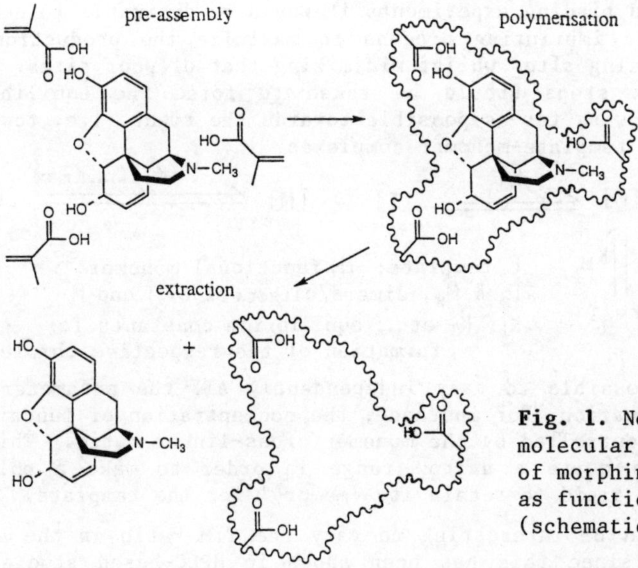

Fig. 1. Non-covalent molecular imprinting of morphine using MAA as functional monomer (schematic).

Published studies on MIP optimization have mostly used HPLC as the means of characterizing the performance of the polymers [7-9]. A number of 'rules of thumb' have been established, which allow new template molecules to be imprinted with some confidence that the products will have reasonable binding characteristics. The same rules have been applied to polymers synthesized for RLB studies, implicitly assuming that the required properties are the same. When considering these two systems, however, it is clear that there are several very significant differences. In HPLC, compounds are usually resolved by kinetic separation, with many rapid exchanges between mobile phase and solid support as the material flows through the column. In order to work efficiently, diffusion kinetics and binding kinetics must both be rapid. In practice, most of the binding events probably occur at highly accessible and relatively low affinity sites. This is confirmed by frontal analysis, whence the K_d is typically estimated to be in the mM range [10, 11]. The characteristic tailing of the peak for the imprinted analyte probably arises from a smaller number of high-affinity interactions, coupled with slow exchange kinetics and the low number of such sites [12].

The situation is quite different when considering RLBA's: since these are usually incubated till they appproach thermodynamic equilibrium, the kinetics are less important. There is time for the analyte to 'explore' the binding site population through multiple binding and release events, whereby the analyte binds to the thermodynamically most favourable high-affinity sites; the presence of numerous sites is undesirable and optimal performance calls for a limited number of high-affinity sites and no low-affinity sites.

In ligand binding experiments it would be desirable to design the molecular imprinting process to maximize the production of the best binding sites whilst minimizing that of poor sites. This implies that steps should be taken to force the equilibrium depicted below as far as possible towards the right, i.e. towards fully formed template-monomer complexes.

$$[T] + [M] \underset{}{\overset{K_1}{\rightleftharpoons}} [T] + [M] \overset{K_2 \ldots\ldots K_{max}}{\rightleftharpoons} [TM_{max}]$$

$$[T] \overset{K_T}{\rightleftharpoons} [T_x] \qquad [M] \overset{K_M}{\rightleftharpoons} [M_x]$$

T, template; M, functional monomer
T_x & M_x, dimers/oligomers of T and M
K_1, K_T etc., equilibrium constants for formation of the respective complexes

It is not possible to vary independently all the parameters in the above equation. For instance, the concentration of functional monomer is controlled by the monomer/cross-linker ratio. This is fixed within quite a narrow range in order to make a polymer sufficiently rigid to retain its 'memory' for the template.

It would be interesting to vary the T:M ratio in the above equilibrium since this has been shown in HPLC-based studies to alter the recognition properties. Similar studies have not been reported for RLB experiments. Since [M] is constrained by the need for a high degree of cross-linking, the way to alter T:M would be to *reduce* [T]. This appears counter-productive, since it will tend to move the equilibrium back towards the left. As discussed above, however, only a small number of high-affinity sites are required for RLB assays, with as few lower-affinity sites as possible. In this case the fraction of the total template added which is fully complexed will be more important than the absolute concentration of fully complexed template, hence to have a large excess of M should be beneficial.

In order to predict accurately the outcome of such changes in T:M ratios, the equilibrium equations would need to be solved mathematically. Whilst this is in principle relatively straightforward, it is in practice rather complex due to the lack of reliable equilibrium constant values for the various interactions measured under realistic imprinting conditions. It is, moreover, unclear how many of the higher order complexes possible within the pre-polymerization mixture might have to be included in order to make meaningful predictions. For instance, the dimerization/oligomerization of carboxylic acids in solution is well documented [13, 14] and will have a profound effect on the position of the above equilibrium. The association constant for the various complexation events can be varied to some extent by altering physical conditions, e.g. temperature or solvent polarity. This has been shown to have some effect on the imprinting process [15, 16], but unfortunately it will usually affect unwanted interactions, e.g. MAA oligomerization, similarly to those desired.

In the absence of a good mathematical model for predicting the outcome of changing the imprinting recipe, a pragmatic evaluation was undertaken, and some results are presented below together with discussion of the implications of the findings.

MATERIALS AND METHODS

Morphine, codeine, heroin, dihydromorphine and pholcodeine were gifts from Dr N. Bruce of our Institute. Morphine was heated at 120° for 4 h to remove water before use. ^{3}H-Morphine (84.5 Ci/mmol in ethanol) came from NEN, Stevenage, UK. Ethylene glycol dimethacrylate (EDMA) and MAA were obtained from Aldrich (Poole) and were redistilled under reduced pressure prior to use. Fisher (Loughborough) supplied analytical or HPLC grade solvents. Solvents for use as porogens were dried either with sodium wire (xylene, cyclohexane) or with molecular sieves (MeCN, amylene-stabilized chloroform). Toluene and acetic acid were used as received for RLBA's. 2,2'-Azobis(dimethylvaleronitrile) (ABDV) was purchased from Wako Pure Chemical Industries, Osaka, Japan). Waterproof silicon carbide abrasive paper (grade P600A from English Abrasives) was purchased from a local hardware store. Sigma (Poole, UK) supplied Sigma-Fluor liquid scintillation cocktail for non-aqueous samples.

Polymer synthesis was by well-established methods. In a typical run (to give T:M 1:50), 103.2 µl (1.2 mmol) MAA was added to 6.84 mg (24 µmol) morphine in a small screw-top vial. After mixing, dry MeCN (1.2 ml) was added and mixed to ensure complete dissolution of the template. ABDV (11.2 mg) and EDMA (952 µl, 4.8 mmol) were added and the solution sparged with N_2 for 5 min on ice to remove dissolved O_2. The vial was quickly sealed with a screw top and parafilm and left in an oven at 45° for 24 h to polymerize. The vial was then gently broken. The removed rod of polymer was rendered into fine particles by rubbing it on a piece of abrasive paper in the bottom of an ethanol-containing glass dish. The particle suspension was filtered through a 50 µm sieve to remove large fragments, then placed in a centrifuge tube and spun (300**g**, 30 min) to sediment the polymer. The template was extracted by re-suspending the polymer in methanol/acetic acid/water (1:1:1) and leaving for some time with occasional mixing (room temp.) before centrifuging and removing the supernatant. This process was repeated 10 times over 5 days. The polymer was then washed (×3) with pure methanol and left to dry in air. Other polymers were similarly made, with more or less morphine (nil if non-imprinted, N/I; code #A-R). The T:M range was:

1:4, #A-4; 1:10, #A-10; 1:50: #A-50; 1:150, #A-150; 1:500, #A-500

Also prepared, using solvents other than MeCN			
	chloroform:	N/I, #C-R;	T:M 1:50, #C-50
	xylene:	#X-R;	1:50, #X-50
	cyclohexane:	#H-R;	1:50, #H-50

The only variables were the T:M ratio and the porogenic solvent.

Binding assays were performed in Eppendorf tubes. To each was added 100 µg polymer in 300 µl solvent, 500 µl solvent having the required concentration of displacer, and 200 µl solvent which contained 6×10^{-13} moles ^{3}H-morphine. The solvent was toluene/acetic acid, 49:1 v/v. The tubes were incubated overnight (room temp., orbital shaker), then the polymer was spun down (16000**g**, 5 min). Duplicate 200 μl aliquots of each supernatant were counted (Beckman LS3801 scintillation counter; 2 ml cocktail).

RESULTS AND DISCUSSION

All the mixtures specified above produced whitish polymers with textures varying from hard, glassy and slightly translucent when chloroform was used as porogen, through hard and opaque for MeCN and xylene as porogen to chalky for cyclohexane. The latter polymer was very soft and ground up very easily; it could even be crushed to a fine powder using finger pressure. The others were much harder. Those made using a large amount of template had a pale brown coloration, but this disappeared after extraction of the template. Grinding on abrasive paper, seemingly novel, is simple, inexpensive and relatively rapid when only small amounts need to be processed. A small amount of silicon carbide particles becomes detached, but they spin down rapidly and can easily be removed from the bottom of the pellet upon re-suspension. Any slight contamination remaining seemed experimentally innocuous.

Fig. 2 shows ^{3}H-morphine displacement by unlabelled morphine from polymers imprinted in MeCN at different T:M ratios. Both total and fractional binding are depicted, to show the effect of altering T:M. As T:M decreases the fractional binding also falls, suggesting a reduction in the number of available binding sites - especially in A-150; A-4, A-10 and A-50 do not differ greatly. For A-500 the binding is essentially at background levels. It is notable, however, that the A-500 binding curve differs in form from that for the non-imprinted reference polymer. This suggests that the template's presence has somehow altered the distribution of carboxyl groups within the structure, affecting the pattern of interaction. The curve has the same form as the displacement curves for higher ratio polymers; so it might be a matter of a little specific binding rather than the low-affinity non-specific binding displayed by the reference polymer. Thus it is probably inappropriate to merely subtract the latter in order to determine 'specific' binding; the B/B_0 data have not been thus corrected.

Transforming the binding curves into B/B_0 form helps compare the curves for different T:M's. The curve differences are quite small, suggesting that the best binding sites produced are essentially the same in all polymers, whatever the T:M used. The IC_{50} μM values as calculated ([17]; Fig. 2 legend) were:- A-4, 3.1; A-10, 1.5; A-50, 1.5; A-150, 0.6; A-500, 0.8. The apparent slight reduction in going from A-4 to A-500 accords with the reduction in receptor concentration which the B/T plots show and which, as B/B_0 plots show, theoretically increases assay sensitivity [18].

Selectivity of A-50 was more closely investigated so as to evaluate the effect of using a below-normal T:M. Curves for the displacement of ^{3}H-morphine by structural analogs were measured.

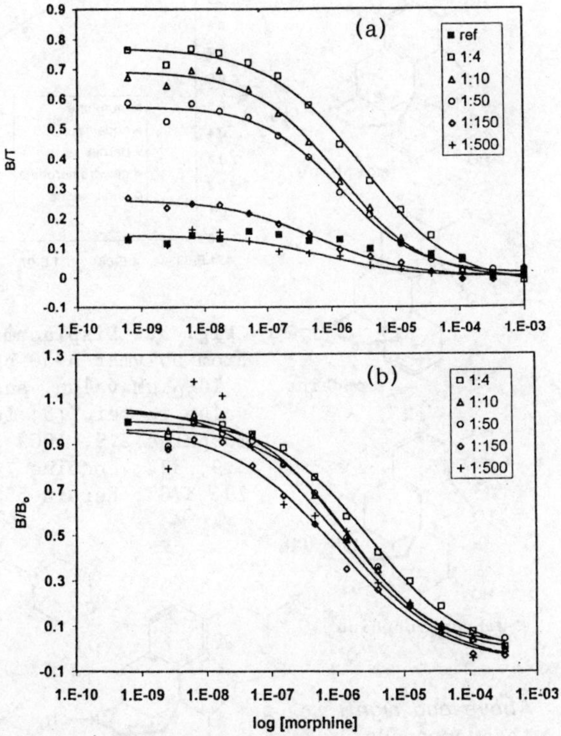

Fig. 2. Dose-response curves for the displacement of ^{3}H-morphine by unlabelled morphine with a range of imprinted polymers. (**a**) Data plotted as cpm bound/cpm total, to show the absolute levels of binding. (**b**) Normalized curves, to compare curve shapes: (cpm bound)/(cpm bound in absence of displacer). The curves represent fits of each data set to a four-parameter logistic function [14].

The results appear in Fig. 3 along with the analyte formulae and (in the legend) IC_{50} values and cross-reactivities. Polymer A-50 is highly selective. Even the most subtle change in compound structure - saturation of the 7-8 double bond in dihydromorphine, which induces a slight twist to the bottom ring, caused a 3-fold reduction in affinity. Codeine, where the 3'-hydroxyl group is blocked, have only 4% cross-reactivity. Heroin, where both the 3'- and 6'-hydroxyls are blocked, was a notably poor displacer; the cross-reactivity was <0.5%. Pholcodeine, where the 3'-hydroxyl is blocked by a more bulky group than in codeine, had an almost identical displacement curve to codeine, suggesting that it has similar difficulty in binding to the sub-set of sites where the 3'-hydroxyl is accurately recognized. It seems that the exact nature of the blocking group is unimportant in this case.

For two of the compounds cognate to morphine there was accord (Fig. 3 legend) with values previously reported using a polymer whose T:M was 1:4 [5]. This points to robustness of the technique.

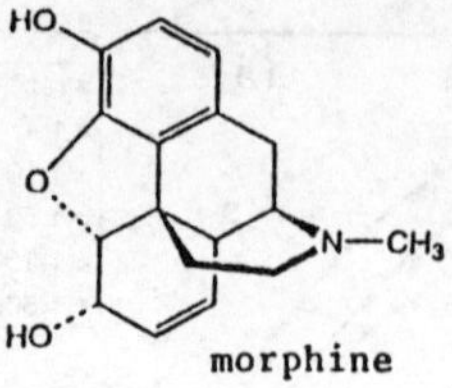

morphine

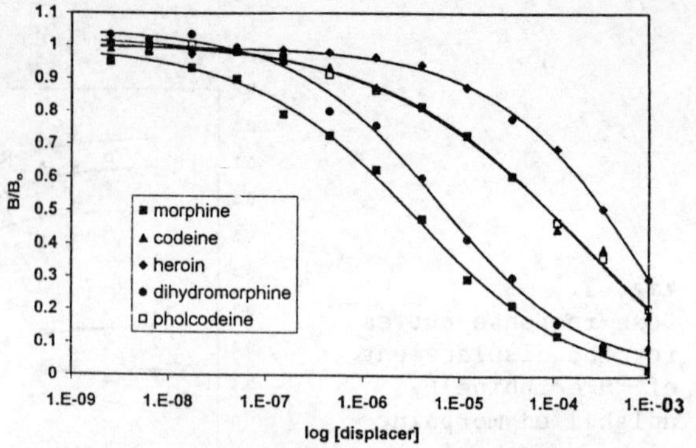

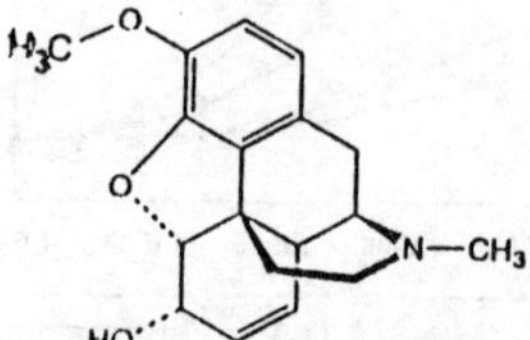

codeine

Fig. 3. Displacement of ^{3}H-morphine from polymer A-50 by cognate compounds.
IC_{50} μM value, and cross-reactivity (cf. value in ref. [5] for a T:M 1:4 polymer):- morphine 2.9, 100% (100%); dihydromorphine 8.9, 32%; codeine 72, 4%(4.7%); pholcodeine 79, 3.6%; heroin 550, 0.5% (2.3%).

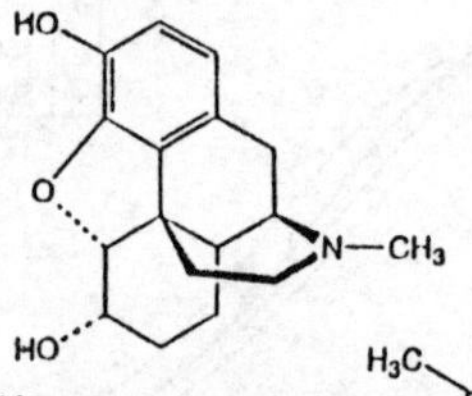

dihydromorphine

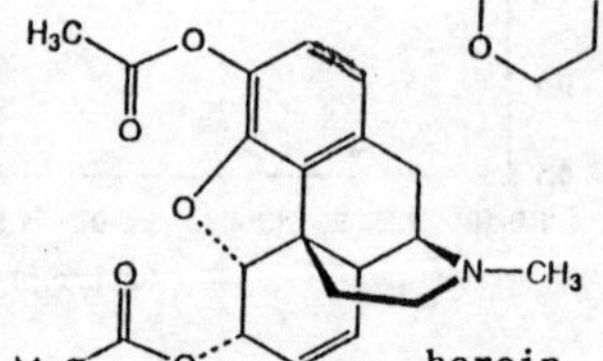

heroin

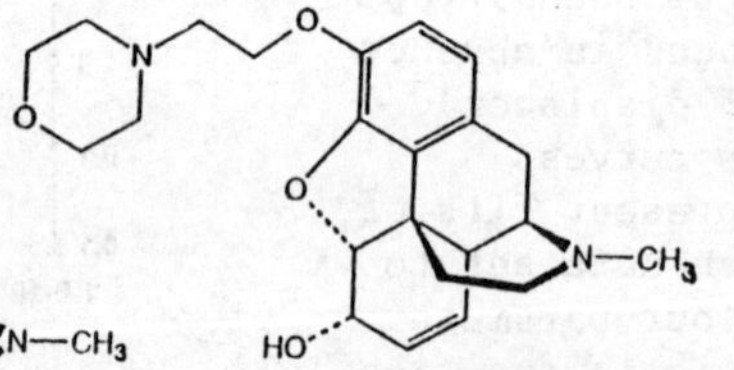

pholcodeine

Above and right: the compounds compared.

It seems from the available data that different polymers made to different recipes by different researchers in different laboratories had very similar affinities and specificities.

Template solubility.- Reducing the amount of template used to synthesize the MIP confers another benefit. With a conventional 1:4 ratio, the solubility of morphine is marginal, necessitating the use of the rather polar solvent MeCN in order to maintain the template in solution in the imprinting mixture prior to polymerization. As the T:M is reduced, the solubility improves since the excess MAA acts as a good solvent for morphine. This allows solvents with much lower polarity to be used than would otherwise be the case for morphine. At 1:50, even very apolar solvents such as cyclohexane could be used. To ascertain any influence on the recognition properties, binding curves (Fig. 4; B/T and B/B_0 plots) were compared for MIP's listed earlier. Those prepared in MeCN, chloroform or xylene gave curves of similar shape, but with cyclohexane no binding was apparent. Specific binding was highest for chloroform, since its reference polymer bound morphine much

Complementing the Key in this Fig., refer 4 pp. back (where solvents are specified)

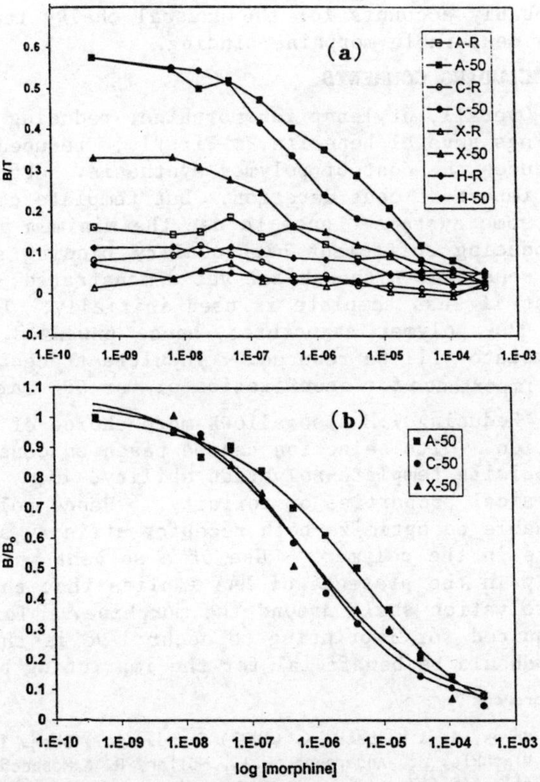

Fig. 4. Dose-response curves for ^{3}H-morphine displacement by unlabelled morphine from imprinted and reference (R) polymers. (**a**) Data as bound/total cpm, to show the absolute levels of binding. (**b**) Normalized curves to compare curve shapes: (cpm bound)/(cpm bound in absence of displacer). Curve-fitting as in Fig. 2 legend.

less than did the MeCN reference. This suggests that there is a higher concentration of accessible high-affinity binding sites with imprinting in chloroform compared with MeCN or xylene. The xylene and cyclohexane reference polymers showed rather little binding. The reason for the variable levels of binding to the different reference polymers is currently unclear and requires further investigation.

The IC_{50} µM values were 4.7 with MeCN, 0.7 with chloroform and 0.6 with xylene. This is consistent with a higher receptor affinity when polymerization-solvent polarity decreases since the strength of the bonds between template and functional monomers during polymerization becomes stronger in a less polar solvent. Clearly more data will be required before definite conclusions can be drawn about the reason for this increased affinity, since solvents have effects other than modulating template-functional monomer interactions. This is clearly shown by the polymer synthesized in cyclohexane - a very poor solvent for growing methacrylate polymer chains. It probably causes polymer precipitation before a continuous polymer network can form. This would lead to a dispersion polymerization, giving non-porous particles, rather than formation of a continuous macroporous network, and probably accounts for the unusual chalky texture and the lack of any measurable morphine binding.

CONCLUDING COMMENTS

Overall, at least for morphine, reducing T:M during imprinting brings several benefits. Firstly, reduced template consumption reduces the cost of polymer synthesis. For morphine this is not an important consideration, but template cost could be important in some systems: hence to use the minimum amount consistent with producing sufficient high-quality binding sites for assays must be sensible. Although not yet demonstrated, one might also expect that if less template is used initially, less will be entrapped in the polymer structure; hence unwanted leaching of residual template will be reduced. In ultra-trace analysis this could be of importance for quantitation or for SPE sample clean-up.

Reducing T:M also allows more choice of solvent for polymerization. Its selection can be based on considerations other than requisite template-solvation ability, e.g. porogenic properties, physical properties or toxicity. Hence polymers should be fine-tunable to optimize both receptor affinity and, e.g., diffusion rate in the polymer. Use of a solvent which dissolves morphine only in the presence of MAA implies that the MAA must be forming a solvation shell around the morphine. This is exactly what is required for imprinting to occur; so if this situation obtains, it should be beneficial for the imprinting process.

References

1. Mayes, A. & Mosbach, K. (1997) Trends Anal. Chem. **16**, 321-332.
2. Vlatakis, G., Andersson, L.I., Müller, R. & Mosbach, K. (1993) Nature **361**, 645-647.
3. Mayes, A.G., Andersson, L.I. & Mosbach, K. (1994) Anal. Biochem. **222**, 483-488.
4. Andersson, L.I. (1996) Anal. Chem. **68**, 111-117.
5. Andersson, L.I., Müller, R., Vlatakis, G. & Mosbach, K. (1995) Proc. Nat. Acad. Sci. USA **92**, 4788-4792.
6. Siemann, M., Andersson, L.I. & Mosbach, K. (1996) J. Agric. Food Chem. **44**, 141-145.
7. Wulff, G., Vietmeier, J. & Poll, H-G. (1987) Makromol. Chem. **188**, 731-740.
8. Sellergren, B. (1989) Makromol. Chem. **190**, 2703-2711.
9. Mayes, A. & Mosbach, K. (1996) Anal. Chem. **68**, 3769-3774.
10. Kempe, M. & Mosbach, K. (1991) Anal. Lett. **24**, 1137-1145.
11. Andersson, H.S., Koch-Schmidt, A-C., Ohlson, S. & Mosbach, K. (1996) J. Mol. Recognit. **9**, 675-682.
12. Sellergren, B. & Shea, K.J. (1995) J. Chromatog. **A 690**, 29-39.
13. Kim, Y. (1996) J. Am. Chem. Soc. **118**, 1522-1528.
14. Stökli, A., Meier, B.H., Kreis, R., Meyer, R. & Ernst, R.R. (1990) J. Chem. Phys. **93**, 1502-1520.
15. O'Shannessy, D.J., Ekberg, B. & Mosbach, K.(1989) J. Mol. Recognit. **2**, 1-5.
16. Sellergren, B. & Shea, K.J. (1993) J. Chromatog. **A 635**, 31-49.
17. Christopoulos, T.K. & Diamandis, E.P. (1996) in 'Immunoassay' (Diamandis, E.P. & Christopoulos, T.K., eds.), Academic Press, San Diego, pp. 40-46.
18. Davies, C. (1994) in 'The Immunoassay Handbook' (Wild, E., ed.), Stockton Press, New York, pp. 25-32.

#A-5

A HIGH-THROUGHOUT SCREENING TECHNIQUE EMPLOYING MOLECULARLY IMPRINTED POLYMERS AS BIOMIMETIC SELECTORS *

M.A.E. Bowman, C.J. Allender, K.R. Brain & C.M. Heard†

Regio and Enantio Selective Interactions,
Science & Technology Group, Welsh School of Pharmacy,
University of Wales Cardiff, Redwood Building,
King Edward VII Avenue, Cardiff CF1 3XF, U.K.

A novel LC-based method is described for the simultaneous high-throughput biomimetic screening and separation of complex mixtures, including combinatorial libraries. The method utilizes the molecular imprinting technique to create highly specific chromatographic materials capable of identifying and isolating potential compounds for further biological evaluation. A model system, targeting compounds capable of β-adrenergic receptor antagonism, was prepared and subsequently evaluated using a small library of structurally diverse compounds.

The advent of combinatorial synthesis has resulted in the production of massive compound libraries [1] needed in the pharmaceutical industry for evaluating biological activity. To obviate bottlenecks in the drug discovery cycle, high-throughput methods for screening/separation must be developed. This article describes such a method which is based upon a novel extension of the existing technique of molecular imprinting [2, 3].

Natural molecular recognition events between ligands and receptor surfaces rely upon the simultaneous cooperative interaction of numerous, relatively weak, non-covalent bonds [4]. Hence such systems are extremely sensitive to the correct ligand shape and functional group orientation in 3-D space. Attempts have recently been made to re-create such selective binding sites by the molecular imprinting technique.- A template molecule (e.g. a natural receptor agonist) is allowed to form a solution complex with acrylate-based monomers possessing complementary chemical functionalities (the non-covalent approach [2]); a solvent, often chloroform, is chosen so as to maximize the strengths of the non covalent interactions (especially hydrogen bonding) involved. The addition of an excess of a suitable cross-linker and a catalytic quantity of an initiator is followed by UV-induced polymerization.

---

†addressee for any correspondence

**Abbreviations*.- MIP, molecularly imprinted polymer (atenolol: ATMIP; propranolol: PRMIP; S-timolol: TIMIP; NIP: non-imprinted; HYMIP, hybrid. k', capacity factor; α, separation factor.

This process results in a matrix which, following removal of the template, possesses receptor-like recognition sites (Fig. 1). Suitable grinding and processing of these polymers gives powders which can be slurry-packed into LC columns. With such columns, the affinity/selectivity of these artificial receptors has been studied [5, 6].

As for Ab's, MIP's exhibit cross-reactivity [7]. Thus, just like their natural counterparts, they can bind molecules that are similar but not identical to the original template in architecture and functionalities. This phenomenon was recently investigated by us [8] using MIP's generated from Boc-protected amino acids (Fig. 2). We concluded that MIP cross-reactivity was exploitable for developing a novel high-throughput screening technique. Such an LC-based method has the added advantage of separating simultaneously the active molecules from the other components of the library. The β-adrenergic receptor [9] was the chosen target for study (cf. the reported preparation of MIP's using commercially available β-blockers as templates [10]). In the first stage, three arbitrarily selected, high-affinity ligands specific for this receptor were chosen as templates: ±-atenolol, ±-propranolol and S-timolol. The individual MIP's obtained were dry-mixed in order to create a hybrid MIP, thus providing a better overall representation of the natural receptor. The screening/separation process then consisted of eluting compounds through an LC column containing the hybrid MIP, retention being longest with candidates for further biological evaluation (Fig. 3).

The selectivity of the hybrid column was initially evaluated by elution of 9 commercially available β-blockers. Its properties were compared to those of columns prepared from each individual β-blocker MIP. A column containing a non-imprinted polymer (i.e. one prepared in the absence of a template molecule; NIP) was also evaluated as a control. The mixed-MIP and NIP columns were then evaluated using a combinatorial bookshelf of some 27 compounds with diverse structures and biological actions [see Merck Index].

MATERIALS AND METHODS

Materials.- Monomers and the 'bookshelf' compounds came from Sigma-Aldrich (Gillingham, UK), azoisobutyronitrile (AIBN) from Janssen Chimica (Geel, Belgium) [all of analytical grade], and solvents [HPLC grade] from Fisher (Loughborough, UK).

Polymerization [method: 11].- The reaction mixture comprised 2 mmol of the imprint molecule, 12 mmol methacrylic acid, 50 mmol ethylene glycol dimethacrylate, 0.56 mmol AIBN and 20 ml chloroform. Polymerization was done in crimp-sealed borosilicate vials after ultrasonic/vacuum degassing and N_2 sparging. The sealed vials were chilled and irradiated (366 nm) overnight at 5°. The polymers were removed from the vials, coarsely ground by hand and allowed to dry overnight under vacuum. Each was wet-rotary ball-milled for 4 h at 90 rpm in a 500 ml porcelain vessel containing

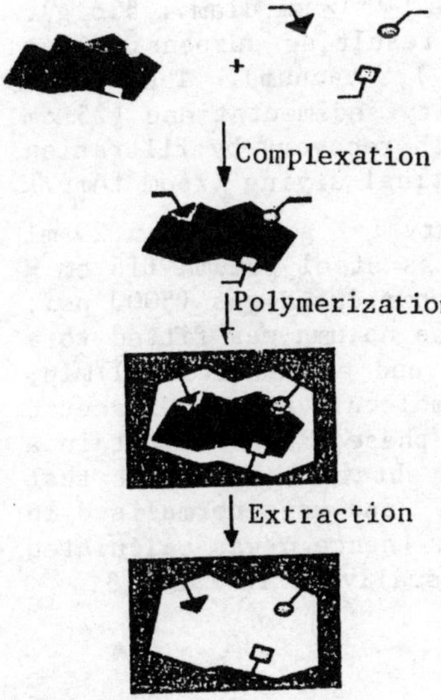

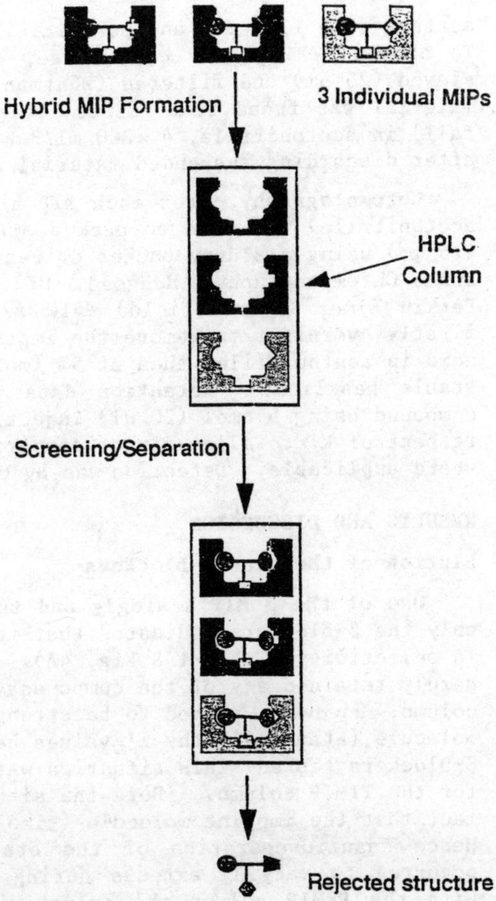

Fig. 1, *above.* The molecular imprinting process.

Fig. 2, *below.* Scatchard plots illustrating cross-reactivity in MIP's generated from Boc-protected amino acids.

Fig. 3, *right.* The screening process.

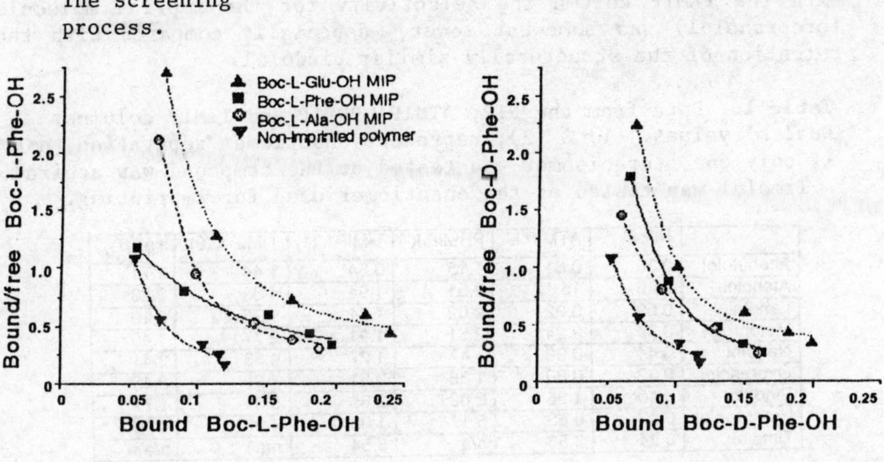

acetonitrile (150 ml) and porcelain balls (10-12 mm diam.; 536 g). To obtain MIP apt for column use, the resulting suspension was sieved (25 µm) and filtered (Whatman No. 1, vacuum). The <25 µm material was freed from 'fines' by gravity sedimentations (25 cm fall) in acetonitrile, 4 ×200 ml/3 h, with recovery by filtration after discarding suspended material, and final drying (room temp.).

Chromatography.- For each MIP a slurry (~1 g powder in 20 ml acetonitrile) was used to pack a stainless steel column (15 cm × 4.6 mm) using a slurry packer driven by compressed gas (5000 psi; Jones Chromatography, Hengoed, UK). The column was fitted to a Perkin Elmer (Beaconsfield) #410 system and eluted at 1 ml/min, firstly overnight to remove the imprint molecule using 20% acetic acid in acetonitrile, then at 5% (mobile phase; <2 h) to attain a stable baseline. Retention data were obtained for each test compound using 5 nmol (20 µl) injections, and were normalized in respect of k' to allow direct comparison; thence α was calculated where applicable. Detection was by UV, usually as in Table 3.

RESULTS AND DISCUSSION

Elution of the nine β-blockers

Use of the 3 MIP's singly and the NIP.- Initial results using only the β-blockers indicated that the MIP's differed considerably in selectivity (Table 1 & Fig. 4A). As expected, the NIP column hardly retained any of the compounds eluted through it. The ATMIP column, however, proved to be strongly selective for the imprint molecule (atenolol), the k' values being much lower for the other β-blockers tested. This situation was mirrored to a lesser extent for the TIMIP column. Here the situation was complicated by the fact that the imprint molecule (timolol) was a single enantiomer. Hence enantioseparation of the other test compounds (racemic) occurred to varying extents during elution through the column. With the PRMIP column the selectivity for the imprint molecule (propranolol) was somewhat lower, especially compared with the retention of the structurally similar pindolol.

Table 1. Data from the NIP, ATMIP, PRMIP and TIMIP columns. Dual k' values, (1) & (2), represent enantiomer separation, n/a if only one stereoisomer was tested or the compound was achiral. Timolol was tested as the enantiomer used for imprinting.

	NIP k'	ATMIP k'	PRMIP k'	TIMIP k'(1)	TIMIP k'(2)	TIMIP α
Acebutolol	0.36	0.61	1.15	0.84	1.44	1.71
Atenolol	0.48	15.12	2.51	1.94	4.33	2.23
Labetalol	0.59	0.92	2.02	1.71	3.75	2.19
Metoprolol	0.39	2.24	2.51	1.91	2.77	1.45
Nadolol	0.44	0.90	1.47	1.27	4.33	3.41
Oxprenolol	0.32	0.80	1.36	0.51	1.25	2.45
Pindolol	0.40	1.66	2.83	2.85	5.04	1.77
Propranolol	0.42	0.63	5.17	1.03	4.22	4.10
Timolol	0.34	3.56	n/a	8.54	n/a	n/a

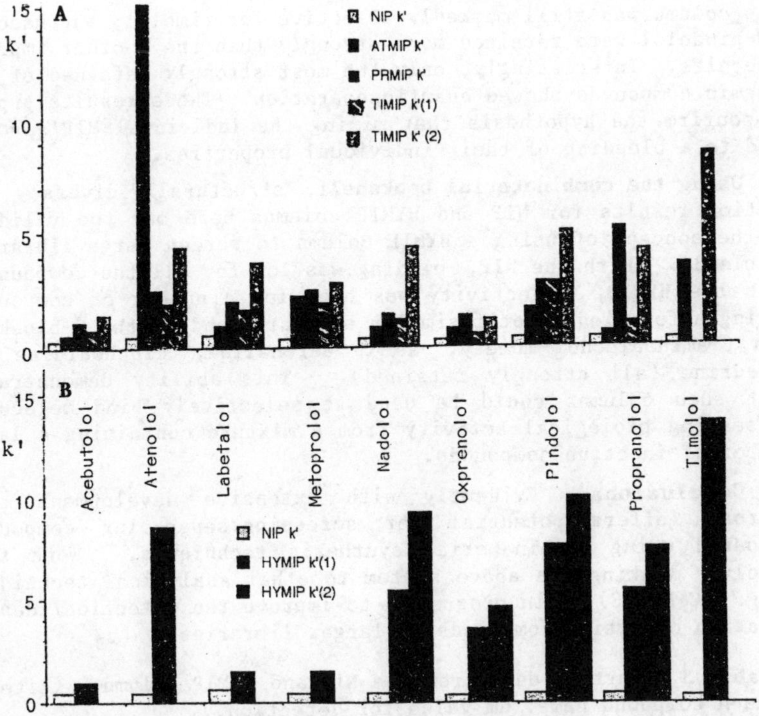

Fig. 4. Representations of data from (A) Table 1 and (B; hybrid MIP) Table 2, for 9 β-blockers (racemic except for timolol). In each set: first (grey) bar, NIP column; then (A): ATMIP, PRMIP and (dark grey; twin if enantiomers separated) TIMIP; (B): hybrid MIP. Detection was at 275 nm.

Table 2. As for Table 1, but NIP and hybrid (HYMIP) columns.

	NIP k'	HYMIP k'(1)	HYMIP k'(2)	HYMIP α
Acebutolol	0.36	0.94	n/a	n/a
Atenolol	0.48	8.64	n/a	n/a
Labetalol	0.59	1.46	n/a	n/a
Metoprolol	0.39	1.45	n/a	n/a
Nadolol	0.44	5.39	9.33	1.73
Oxprenolol	0.32	3.51	4.63	1.32
Pindolol	0.40	7.70	10.15	1.32
Propranolol	0.42	5.54	7.61	1.37
Timolol	0.34	13.82	n/a	n/a

Use of hybrid MIP (HYMIP) and NIP columns.- The k' values for the HYMIP column (Table 2 & Fig. 4B) were notably interesting since they appeared to vary in a highly complicated fashion when compared to the results obtained with the individual MIP columns.

This column was still markedly selective for timolol; but nadolol and pindolol were retained more strongly than the 2 other imprint molecules. Interestingly, only the most strongly retained of the racemic compounds showed enantioseparation. These results appear to confirm the hypothesis that mixing the individual MIP's would lead to a blending of their individual properties.

Using the combinatorial bookshelf, structurally diverse, the elution results for NIP and HYMIP columns bore out the validity of the concept of using a HYMIP column to screen large libraries (Table 3). With the NIP, binding was low for all the compounds. With the HYMIP, selectivity was high for a number of compounds having a functional motif similar to that found in the β-blockers - a β-aminoalcohol moiety, as in adrenaline, cinchonidine and ephedrine (all strongly retained). This ability demonstrates that such columns could be used to selectively bind molecules possessing biological activity from a mixture containing a large number of inactive compounds.

Conclusions.- Evidently with extensive development this approach offers potential for screening/separating compounds prepared using combinatorial synthesis techniques. Work that involves linking the above system to other analytical techniques (e.g. LC-MS-MS) is in progress, to improve the detection/identification of active compounds in larger libraries.

Table 3. Further data from the NIP and HYMIP columns. (After each compound name: nm value for detection.)

	NIP k'	HYMIP k'(1)	HYMIP k'(2)	HYMIP α
Acebutolol (275)	0.36	0.94	n/a	n/a
Adrenaline (280)	0.48	5.39	n/a	n/a
Atenolol (275)	0.48	8.64	n/a	n/a
Atropine (258)	0.19	0.93	n/a	n/a
Chlorpheniramine (265)	0.23	1.12	n/a	n/a
Cinchonidine (235)	1.01	4.27	n/a	n/a
Cinchonine (236)	0.97	3.74	n/a	n/a
Ephedrine (257)	0.31	3.30	n/a	n/a
Hydrocortisone Acetate (240)	0.42	0.37	n/a	n/a
Ibuprofen (265)	0.19	0.13	n/a	n/a
Indapamide (242)	0.28	0.25	n/a	n/a
Ketoprofen (260)	0.21	0.19	n/a	n/a
Labetalol (275)	0.59	1.45	n/a	n/a
Metoprolol (275)	0.39	1.45	n/a	n/a
Nadolol (275)	0.44	5.39	9.33	1.73
Oxprenolol (275)	0.32	0.51	4.63	1.32
Pindolol (275)	0.40	7.70	10.15	1.32
Proguanil (259)	0.38	1.23	n/a	n/a
Promethazine (249)	0.32	1.29	n/a	n/a
Propranolol (275)	0.42	5.54	7.61	1.37
Quinidine (250)	0.90	2.56	n/a	n/a
Quinine (250)	0.88	2.52	n/a	n/a
Theophylline (270)	0.40	0.38	n/a	n/a
Timolol (275)	0.34	13.82	n/a	n/a
Boc-Tryptophan (275)	0.23	0.15	n/a	n/a
Boc-Tyrosine (275)	0.17	0.20	n/a	n/a
Verapamil (278)	0.05	1.10	n/a	n/a

Acknowledgement- The authors thank the Wellcome Trust for providing a Sir Henry Wellcome Commemorative Award for Innovative Research.

References

1. Floyd, C.D., Lewis, C.N. & Whittaker, M. (1996) Chem. Br. **32**(3), 31-35.
2. Mosbach, K. & Ramström, O. (1996) Biotechnology **14,** 163-170.
3. Wulff, G. (1995) Agnew. Chem. Int. Ed. Eng. **34,** 1812-1832.
4. Alberts, B., Bray, D., Lewis, J., Raff, M., Roberts, K. & Watson, J.D. (1994) in 'Molecular Biology of the Cell', 3rd edn. (Alberts, B., ed.), Garland, New York, pp. 89-98.
5. Cheong, S.H., McNiven, S., Rachov, A., Levi, R., Yano, K. & Karube, I. (1997) Macromolecules **30,** 1317-1322.
6. Matsui, J., Nicholls, I.A. & Takeuchi, T. (1996) Tetrahedron Asymmetry **7,** 1357-1361.
7. Vlatkis, G. (1993) Nature **361,** 645-647.
8. Allender, C.J., Brain, K.R. & Heard, C.M. (1997) Chirality **9,** 233-237.
9. Main, B.G. (1990) in 'Comprehensive Medicinal Chemistry', Vol. 3 (Hansch, C., ed.), Pergamon, Oxford, pp. 187-228.
10. Fischer, L., Müller, R., Ekberg, B. & Mosbach, K. (1991) J. Am. Chem. Soc. **113,** 9358-9360.
11. O'Shannessey, D., Ekberg, B. & Mosbach, K. (1989) Anal. Biochem. **177,** 144-149.

#A-6

NEW IMPRINTING STRATEGIES

[1]Evgeny N. Vulfson, [1]Michael J. Whitcombe,† [1]Markus Lübke, [1]Lourdes Martin, [1]Ixiar Esteban, [2]Jae-Yeong Ju, [2]Chul Soo Shin & [3]Jens-Uwe Klein

[1]Institute of Food Research, Earley Gate, Whiteknights Road, Reading RG6 6BZ, U.K.

[2]Dept. of Food & Biotechnology, College of Engineering, Yonsei University, Seoul, South Korea

[3]Institute of Organic Chemistry & Biochemistry, Czech Academy of Sciences, Flemingovo 2, 16610 Prague 6, Czech Republic

Improvements in the performance of imprinted polymers are critically dependent on the design of efficient chemical strategies. Here we describe how new methods involving 'sacrificial spacers' can be used to introduce a range of chemical functionalities in a polymer recognition site. In addition, the use of novel monomers in non-covalent imprinting, designed to interact with specific template functional groups, has been demonstrated for a range of template structures.

Molecular imprinting as a methodology for the preparation of highly selective polymeric adsorbents has received much attention in recent years [1]. Most of the research has focused on the synthesis of polymers capable of resolving racemic mixtures [2]; but other applications have been reported, notably in SPE [3], catalysis [4], sensors [5], removal of undesirable components from complex mixtures [6] and separation and concentration of proteins [7] and micro-organisms [8].

Conceptually, molecular imprinting relies upon the template-directed self-assembly of functional monomers in the polymerization mixture. This enables the positioning of complementary functionality in what, after the polymerization, becomes the polymer's recognition site in the precise spatial arrangement to ensure specific interactions with the functional groups of the template on rebinding. Hence it is the strength of the initial interactions between template and the polymer-forming components which largely determines the recognition properties of the imprinted polymer [9]. So progress in this field and in our ability to produce ever more selective materials is crucially dependent on the design of new, superior monomers, capable of forming stronger complexes with the template, and on the development of novel imprinting strategies for positioning complementary functionalities in the recognition sites. Both approaches have been actively investigated in our laboratory in the last few years. Some preliminary results and conclusions appear below.

†correspondent; e-mail: michael.whitcombe@bbsrc.ac.uk

RESULTS AND DISCUSSION

Traditionally imprinted polymers have been prepared either by the covalent method, pioneered by Wulff's group [10], or through non-covalent assembly of the template-polymer complexes. The later methodology was introduced by Mosbach's group [11] and, due to its simplicity and versatility, is most commonly used for preparing imprinted polymers. The relative merits of the two methodologies have been discussed in detail in numerous reviews; but the major advantages and drawbacks still warrant mention. The great attraction of the non-covalent method is that there is no requirement for organic synthesis. However, the binding sites obtained are usually diverse, in terms of their affinity for the ligand, due to the formation of a multitude of complexes between monomers and template during the initial stages of polymerization. For some practical applications this may pose a serious problem. Also, the 'quality' of non-covalently imprinted polymers is very dependent on the polymerization solvents as it is necessary to shift the equilibrium towards complex formation. Although the two imprinting methods rely on very dissimilar monomers, and the classes of template that are targeted are often very different too, they share one common feature: they strive to produce an 'exact fit' recognition in the sense that the ligand, when it binds to the polymer, reconstitutes exactly the same molecular interactions with the functional groups in the site that had been used for its creation.

The 'sacrificial spacer' strategy

Recently we have developed an alternative imprinting strategy that relies on the use of a covalent template monomer to create a non-covalent recognition site. The feasibility of this approach was first demonstrated by the synthesis of cholesterol-specific polymers bearing a phenolic residue in the active sites, using the 4-vinylphenylcarbonate ester of cholesterol as a template [12]. The labile carbonate ester bond can be easily and efficiently cleaved hydrolytically with the loss of CO_2 which acts as a sacrificial spacer. This results in liberation of a phenolic residue capable of interacting with cholesterol through hydrogen bonding. The polymers prepared by this method were shown to bind cholesterol with a single dissociation constant of 0.6 mM [12]. The principal advantage of this 'sacrificial spacer' methodology is that it is applicable to a wide range of templates, and diverse functionalities such as amino-, carboxyl-, hydroxyl- and amido-groups can be incorporated in the polymer's recognition site, as illustrated by the following examples.

There are several reasons why polychlorinated aromatic compounds are interesting targets for imprinting. Firstly, they possess functionality which is extremely difficult to engage using conventional non-covalent methods and monomers. Indeed, the aromatic chlorines, although capable of forming exceedingly weak hydrogen bonds in non-polar solvents [13], are unlikely to participate in complexation with (meth)acrylic acid or (meth)-acrylamide. Secondly, the presence of polychlorinated aromatic compounds (polychlorodibenzodioxins, polychlorodibenzofurans and polychlorobiphenyls) in the environment and in the food chain is of great concern, and the development of a generic approach to the imprinting of these molecules would be of great practical value. In our work, 2,3,7,8-tetrachlorodibenzodioxin (TCDD) was chosen as the target, representative of this class of compounds. It is the most prominent member of a large family of polychlorinated dibenzodioxins which are currently the subject of extensive monitoring due to their extraordinary mammalian toxicity [14].

The 'sacrificial spacer' methodology was suitably modified such that hydrogen bond-forming monomers were incorporated into the polymer's recognition site to ensure the interactions with aromatic chlorines (Fig. 1). The resulting polymers showed a clear imprinting effect when challenged with radioactively labelled TCDD at pM concentrations. Even better results were obtained with polymers prepared in the presence of specially synthesized cross-linkers to form charge-transfer complexes with the template in the polymerization [15].

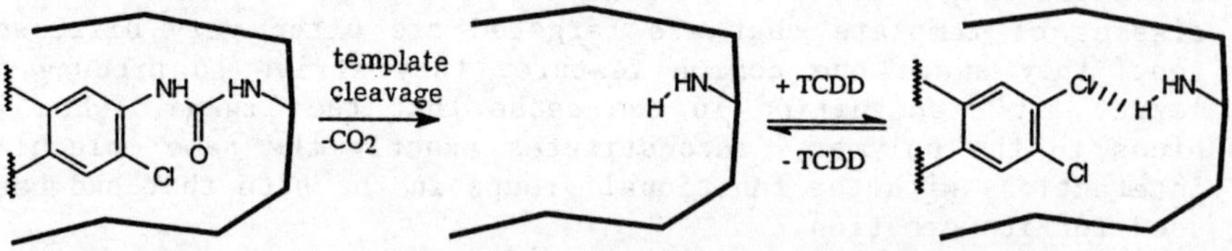

Fig. 1. The 'sacrificial spacer' methodology as applied to the imprinting of TCDD, incorporating amino groups in the polymer recognition site (schematic).

Another example of the utility of the 'sacrificial spacer' methodology when used in conjunction with conventional non-covalent imprinting is the preparation of polymers specific to a particular amino acid sequence. In the simplest case, an oligo-peptide template can be engaged through two types of interactions, covalently on the N-terminus and non-covalently on the C-terminus (or conversely). The presence of additional multifunctional amino acids in the sequence, e.g. lysine, arginine and aspartic or glutamic acid, provide extra points of interaction, using essentially the same chemistry/monomers for targeting the terminal and side-chain NH_2 (or COOH) groups of lysine (or aspartic acid) respectively. Fig. 2 shows a typical arrangement of functionality in the polymer's recognition site.

It is emphasized that although either end of the oligopeptide template can be used for covalent modification *via* a 'sacrificial spacer', in our work we opted for the modification of the amino group. A detailed characterization of the polymers thus prepared is still ongoing but the initial results are very encouraging. It was shown that the polymers display a high degree of selectivity and bind the template well even in aqueous solution. Moreover, the substitution of key amino acids in the sequence, especially those which were engaged covalently, has a dramatic effect on the binding characteristics. Functionally identical non-imprinted polymers show virtually no binding of the template and its structural analogs under the same assay conditions.

Fig. 2. Imprinted site for a generalized tripeptide, incorporating acidic and basic polymer functionality to interact with the target's C and N termini and side-chain functional groups (shown in the neutral form for the sake of clarity).

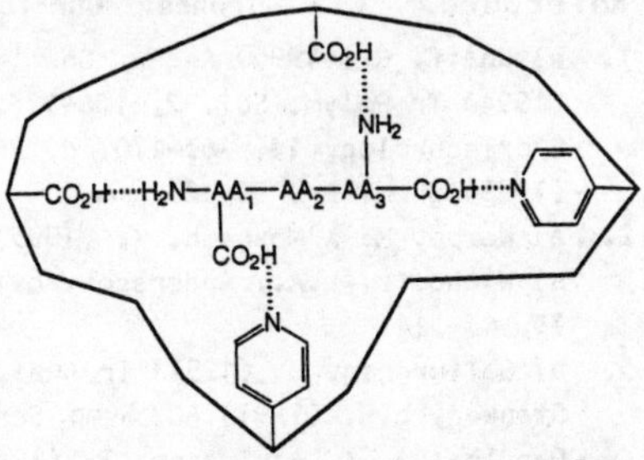

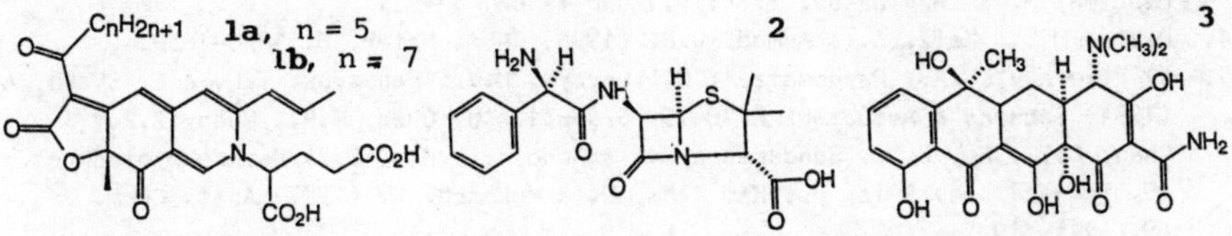

Fig. 3. Structures of some template molecules. Pigments from *Monascus spp.* fungi (**1**), ampicillin (**2**) and tetracycline (**3**).

Functional monomer innovations

An alternative approach to improving the selectivity of imprinted polymers is the design of new functional monomers which are capable of forming stronger non-covalent complexes with the template. In order to demonstrate the validity of this approach we have prepared a range of vinyl derivatives of conventional low mol. wt. receptors used in host-guest chemistry to provide efficient hydrogen bonding. These syntheses yielded several promising monomers which are currently being investigated in the imprinting of natural products; a few are depicted in Fig. 3. So far, carboxyl and amino groups have been targeted and good specificity was observed in batch experiments, even when carried out in the presence of aqueous buffers. For example, polymers imprinted with (L)-CBZ-aspartic acid [CBZ = **N**-carbobenzoxy] bound 20% more of the (L)-isomer than the (D)-amino acid in individual batch experiments (10 mg/ml polymer) at mM concentrations, whereas the (D)-imprinted material showed the opposite specificity. The optimization of these polymers and their potential for preparative resolutions are currently being investigated in detail in our laboratory at Reading.

CONCLUSION

The design of new covalent imprinting strategies, based on the sacrificial spacer approach, and the synthesis of new monomers for non-covalent imprinting, targeted to specific functional groups, has enabled us to imprint a wide range of structurally diverse templates. The resulting polymers show good specificity in batch binding experiments and may prove useful in downstream processing, SPE, racemic resolution, analyses and chromatography.

References [**Tr** = Trends; **Chr'ia** = Chromatographia]

1. a) Wulff, G. (1995) Angew. Chem. Int. Ed. **34,** 1812-1832. b) Shea, K.J. (1994) Tr Polym. Sci. **2,** 166-173. c) Mosbach, K. & Ramström, O. (1996) Bio/Technology **14,** 163-170. d) Vulfson, E.N., Alexander, C. & Whitcombe, M.J. (1997) Chem. Br. **33,** 23-26.
2. a) Kempe, M. & Mosbach, K. (1995) J. Chromatog. **A 691** 317-323 & **694,** 3-13. b) Nicholls, L.A., Andersson, L.I., Mosbach, K. & Ekberg, B. (1995) Tr Biotech. **13,** 47-51.
3. a) Sellergren, B. (1997) Tr Anal. Chem. **16,** 310-320. b) Muldoon, M.T. & Stanker, L.H. (1997) ACS Symp. Ser. **657,** 314-330. c) Andersson, L.I., Paprica, A. & Arvidsson, T. (1997) Chr'ia **46,** 57-62.
4. Davis, M.E., Katz, A. & Ahmad, W.R. (1996) Chem. Mater. **8,** 1820-1839.
5. a) Piletsky, S.A., Parhometz, Y.P., Lavryk, N.V., Panasyuk, T.L. & El'Skaya, A.V. (1994) Sensors & Actuators **B 18-19,** 629-631. b) Chen, G.H., Guan, Z.B., Chen, C.T., Fu, L.T., Sundaresan, V. & Arnold, F.H. (1997) Nature Biotechnol. **15,** 354-357. c) Kriz, D., Ramström, O. & Mosbach, K. (1997) Anal. Chem. **69,** A345-349.
6. Whitcombe, M.J., Alexander, C. & Vulfson, E.N. (1997) Tr Food Sci. Technol. **8,** 140-145.
7. a) Mallik, S., Plunkett, S.D., Dhal, P.K., Johnson, R.D., Pack, D., Schnek, D. & Arnold, F.H. (1994) New J. Chem. **18,** 299-304. b) Hjerten, S., Liao, J.L., Nakazato, K., Wang, Y., Zamaratskaia, G. & Zhang, H.X. (1997) Chr'ia **44,** 227-234.
8. a) Aherne, A., Alexander, C., Payne, M.J., Perez, N. & Vulfson, E.N. (1996) J. Am. Chem. Soc. **118,** 8771-8772. b) Alexander, C. & Vulfson, E.N. (1997) Adv. Mater. **9,** 751-755.
9. Whitcombe, M.J., Martin, L. & Vulfson, E.N. (1998) Chr'ia **47,** in press.
10. Wulff, G. (1982) Pure Appl. Chem. **54,** 2093-2102.
11. Mosbach, K. (1994) Tr Biochem. Sci. **19,** 9-14.
12. Whitcombe, M.J., Rodriguez, M.E., Villar, P. & Vulfson, E.N. (1995) J. Am. Chem. Soc. **117,** 7105-7111.
13. Smith, J.W. (1973) in 'The Chemistry of the Carbon-Halogen Bond' (Patai, S., ed.), Wiley, Chichester, Ch. 5.
14. Safe, S.H. (1986) Ann. Rev. Pharmacol. Toxicol. **26,** 371-399.
15. Lübke, M., Whitcombe, M.J. & Vulfson, E.N. (1998) J. Am. Chem. Soc., in press.

Editor's add-on to the foregoing broad survey

Conditions in refs. 1d & 12 for use of a cholesterol ester as template monomer: 24 h at 65° in the presence of AIBN (Fluka) and ethyleneglycol dimethacrylate (EGMDA); hydrolysis by reflux with 1 M NaOH/methanol, leaving a 4-vinylphenol moiety behind.

#A-7n

A Note on

MOLECULAR IMPRINTED POLYMER FOR SOLID PHASE EXTRACTION OF TAMOXIFEN

D. Stevenson†, R.J. Briggs, ▫J. Hay & B. Rashid

School of Biological Sciences & ▫School of Physical Sciences, University of Surrey, Guildford GU2 5XH, U.K.

Although analytical instrumentation continues to develop, the analysis of trace organics in biofluids usually requires some form of sample preparation. SPE is becoming the method of choice, partly due to the availability of automated equipment. Recently there has been interest in the development of SPE procedures using immobilized Ab's* as the sorbent [1, 2, & following art.]. These can give highly selective extraction and clean HPLC traces. The major disadvantage of this approach is that it requires a supply of Ab to the analyte of interest, with unpredictability of time needed to produce it, its quality or even the success rate.

For highly selective extraction, an alternative to Ab use is the MIP approach, surveyed in reviews [e.g. 3] and accompanying articles. The polymer, freed from the analyte template and then ground, serves for packing HPLC columns or, more recently, SPE columns. Drugs that have furnished MIP's include propranolol [4], morphine [5], tamoxifen [6] and naproxen [7]. Here we describe trial study of a tamoxifen MIP for use as a selective SPE column.

EXPERIMENTAL

Preparation of the MIP.- A solution, in MeCN, of tamoxifen (0.56 g, 1 mmol) [omitted for control polymer], methacrylic acid (0.68 g, 8 mmol) and ethylene glycol dimethacrylic acid (8.0 g, 40 mmol), in a headspace sampling vial whose septum cap had gas inlet and outlet tubes, was purged for 5 min with N_2. Then 0.08 g (0.5 mmol) 2,2'-azobis-(2-methylpropiononitrile) initiator was added, and the capped mixture polymerized under N_2 on a heating block (60°, 16 h). The polymer obtained after breaking the bottle was ground (using a mortar and pestle), washed with MeCN (2 × 250 ml), freed from fines by sedimentation, and air-dried (room temp., fume cupboard). The MIP or control polymer (500 mg) was packed into 70 × 10 mm SPE columns, of plastic syringe type, for recovery evaluation which typically involved loading 1 ml tamoxifen solution in different solvents, washing, and eluting with different solutions.

Tamoxifen analysis on dried-down (N_2) eluates redissolved in 200 µl mobile phase was done by HPLC on a Spherisorb ODS-2 column (150 × 4.6 mm; 5 µm), using 80:20 methanol/pH 3 phosphate buffer containing 10 mM sodium heptanesulfonate (1.5 ml/min). The limit of detection (240 nm) was ~50 ng/ml.

†correspondent

**Abbreviations used.*- Ab, antibody; MeCN, acetonitrile; MIP, molecularly imprinted polymer.

Specificity experiments.- Each compound (1 µg in 1 ml MeCN) was loaded onto the MIP and control columns. After washing with MeCN (3 × 5 ml) and elution (3 × 5 ml) with 4:1 MeCN/glac. acetic acid, fractions were collected. These and the analyte-loading breakthrough were analyzed after the drying down/redissolution. The following model compounds were tested:

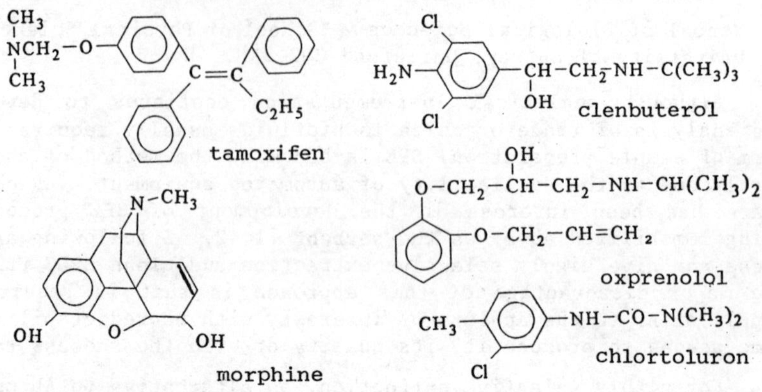

RESULTS AND DISCUSSION

Polymer preparation was very simple and essentially completed within 2 days. The combined cost of the reagents was <£50. This contrasts with the difficulty in obtaining immunoadsorbents where Ab's are unavailable. We did, however, experience considerable difficulty in washing all the tamoxifen template from the MIP column. After the MeCN washes specified above the MIP was washed with 2 × 250 ml 9:1 MeCN/acetic acid and then 9 × 100 ml methanol/ acetic acid (4:1). The washes were monitored by HPLC until no trace of tamoxifen could be found. Yet our early experiments gave total tamoxifen recoveries much greater than 100% from the MIP column but close to 100% from the control polymer, suggesting that not all the template could be removed from the MIP column despite the extensive washing. There was anecdotal evidence to suggest that if the MIP was used immediately after washing, tamoxifen leaching from the column was much less significant than if the MIP was used a few days later.

In our preliminary report on the tamoxifen MIP [6], careful optimization of elution solvent was necessary in order to retain the drug on the polymer and then elute. MeCN was a suitable wash solvent and 4:1 MeCN/acetic acid a suitable eluent. With the control polymer, 80% of the tamoxifen eluted in the breakthrough and wash fractions, clearly demonstrating a difference between the control and the MIP. In contrast with our immunoaffinity work using Ab's, 15 ml of elution solvent was needed to desorb the tamoxifen from the MIP. In all our immunoaffinity work we have been able to elute in as little as 1 ml of elution solvent.

Specificity experiment (Fig. 1; all 3 wash and elution fractions aggregated).- Evidently, even though for tamoxifen the best result is shown, the MIP retains structurally unrelated compounds. Although we have not yet prepared a tamoxifen immunoabsorbent for direct comparison, all our other work with immunoabsorbents suggests they are much more specific than the MIP.

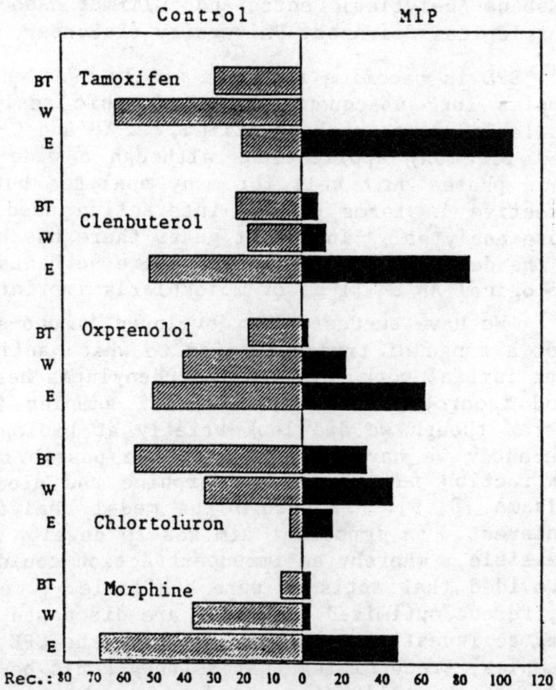

Fig. 1. Recovery (%) of model compounds from control and MIP column. BT, breakthrough; W, washes; E, eluate.

CONCLUSIONS

1. MIP's are easier to obtain than biological Ab's.
2. MIP's are more stable to organic solvents and pH extremes.
3. Difficulty in removal of analyte template could cause analytical errors.
4. Careful selection of the elution solvent is needed.
5. Specificity is not so good as for biological Ab's.
6. As newer methods of synthesis are optimized, these may well produce MIP's with better specificity characteristics.

References

1. Rashid, B.A., Kwasowski, P. & Stevenson, D. (1996) in 'Biofluid Assay for Peptide-Related and Other Drugs' [Vol. 24, this series] (Reid, E., Hill, H.M. & Wilson, I.D., eds.), Royal Society of Chemistry, Cambridge, pp. 219-223.
2. Rashid, B.A., Kwasowski, P. & Stevenson, D. (1996) Pharm. Sci. **2**, 115-116.
3. Sellergren, B. (1997) Trends Anal. Chem. **16**, 310-320.
4. Andersson, L.I. (1996) Anal. Chem. **68**, 111-117. [Cf. art. #A-1.]
5. Andersson, L.I., Muller, R., Vlatkis, G. & Mosbach, K. (1995) Proc. Nat. Acad. Sci. USA **92**, 4788-4792.
6. Rashid, B.A., Briggs, R.J., Hay, J.N., Stevenson, D. (1997) Anal. Comm. **34**, 303-305.
7. Kempe, M. & Mosbach, K. (1994) J. Chromatog. **A 664**, 277-279.

#A-8n

A Note on

ILLUSTRATIVE IMMUNOAFFINITY PROCEDURES

D. Stevenson†, B. Rashid, M.F. Katmeh & ᴾP. Kwasowski

Robens Analytical Centre and ᴾClifmar Associates, School of Biological Sciences, University of Surrey, Guildford GU2 5XH, U.K.

SPE is becoming the most widely used method to prepare bio-samples for subsequent chromatographic analysis. Commercially available phases such as silica, C-18 and C-8 bonded silicas are used for many applications although a wide range is available. These phases suit well for many analytes but are generally non-selective in terms of the interaction used to retain and then elute analytes. In recent years there has been growing interest in the development of more selective sorbents, e.g. those using biological Ab's* [1-3] or molecularly imprinted polymers [4-7].

We have successfully developed immuno-extraction procedures for a range of trace organics to which antisera were available. Our initial work involved the phenylurea herbicides chlortoluron and isoproturon in environmental samples [1, 2], particularly water though we did look briefly at biological samples. More recently we have investigated the possibility of using immuno-extraction for the drugs morphine and clenbuterol in urine and plasma [8, 9]. Although the model analytes were of intrinsic interest, an important aim was to develop a generic protocol if feasible, whereby an immunoextraction could be developed rapidly provided that antisera were available. Comparisons between the different optimized protocols are discussed below. A further aim was to investigate the stability of the SPE columns to biological samples, recognizing that as they could be reckoned expensive to prepare, ideally they should be usable several times. Immuno-sorbent deterioration has been monitored by measuring column capacity before and after subjecting the columns to biosamples. To determine analyte content, SPE eluates were subjected to HPLC.

EXPERIMENTAL

HPLC conditions for the analytes are fully described elsewhere [8]. Briefly, morphine was run on a Hypersil CPS column, 5 µm, 250 × 5 mm, using 13% (v/v) acetonitrile in pH 2.5 65 mM phosphate buffer containing 1.5 mM sodium lauryl sulfate (flow rate 1 ml/min). Detection was by a Coulochem ESA electrochemical detector at a potential of +0.45 V. Clenbuterol was run on a Spherisorb ODS-2 column, 5 µm, 150 × 5 mm, using 25% acetonitrile in 50 mM phosphate buffer, pH 2.5, containing 1% triethylamine (1 ml/min). Detection was by UV at 230 nm.

---

†addressee for any correspondence

**Abbreviations*.- Ab, antibody; PBS, phosphate-buffered saline.

Immunosorbents*.- The solid support was aldehyde-activated controlled-pore (10 μm) glass beads (100 μm diam)*. Typically the antiserum (morphine, 1 ml; clenbuterol, 0.5 ml) was added to 2 g of support (columns prepared by Clifmar Associates). Typical experiments involved loading 0.1-1 ml of sample onto the sorbent, washing with PBS at neutral pH, and then eluting with PBS at low pH containing 40-50% organic modifier (methanol or ethanol).

Binding capacity was determined in two ways. The frontal method entailed adding 1 ml drug solution of known concentration to the immunocolumn until breakthrough was observed; fractions were collected and analyzed by HPLC. Alternatively, the immunosorbent was loaded with excess drug and washed, then eluted.

RESULTS

Optimization of immuno-extraction (details in [8-10]) was done successfully with all immunocolumns (for four analytes) that we have prepared and evaluated. For the washing, pH 7 PBS was always used. For chlortoluron and isoproturon the adopted eluant was 50% ethanol in pH 2 PBS. For morphine it was 40% ethanol in pH 4 PBS, and for clenbuterol it was 50% methanol in pH 2 PBS. Notably the protocols for analyte retention and elution are evidently near-identical; in all cases loading is done close to neutral pH, removal of interferences by washing (if needed) is with neutral PBS, and elution is achieved in a 1 ml fraction by lowering the pH and adding ethanol or methanol. Clean HPLC traces were obtained for each analyte, whether from environmental waters in the case of the herbicides or from plasma and urine in the case of the two drugs. (For details see refs. [1, 2, 10].)

Column capacities and survival.- Gentle elution conditions are favoured as this allows columns to be used many times, as is shown by the results in Table 1 for 4 columns, two for each drug.

Table 1. Capacities of columns for morphine and clenbuterol.

Drug	Day no.	No. of extractions	Capacity, ng	
Morphine	0 (new)	1	Col. #1: 150;	#2: 150
	57	13 (cumulation)	126	117
	123	25 (")	73	57
Clenbuterol	0 (new)	1	Col. #1: 286	#2: 316
	14	13 (cumulation)	214	221
	55	48 (")	123	95

* Beads from Sigma (activation by Clifmar [10])

Evidently the immunocolumns can be used many times before capacity drops significantly. It is also notable that although quite low column capacities pertain in the field of trace-organic analysis, almost always it is low concentrations of analyte that cause difficulties, not high levels. The maximum capacities in Table 1 are certainly compatible with modern practice of HPLC for drug bioanalysis.

CONCLUDING COMMENTS

In our experience the development and optimization of immunoaffinity SPE procedures is certainly achievable. These can be based on a simple retention and elution protocol using aqueous-based solvents. Although Ab's are currently expensive to produce, the immunosorbents can be re-used. If suitable enzymic, fluorescence or radiolabel means of detection are available, immunoaffinity SPE may not need the subsequent chromatographic step before determination. Where some confirmation of identity is required, the coupling of immunoextraction with instrumental chromatography offers a powerful combination of techniques. The approach may prove the method of choice in future studies where analytes are water-soluble high-mol. wt. compounds difficult to extract by other procedures.

References

1. Shahtaheri, S.J., Katmeh, M.F., Kwasowski, P. & Stevenson, D. (1995) J. Chromatog. **A 697**, 131-136.
2. Martin-Esteban, A., Kwasowski, P. & Stevenson, D. (1997) Chromatographia **45**, 364-368.
3. Lingeman, H., van de Merbel, N.C. & Brinkman, U.A.Th. (1994) in 'Biofluid and Tissue Analysis for Drugs including Hypolipidaemics' [Vol. 23, this series] (Reid, E., Hill, H.M. & Wilson, I.D., eds.), Royal Society of Chemistry, Cambridge, pp. 315-326.
4. Martin, P., Wilson, I.D., Morgan, D.E., Jones, G.R. & Jones, K. (1997) Anal. Comm. **34**, 45-47. [Cf. art. #A-3, this vol.]
5. Andersson, L.I. (1996) Anal. Chem. **68**, 111-117. [Cf. art. #A-1, this vol.]
6. Rashid, B.A., Briggs, R.J., Hay, J.N. & Stevenson, D. (1997) Anal. Comm. **34**, 303-305. [Cf. preceding art., this vol.]
7. Kempe, M. & Mosbach, K. (1994) J. Chromatog. **A 664**, 277-279.
8. Rashid, B.A., Kwasowski, P. & Stevenson, D. (1996) in 'Biofluid Assay for Peptide-Related and Other Drugs' [Vol. 24, this series] (Reid, E., Hill, H.M. & Wilson, I.D., eds.), Royal Society of Chemistry, Cambridge, pp. 219-223.
9. Rashid, B.A., Kwasowski, P. & Stevenson, D. (1996) Pharmaceut. Sci. **2**, 115-116.
10. Rashid, B.A., Aherne, G.W., Katmeh, M.F., Kwasowski, P. & Stevenson, D. (1998) J. Chromatog. **A 797**, 245-250.

#A-9

MEMBRANE-BASED SAMPLE-PREPARATION PROCEDURES

Henk Lingeman

Dept. of Analytical Chemistry, Free University,
De Boelelaan 1083, 1081 HV Amsterdam, The Netherlands

For sample preparation, on-line and/or automated systems that typically precede LC and entail coupling [1] are becoming prevalent. Some are of flexible type, e.g. robotic systems. Others, with an autosampler, are fixed (inflexible) and can perform only one or two tasks, e.g. column switching. Semi-flexible types typically have an X-Y-Z liquid dispenser and can perform a limited number of tasks (e.g. mixing, pipetting, injection); those having continuous-flow operation are embraced by this follow-up of a previous survey [2].•

Consideration is first given to membrane types - porous or non-porous - and to transport processes and aspects of flux, in the context of assemblies for on-line dialysis. Non-porous (selective) membranes of the SLM type allow LLE and dialysis to be combined. The dialysis pH must be appropriate for the molecules concerned. To overcome dilution problems and maybe aid selectivity, a TEC is put between the dialysis system and the LC column. Use of a membrane for microdialysis or gas extraction is touched on. Consideration is given to electrodialysis (EDIST), and to pressure-aided dialysis (ultrafiltration) which can help in monitoring fermentations. The various dialysis modes have been applied in biotechnological and environmental areas. The examples also include bioanalytical uses.*

The coupling of membrane-based systems with LC offers sample preparation procedures that are automatable, usable on-line with the separation unit, simple, fast, robust, and give high recoveries. Membranes can be defined as *selective barriers between two phases* [3]. They are used for many industrial processes. There is growing recognition of their usefulness in analytical laboratories for sample preparation, where separations are achieved because some components pass through the membrane more easily then others. The phases divided by the membrane are called the *feed or donor* and the *permeate or acceptor* phases; molecular *flux* occurs between them. Two parameters characterize membrane-based sample-preparation systems: the membrane type and the force that furnishes a flux.

•Membrane disc supports for sorbents are not pertinent.- *Ed.*
**Abbreviations*.- EDIST, electrodialytic sample treatment; EF, enrichment factor; MWCO, molecular wt. cut-off; oPA, *o*-phthaldialdehyde (*Author prefers* OPA); SLM, supported liquid membrane; TEC, trace-enrichment column/cartridge [particular manufacturer not implied]. Extraction: LLE, liquid-liquid; SPE, solid phase.

Membrane types fall into two groups: *porous* (non-selective) and *non-porous* (selective). The latter consist of a liquid or polymer film in which molecules must be able to dissolve to allow transport, in respect of which the different partitioning coefficients (K_D) of analytes between the liquid phase and the membrane confer an enhancement of selectivity. Porous membranes, on the other hand, act as simple sieves: molecules small enough to pass the pores of the membranes will go from the donor to the acceptor phase, whereas larger molecules will not pass the membrane. Thus porous membranes are generally ranked according to the molecular mass of the smallest compound that is >90% retained in the donor phase (MWCO) [3]. The exact composition of the membrane is less important, although every porous membrane shows some non-porous characteristics (some solutes will penetrate it).

Membrane transport forces usable to foster migration include osmosis and, for solutes, dialysis in which a *concentration difference* (ΔC) gives the force causing the *molecular flux* (J_m; mol/sec). In osmosis (generally with porous membranes) only the solvent migrates, the donor phase becoming concentrated; in dialysis (either membrane type) solute molecules can also be transported. The force exerted by the concentration difference can be supported or entirely replaced by application of an *electric potential* (ΔE) over the membrane, giving an *electrical flux* (J_e; C/sec). In these membrane processes (electrodialysis, electro-osmosis), separation based on analyte charge is added to that based on molecular size and shape. The exact charge of weakly acidic or basic compounds can be altered by changing the pH, with benefit to selectivity.

Migration can also be forced (generally with a porous membrane) by means of a *pressure difference* (ΔP; 'filtration'), giving J_v, *volume flux* (unit: m^3/sec). Filtration using between 50 kPa and 1 MPa pressure ('ultrafiltration') is widely used in industry; in general it serves to separate low- and high-mol. wt. compounds.

Donor and acceptor phases can each be *static* or (continuous pumping over the membrane) *dynamic*. With a dynamic donor phase, substances are transferred faster and large sample volumes can be processed. If the acceptor phase is dynamic a high concentration gradient prevails and recovery is better; the drawback is sample dilution, circumventable by having a TEC on-line with dialysis.*

CONCENTRATION-GRADIENT BASED SYSTEMS

Dialysis can be performed using porous membranes (on-line dialysis) as well as by using non-porous membranes (supported liquid membranes - SLM's). Principles and applications of both techniques are now discussed. In addition, use of microdialysis† and of gas extraction are given some consideration.

---

*#A-10n and a Vol. 23 art. (cf. [2]; Rasmussen's lab.) concern such a system ('ASTED'; Gilson); cf. 'Dialysis' indexing,- *Ed.*
†in 'true' sense; done with a probe *in vivo; cf. Preface. - Ed.*

One of the main problems to be solved in sample preparation is the removal of macromolecules (notably proteins) from the sample. One option is to use an on-line dialysis system [2-8]. The steps in such a fully automated procedure are: injection, dialysis, trace-enrichment, desorption, separation and detection (Fig. 1) [5-9]. The heart of the system is a dialysis block, made of Perspex or PTFE. It comprises two parts with an intervening membrane (e.g. cellulose acetate having 10-15 kDa MWCO). Each part has grooves forming the donor or acceptor channel; the membrane functions only to separate macromolecules from small molecules. Contrary to traditional equilibrium dialysis, on-line dialysis is relatively fast (10-20 min) and gives high recoveries (70-90%). This is possible since, normally, the acceptor solvent (water) is continuously flowing, so giving a high concentration gradient over the membrane during the whole dialysis process. In consequence, the sample (0.1-1 ml) is diluted enormously. To counteract this effect, and to provide additional selectivity, a re-concentrating TEC 10 mm long is incorporated into the system; it can be packed with any SPE sorbent. Desorption is performed by the LC eluant; then the analytes are separated and detected.

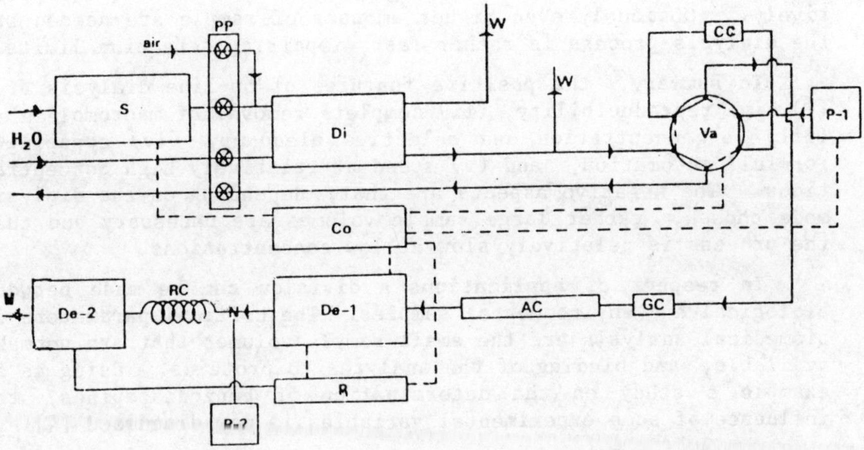

Fig. 1. Set-up for continuous-flow dialysis, then HPLC (with a post-column reaction). *From ref. [7], courtesy of Elsevier.*

S, sampler; Di, dialyzer; Va, 6-port valve; PP, peristaltic pump; CC, concentration column (TEC); P, HPLC pump; GC, guard column; AC, analytical column; De, detector; R, recorder; Co, controller; RC, reaction coil; W, waste.

The number of molecules transported per unit time (flux; J) depends on several parameters, described by Fick's law:

$$J = -\frac{A.D}{\tau}\frac{\delta c}{\delta x}$$

where A is the membrane area, D the diffusion coefficient, τ the tortuosity and $\delta c/\delta x$ the concentration gradient over the membrane.

D depends on the temperature (T), the viscosity of the sample (η) and the radius (r) of the analyte: $D = k.T/(6.\pi.\eta.r)$ [k is the Boltzmann's constant]. Manipulation of these two equations can generate a greater flux. This implies either a higher recovery in the same time or the same recovery in a shorter time period.

On-line dialysis can be performed in several modes. In the *static (stagnant-flowing) mode*, the donor compartment of the block is filled with the sample and the acceptor side is dynamic. The advantage of this mode is that only a small amount of sample is needed at a medium speed, providing medium detection limits compared with the other modes of dialysis. The second mode is the *pulsed (pulsed-flowing) mode*. This is a special form of the static mode. The donor side of the block is filled not once but several times. This results in significantly improved detection limits, but the drawbacks are that higher amounts of sample are needed and that the process is slower. A third option is the *continuous (flowing-flowing) mode*. Now both the donor and the acceptor side are continuously flushed with sample or recipient solvent respectively. Obviously even higher amounts of sample are needed but the dialysis process is rather fast, impairing detection limits.

In summary, the positive features of on-line dialysis are: (i) high reproducibility, (ii) complete removal of macromolecules, (iii) re-concentration and selective clean-up, (iv) capability for full automation, and (v) speed at relatively high concentrations. The negative aspects are that, depending on the dialysis mode chosen, rather large sample volumes are necessary and that the process is relatively slow at low concentrations.

In respect of **applications** a division can be made between biological and environmental samples. The critical parameters in biomedical analysis are the small sample volumes that are normally available, and binding of the analytes to proteins. Using as an example a study on the determination of benzodiazepines, the influence of some experimental variables is now discussed [4].

The first parameter of importance is the flow rate of the acceptor phase. The result of a higher flow rate is that the maximum recovery will be obtained more quickly because of a higher concentration gradient across the membrane. The hydrophobicity is another important parameter. The more hydrophobic the analyte, the lower is the maximum recovery. This is probably due to hydrophobic and electrostatic interactions between the analyte and the backbone of the membrane. Another parameter of major importance is the pH. For weak acids and bases the highest recovery is obtained at pH values where the analytes are neutral, while for ampholytes (e.g. amino acids, peptides, sulfonamides) the pH should be near the i.e.p. of the analyte. Higher temperatures (up to 50°) can be used to speed up the analysis because of larger diffusion coefficients at increased temperatures.

In bioanalysis, ionic strength and viscosity are important parameters. This is because method development is done using standard (aqueous) solutions, while real samples have significantly higher ionic strength and viscosity. Fortunately recovery is independent of both parameters if aqueous standards are compared with physiological conditions. The most important parameter in bioanalysis, however, is analyte-protein binding. In principle, only non-bound analytes can pass the membrane. Hence for highly protein-bound solutes the recovery will be rather low unless disruption techniques are applied. Possible ways include altering the conformation of the protein or adding a displacer* [10-15].

The choice of method depends strongly on which protein and which binding site is involved. Thus nitrazepam recovery is ~85% from water and only 40% from plasma, increasable to 70% using octanoic acid as a displacer [4]. For several NSAIDS's (e.g. ketoprofen, ibuprofen, flurbiprofen) recovery can be increased from 0-1% to 40-65% by adding acetonitrile (to 20%) and n-decanoic acid (to 8 mM) to the sample [16]; the dialysis and reconcentration take ~20 min (~25 min for total analysis). For nitrazepam the membrane and TEC survive >100 analyses, and the repeatability at the detection limit (25 ng/ml) is better than ±4% [4].

Different membrane materials have been tested, and configurations other than planar - notably hollow-fibre membranes [7; immunoaffinity mode], as used for various biosamples. Sample solubilization may be needed to obviate dialysis-block clogging.

Automated systems with on-line dialysis coupled to LC can embrace a reaction step to give a good fluorescence signal from the analyte. Oxytetracyclines in biofluids and salmon tissues were run by RP-LC followed by alkalinization and irradiation at 366 nm [18; cf. 17]. For amino acids in serum and urine a derivatizing agent (oPA or, for secondary amines, 9-fluorenylmethylchloroformate) was added to the sample and 100 µl put into the donor channel of the dialysis block; 25 µM levels were measurable [19].

Supported liquid membranes (SLM's)

Dialysis using non-porous membranes has been used in bioanalysis, but so far mainly environmental applications have been described. Some applications have entailed use of silicone rubber or other non-porous polymer membranes [20], but normally SLM's are used. They comprise a porous PTFE membrane impregnated with a water-immisicible organic solvent such as n-undecane or dihexyl ether [21]. The choice of immobilized solvent is ratner important,

*See index entry 'Protein binding' in this and earlier vols. and, in Vol. 23 [cf. 2], pp. 237-244 (Krogh & Rasmussen).- *Ed*.

because the distribution ratio of the analyte between the donor and the acceptor phase can vary significantly. For example, the EF for the herbicide metsulfuron is 0.5 in n-undecane and 3 in dihexyl ether [22]. Important properties of the solvent are:

- viscosity: greater viscosities will provide higher diffusion coefficients;
- water-solubility: if low, system stability will be high;
- distribution ratio between donor phase and solvent: it should be high for the analytes and low for interferants.

In principle, this set-up enables LLE and dialysis to be combined in a single process. Firstly all the analytes are extracted from the donor phase into the liquid membrane acting as a separate phase, and then transferred to the acceptor phase (Fig. 2) [21-23]. Accumulation of the analytes in the acceptor phase is achieved by having it stagnant whereas the donor phase is flowing. The concentration in the acceptor phase will vary linearly with the concentration in the donor phase and with the sample volume. Clean-up is possible by pumping a sample-free donor phase after completing sample injection, thereby washing back compounds that are not ionized in the acceptor phase [23].

A one-directional flow of weakly acidic or basic analytes is obtained by making the donor phase acidic or basic respectively, such that the analytes are in their neutral form; then they can easily dissolve into a non-polar SLM and undergo transfer to the acceptor phase. When the latter is neutral, the acids or bases will regain their ionic form and cannot be back-extracted through the non-polar SLM. Hence neutral compounds, which may pass freely from one phase to the other and back again, can be washed out of the acceptor phase by passing a large volume of pure water over the membrane on the donor side. Depending on the sample size available, flow rates up to 7 ml/min can be used. We could detect 1 ng/ml triethylamine in 10 ml urine (acceptor vol. 56 µl; 2.5 µl injected for GC).

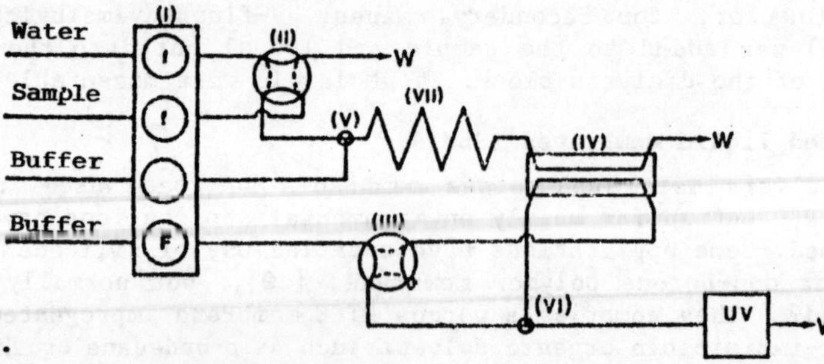

Fig. 2.
SLM flow set-up (version with outflow directly measured).
Cf. Fig. 1 in ref. [23]; acknowledgement to author and Am. Chem. Soc.
I, peristaltic pumps; II, III: 4-way valves (sample inlet; switching); IV, membrane separator (grooved plastic blocks); V, mixing point; VI, bypass loop coupling; VII, mixing coil.

Microdialysis

This sample preparation technique entails inserting a small probe into living systems to perform *in vivo* observations without disturbing them. Firstly this technique was used mainly for the determination of catecholamines, but recently it has been used to monitor glucose, lactate and glycerol levels in patients. Microdialysis has also been used for measuring oligosaccharides in plants and fermentation broths [24].

Gas extraction [cf. index entry 'Gas...' - *Ed.*]

Hollow-fibre membranes can serve for extraction of volatile organic compounds by gas extraction. A sample is passed through a single hollow-fibre membrane (porous polypropene, or non-porous silicone rubber), traversed by a carrier gas as the acceptor phase. The method has been used to extract 1,1,1-trichloroethane, trichloroethene and tetrachloroethene from water, the analytes being trapped by cryofocusing and analyzed by LC or GC [25].

ELECTRIC-POTENTIAL BASED SYSTEMS (earlier survey: [2])

EDIST entails applying an electrical potential difference (5-10 V) over the membrane to create an *additional driving force (electromigration)*. It is a special form of electrodialysis, dating back some 70 years [26], with a block [27] (illustrated in [2]) that comprises two electrode vessels with electrode compartments and, shielding the electrodes from them, two ion-exchange membranes. PTFE spacers with flow channels are used to provide the donor and acceptor channels, separated by a neutral or ion-exchange membrane [28]. The two ion-exchange membranes are necessary to prevent electrochemical degradation of the analytes. The flux in EDIST can be expressed by the Nernst-Planck equation:

$$J = J_m + J_e = -A.D.\frac{\delta c}{\delta x} . \frac{D.A.c.z.F}{RT} . \frac{\delta E}{\delta x}$$

z, analyte valency;
F, Faraday constant;
R, gas constant.

The total flux is in this case the sum of the molecular flux, due to a concentration gradient, and the electrical flux, due to an electrical potential gradient. Theoretical models for EDIST in trace analysis by LC are discussed by Debets & co-authors [29]. The one difference from on-line dialysis is that EDIST needs no TEC, as the applied potential obviates possible back-transport of analyte molecules from acceptor to donor phase. So, rather than dilution, analyte enrichment occurs. A steady state is reached when the back-diffusion flux equals the electromigration flux, allowing no further enrichment (factor: EF = c_a/c_s [30] where c_a refers to analyte in the acceptor phase after EDIST and c_s to the start). Using a flowing donor and stagnant acceptor phase, EF in conventional dialysis will always be <1 but in EDIST up to 10-20. EF depends on the voltage; if >10 V, gas forms at the electrodes.

A problem in EDIST is that as the pH of the phases changes during dialysis due to membrane transport of protons, weak bases especially may change from charged to neutral form, so hampering the dialysis. A second problem is that there can also be concentration of charged macromolecules- proteins and humic substances; the latter are retained if the membrane MWCO is, say, 3.5 kDa.

Only in the last decade have analytical applications of EDIST been reported - for analyte enrichment [27, 30, 31] and matrix normalization [32-34] prior to LC. The field of application has been largely biomedical, similarly to on-line dialysis [3].

PRESSURE-GRADIENT BASED SYSTEMS

Ultrafiltration is a pressure-based membrane system that is particularly well suited to separate compounds with a high molecular mass from those with a low mass. This property is used in sample preparation for off-line or on-line removal of macromolecules, aimed to avoid clogging of the LC column [3]. In ultrafiltration the solvent and solutes of appropriate size are pressure-driven through a porous membrane:

$$J_v = \frac{\varepsilon . A . R^2}{8\eta} . \frac{\delta p}{\delta x}$$

The volume flux (J_v) depends on the porosity of the membrane (ε), the surface area (A), the pore radius (R), η and the pressure gradient across the membrane $\delta p/\delta x$. In pressure-driven systems the main problem is *concentration polarization,* i.e. membrane fouling. The most frequently applied method to minimize it is use of *cross-flow* instead of *dead-end* filtration.

On-line ultrafiltration systems have been used for carbohydrates, for metabolites (acids: [35, 36]; alcohols: [37, 38]) and, with on-line monitoring, for cephalosporins [39, 40] and penicillins [35, 41]. Most applications are biotechnological, typically to fermentation media with unattended monitoring and real-time analyses. Thus, with a recombinant *Escherichia coli,* amino acid metabolism was followed using a hollow-fibre ultrafiltration module, then ion-exchange LC and derivatization by oPA [42]. Advantageously, an entire ultrafiltration system can be steam-sterilized, fully closed, along with the fermentor, so significantly reducing the risk of infection of the broth.

Concluding comment.- The systems now surveyed may warrant more exploitation by drug bioanalysts although, at least in Europe, the 'ASTED' system mentioned earlier is already quite well known besides others unrelated to it. However, principles and possibilities are less well known. It is with this in mind that the present review has been compiled, with some 'classical' citations*, to strengthen understanding and awareness.

**Note by Editor*.- To the author's regret, a few 'non-classical' refs. (mainly late in his text) were excised editorially, as was an 'EDIST' diagram that appeared in Vol. 23 (inserted ref.: [2]; can be regarded as a 'companion' article); terse allusions to 'ASTED' were also inserted. Book-length constraint influenced the editing, and the streamlined layout adopted for the ref. list.

References [J. Chromatog. DENOTED J Chr; for 'home' authors, e.g. Brinkman, only **first** citation gives initials

1. Scott, R.W.P. (1997) 'Tandem Techniques', Wiley, Chichester, 538 pp.
2. Lingeman, H., van de Merbel, N.C. & Brinkman, U.A.Th. (1994) in 'Biofluid and Tissue Analysis for Drugs including Hypolipidaemics' [Vol. 23, this series]: eds. (Reid, E...) & publisher (Roy. Soc. Chem.) AS NOW, pp. 315-326.
3. van de Merbel, Hageman, J.J. & Brinkman (1993) J Chr **634**, 1-29.
4. van de Merbel, Teule, J.M., Lingeman & Brinkman (1992) J. Pharm. Biomed. Anal. [**10**, 225-233.
5. Turnell, D.C. & Cooper, J.D.H. (1987) J Chr **395**, 613-621.
6. van de Merbel, Lagerwerf, F.M., Lingeman & Brinkman (1994) Int. J. Environ. Anal. Chem. **54**, 105-118.
7. Aerts, M.M.L., Beek, W.M.J. & Brinkman (1988) J Chr **435**, 97-112.
8. Agasøster, T. & Rasmussen, K.E. (1990) J Chr **564**, 171-179.
9. van de Merbel & Brinkman (1993) Trends Anal. Chem. **12**, 249-256.
10. Peters, T. (1975) in 'The Plasma Proteins' (Putnam, F.W., ed.), Academic Press, New York, pp. 153- (Chapter).
11. Fehske, K.J., Muller, W.E. & Wolhert, U. (1983) Biochem. Pharmacol. **30**, 687-692.
12. Kragh-Hansen, U. (1988) Mol. Pharmacol. **34**, 160-165.
13. Wanwimolruk, S., Birkett, D.J. & Brooks, P.M. (1981) Mol. Pharmacol. **24**, 458-465.
14. Tillement, J-P., Houin, G., Zini, R., Urien, S., Albengres, E., Barre, J., Lecomte, M., d'Atis, P. & Sebille, B. (1984) Adv. Drug Res. **13**, 60-94.
15. Lindup, W.E. (1987) Progr. Drug Metab. **10**, 141- (Chapter).
16. Herráez-Hernández, R., van de Merbel & Brinkman (1995) J Chr **B 666**, 127-137. [149.
17. Farjam, A., van de Merbel, Nieman, A.A., Lingeman & Brinkman (1992) J Chr **589**, 141-
18. AS FOR 8. (1991) J Chr **570**, 99-107; ALSO (1998) J. Pharm. Biomed. Anal., in press.
19. Vérillon, F. & Qian, F. (1991) Analusis **19**, 271.
20. Melcher, R.G. & Morabito, P.L. (1990) Anal. Chem. **62**, 2183-2188.
21. Jönsson, J.Å. & Mathiasson, L. (1992) Trends Anal. Chem. **11**, 106-114.
22. Nilvé, G., Audunsson, G. & Jönsson, J.A. (1989) J Chr **471**, 151-160.
23. Audunsson, G. (1986) Anal. Chem. **58**, 2714-2723.
24. Westerink, B.H.C. (1992) Trends Anal. Chem. **11**, 176-181.
25. Pratt, K.F. & Pawliszyn, J. (1992) Anal. Chem. **64**, 2101-2106 & 2107-2110.
26. Prausnitz, P.H. & Reitstötter, J., eds. (1931) 'Elektro-osmose, Elektrodialyse in Flüssigkeiten', T. Steinkopff, Dresden.
27. Debets, A.J.J., Kok, W.Th., Hupe, K-P. & Brinkman (1990) Chr'graphia **30**, 361-366.
28. Groenewegen, M.G.M., van de Merbel, Slobodnik, J., Lingeman & Brinkman (1994) Analyst **119**, 1753-1758.
29. Debets, van Wier, P., Hupe, Brinkman & Kok (1994) Chr'graphia **39**, 460-468.
30. Debets, Hupe, Kok & Brinkman (1992) J Chr **600**, 163-173.
31. Cox, J.A. & Carlson, R. (1981) Anal. Chim. Acta **130**, 313-321.
32. Pettersen, J.M., Johnson, H.G. & Lund, W. (1988) Talanta **35**, 245-247.
33. Okamoto, Y. Sakamoto, N., Yamamoto, M. & Kumamaru, T. (1991) J Chr **539**, 221-224.
34. Haddad, P.R., Laksama, S. & Simons, R.G. (1993) J Chr **640**, 135-143.
35 Schügerl, K., Brandes, L., Dullau, T., Holzhauer-Rieger, K., Hotop, S., Hübner, U., Wu, X. & Zhou, W. (1991) Anal. Chim. Acta **249**, 87-100.
36. van de Merbel, Ruijter, G.J.G., Lingeman, Brinkman & Visser, J. (1998) Appl. Microbiol. Biotechnol., in press.
37. van de Merbel, Lingeman, Brinkman, Kolhorn, A. & de Rijke, L.C. (1993) Anal. Chim. Acta **279**, 39-50.
38. Buttler, T.A., Johansson, K.A.J., Gorton, L.G.O. & Marko-Varga, G.A. (1993) Anal. Chem. **65**, 2628-2636. [213-219.
39. Bayer, Th., Herold, Th., Hiddessen, R. & Schügerl, K. (1986) Anal. Chim. Acta **190**,
40. Holzbauer-Rieger, K., Khou, W. & Schürgerl, K. (1990) J Chr **499**, 609-615.
41. Christensen, L., Nielsen, J. & Villadsen, J. (1991) Anal. Chim. Acta **249**, 123-136.
42. van de Merbel, Zuur, P., Frijlink, M., Holthuis. J.J.M., Lingeman & Brinkman (1995) Anal. Chim. Acta **303**, 175-185.

#A-10n

A Note on

SUMATRIPTAN ASSAY BY HPLC WITH TRACE-ENRICHED DIALYSATES§

J.D.H. Cooper† & J.E. Taylor

Clinical Innovations Ltd., Building 28, Stoneleigh Deer Park, Stareton, Kenilworth CV8 2LQ, U.K.

Require-ment *An ASTED¶ procedure for automated assay of sumatriptan (a basic compound) at ng/ml levels in human plasma.*

End-step *HPLC on a high carbon-loaded column (Kromasil C4; Hichrom) with DEA present to block silanols, giving Gaussian peaks; detection at 225 nm. ASTED-coupling by the control system.*

Sample handling *Plasma acidified (MCA) to release protein-bound drug and, with pH 7 buffer added (MCA neutralized), dialyzed and trace-enriched; TEC freed from interferants by 60% MeCN.*

Comments *The assay, as used in clinical pharmacokinetic studies, had good linearity (LLOQ 1 ng/ml), accuracy (+2% bias), precision (C.V. 1.7%) and specificity. Conditions apt for the difficult analyte had to be carefully chosen.*¤

The basic drug sumatriptan (anti-migraine) has previously been assayed in biofluids by HPLC-MS [1] or, with complexity that would hinder automation, by liquid-liquid extraction and HPLC-EC [2]. Our approach for using ASTED-RP-LC to assay basic compounds on-line [3] has now been applied, using mainly Gilson equipment supplied by Anachem (Luton). The control/peak integration system had Microsoft Windows V3.11 software in a 486 PC. (*Note by Editor:* ASTED served for basic analytes in reports by K.E. Rasmussen [4, 5]; * *below* denotes differences from the system he depicts [4].)

The ASTED unit has, besides a Rheodyne 7010 valve for high-pressure HPLC loading, a second one* at low pressure to isolate the membrane from its loading syringe-injector - preceding which there is provision for introducing and mixing reagent* as well as drawing sample from the autosampler vial. The Kel F dialyzer unit (~370 µl donor vol.) has 15 kDa Cuprophan membranes (Enka). The stainless steel TEC, packed in-house with 70 mg 10 µm Hypersil C1 (Shandon), replaces the loop on the high-pressure valve. Plastic tubing used to move liquids is chemically pre-treated [6] to increase flow velocity without disrupting sample integrity.

§This is an edited-down version of a fuller text, omitting the system diagram supplied.

†addressee for any correspondence

¤The index entry 'Basic...' throughout this series bears on HPLC difficulties.- *Ed.*

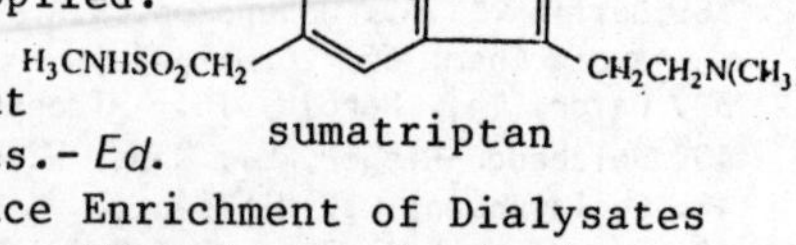

sumatriptan

¶*Abbreviations.*- ASTED, Automated Trace Enrichment of Dialysates (System; Gilson XL); DEA, diethylamine; EC, electrochemical; LLOQ, lower limit of quantitation; MCA, monochloracetic acid; MeCN, acetonitrile; TEC, trace-enrichment cartridge.

Materials.- Chemicals (analytical grade) were from Sigma, and HPLC solvents from Romil; HPLC-grade water came from a Purite (Thame) system. Buffers were phosphate-based (ammonium salts). Drug standards (stock and working) were in 20% v/v methanol.

Method outline.- Plasma and 1 M MCA (400 + 100 µl) mixed; 480 µl delivered into the donor channel, and, by Gilson diluter, 2 mM pH 7 buffer. Recipient solvent (10 mM pH 7 buffer with 5% methanol) applied by the other diluter; 5 ml put through valve and TEC over 5 min, then 200 µl 60% MeCN and 1 ml donor solvent. Elution *via* valve by mobile phase for HPLC (Fig. 1). **Validation** reassuring; the intra-/inter-assay bias and C.V. values were: spiked at 1 ng/ml: +5.0/+6.0% & 5.1/9.0%; at 3, -5.0/-2.0% & 3.0/6.7%; at 20: +0.5/+1.0% & 2.1/1.8%; at 80: +2.3/+1.8% & 1.4/1.05%. The method was sensitive (as with EC [2]), economical and robust.

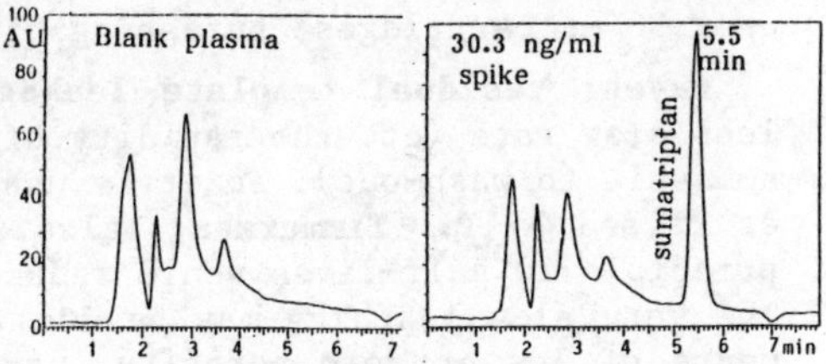

Fig. 1. HPLC traces (column 100 × 4.6 mm; survives thousands of runs -hundreds for the TEC). Mobile phase: 1:9 MeCN/45 mM buffer pH 3.5 (optimal separation) with 10 mM DEA; 1 ml/min. Note absence of blank interferences. No internal std. used (yet good precision).

Purging/regeneration of tubing/channel/TEC/column between samples was thorough (carry-over <2%), by donor and/or recipient solvent as appropriate. **Removal of interferants** hinged on the 60% MeCN TEC wash and use of pH 7 buffer (helpful *re* diffusion of MCA into the dialysate). **Basic analytes** become enriched by ion-exchange interactions with silanols on non-end-capped short-alkyl RP silicas ([5], cf. [9]; breakthrough vol. was >10 ml). **Protein binding** besets ASTED assays (and drugs may adsorb onto membranes [5]); pH-lowering by MCA [3, 8] effects the disruption requisite for reliable total-drug values. Matrix differences did not affect accuracy/precision, tested by analyzing spiked blanks from volunteers. **Drug stability** was good (at -20°, or 24 h at room temp.).

References [J. Chromatog. ABBREVIATED TO J Chr; refs. 4, 5 & 9 inserted by Ed.]

1. Oxford, J.M. & Lant, M.S. (1989) J Chr **B 496,** 137-146.
2. Andrew, P.D., Birch, H.L. & Phillpot, D.A. (1993) J. Pharm. Sci. **82,** 73-76.
3. Cooper, J.D.H., Muirhead, D.C., Taylor, J.F. & Baker, O.R. (1997) J Chr **B 701**, 87-95.
4. Krogh, M. & Rasmussen, K.E. (1994) in present series, V. 23, pp 237-244.
5. Johansen, K., Krogh, M. & Rasmussen, K.E. (1997) J Chr **B 690,** 223-231.
6. Cooper, J.D.H., Buick, A.R. & Fook Sheung, C.T.C.(1994) Lab. Robot. Autom. **6,** 15-20.
7. —, Fook Sheung, C.T.C. & Buick, A.R. (1994) J Chr **B 652,** 15-21.
8. —, Turnell, D.C., Green, B. & Verillon, F. (1988) J Chr **456,** 53-69.
9. Law, B. (1992) in this series, V. 22 (ed. Reid et al.; RSC - as now), pp. 57-64.

For details lacking in Editor's condensate, ask Dr Cooper: fax (0)1203-639568.

Comments (grouped according to aspect) on **MIP contributions** in Sect. #A [see Contents for full names of those concerned]

MIP particle features.- Mayes (reply to Pleasance) reckoned it is immaterial, for template removal, what approach (grinding or crushing) is used to prepare the particles. What is important is ascertainment, perhaps by laser diffraction, of particle size; this influences (e.g.) internal and external abundance of binding site and the amount of trapped drug, pertinent to leakage. Size and flow-rate optimization has been little studied (responses to query by **H.M. Hill**), nor how much to use for SPE: up to 100 mg, but even 5 mg gives excellent capacity and allows low elution volumes (5× 5 ml for tamoxifen - **Stevenson). Andersson** routinely uses 30 mg cartridges, suffering hindrance of flow by fines.

Mayes: residual template leakage occurs at a seemingly endless slow rate (cf. the rapidity of removing template, most being amenable to wash-out); so it is unhelpful to consider 'half-life' as raised by **P. Timmerman, also J. Schmid** who remarked that in principle 10 half-lives would reduce the concentration 1000-fold. The very slow leaching may be due to entrapment in microenvironments of low or zero porosity, such that the molecules have to diffuse through a highly rigid area of polymer structure to reach the 'outside world'. This is probably why 'shocking' the polymer, e.g. by temp. or solvent changes, releases more template each time. Maybe the shocks cause transient re-ordering of packed chain structures in the polymer, allowing diffusion of trapped template at an accelerated rate.

Answers to D. Dell: leakage problems should be circumventable by using as template an analog of the analyte. **S.A. Westwood** commented that MIP specificity would benefit if the analog were a 'cold' (non-radio) labelled version rather than structural. He further advocated that the template be removed by SFE, which with an appropriate modifier could swell the polymer and greatly speed up within-matrix diffusion. **Mayes** mentioned that when trying supercritical CO_2 as mobile phase with MIP columns an unexpected absorbance was found as soon as the CO_2 emerged from the column. This was attributed to washing out of entrapped template. No such absorbance was observed after a 'rest' period; but prolonged elution by CO_2 resulted in loss of stationary phase, presumably by reduction of polymer to ultra-fines which passed through frits and washed out of the system. This also caused severe baseline problems. Conceivably leakage problems in ultra-trace analysis might be solved by briefly treating the polymer with CO_2/modifier and then retrieving the polymer and re-trying the SPE procedure. **M. Bayliss asked (Andersson** replied "At present, yes") whether sensitivity loss due to residual template impairs 'imprint RIA'.

Replies to **K. Borner.- Binding** constants for MIP's can be as high as 10^3 (comparably to polyclonal Ab's) and different sites are feasible, e.g. for analyte and i.s. Remarks on **equilibrium (Mayes)**.- Prediction of its position is being aimed at through modelling work, although complex and fraught with dangers. Even for a simplified case the equations are already quite complex. Thus for morphine we could consider dimerization of methacrylic acid **and** 3 template/MAA complexes, but this already requires iterative solution of a quite complex equation. Moreover, we can only roughly estimate the values of the equilibrium constants from published reports on similar interactions or by attempting NMR titration experiments. Mere simplistic consideration of the equilibrium is, however, useful in that it forces one to think about which factors alter it favourably to enhance the concentration of the fully complexed template species.

Agreed by **Andersson/Heard/Martin: MIP reproducibility** in the data obtained is good between batches - although (**Mayes**) scale-up of the polymerization (even to multi-kg amounts) is so easy that there may be no need to synthesize multiple batches. So far no manufacturer of SPE or affinity-chromatography packings has become involved with MIP's (query by **D. Dell**).

Citations contributed by Senior Editor **concerning MIP's** (see also p. 70)

For journals, Chr'y DENOTES Chromatog.

Surveys of molecular imprinting have appeared (1997): Anal Chem **323A**, 1973-2224; also (non-covalent mode; B. Sellergren) TRAC **16**, 310- 320.

A uniform-size MIP, effective for enantiomers as shown with *N*-(3,5-dinitrobenzyl)-α-methylbenzylamine, has been prepared by a 2-step swelling procedure wherein it is useful, not obligatory, to include methacrylic acid as well as a cross-linking agent that can be chiral.- Hosoya, K. et al. (1996) J Chr'y **A 728**, 139-147.

MIP's for CE separation of amino acid enantiomers were synthesized on the actual capillary wall.- Lin, J.M. et al. (1997) J Anal Chem **357**, 130-132. **A novel protein-imprinted support** was selective for whatever protein (e.g. Hb) had been used for the imprinting.- Liao, J-L. et al (1996) Chr'ia **42**, 259-262. **An MIP for 1,3,5-triazines** has been prepared using atrazine.- Matsui, J. et al. (1995) Anal Chem **67**, 4404-4408.

Comments on non-MIP contributions, late in Sect. A (see Contents)

Remark by **R.F. Pemberton (to Stevenson**, who hadn't done the comparison).- It might be advantageous to use an mAb rather than a polyclonal antiserum (with a population of Ab molecules of the same affinity), so that all antigen molecules are simultaneously eluted in a minimized volume. **Lingeman** had found for a drug that an mAb indeed gave a more concentrated eluate (×4).

Replies by **Stevenson.- To Mayes**: falling cartridge capacity could indeed be due to Ab leaching and/or clogging by non-specific binding, besides Ab denaturation with low pH and organic eluants. To **Gilbert**: with a specific Ab, measurement need entail only, say, ELISA; but identity verification is more assured with HPLC, especially if coupled to diode array or MS.

Lingeman, answering Venn.- Continuous constant-flow renewal of the recipient fluid - traversing the trace-enrichment cartridge (TEC) also - drives the equilibrium, and eventually even highly protein-bound compounds are fully recovered.

Cooper, asked by W. Herron whether forward- or back-flushing of the TEC gave better results, replied that the system was set up for back-flushing. **K-H. Lehr** remarked that the dialysis rate for an analyte prone to protein binding might differ in different patients; so a calibration done with one plasma matrix might not be valid for another plasma.

Citations contributed by Senior Editor
relevant to non-MIP Sect. #A topics; often with 'model' compounds (cf. some * entries in #ABC; SEE end of #C for HPLC packings)

Affinity approaches (not imprinting)

'Chromatographic Separations Based on Molecular Recognition' - a Wiley book (1997) by K. Jinno.

Immunoaffinity packings have been used to clean up animal biosamples such as liver extracts for assay of (1) ivermectin, pre-HPLC; (2) synthetic corticoids, pre-GC-MS; (3) clenbuterol, pre-HPLC.- (1997) J Chr'y **B 696**: (1) Li, J.S. et al., 166-171; (2) Delahaut, P. et al., 203-215; (3) Lawrence, J.F. et al. 291-297.

Insulin secretion from single islets (rat; UV detection) was measured using C-8 microbore HPLC columns after an Ab-modified protein G step.- Shen, H. et al. (1997) J Chr'y **B 689**, 295-303.

'Live' **micro-dialysis** (term ambiguous! cf. Preface remark)

***In vivo* microdialysis samples** from rat brain were analyzed off-line (C-18 cols., with switching; fluor) for 5HT and simultaneously for (1) fluoxetine & norfluoxetine, or (2) paroxetine.- Ramaiya, A. et al. (1997): (1) J Chr'y **B 691**, 119-129; (2) J Pharm Biomed Anal **15**, 729-738.

***In vivo* microdialysis with on-line transfer** every 10 min to a microbore HPLC(-CN)-EC has been done, using a brain probe, in rats given apomorphine; 0.1 ng/ml was detectable.- Tsai, T.H. et al. (1997) J Liq Chr'y **20**, 481-490. **Off-line** HPLC (not microbore) was done with probe samples of rat extracellular fluid - cortical (also CSF, blood and plasma ultrafiltrates; dosage with 'EAB 15') [Malhotra, B.K. et al. (1996) J Chr'y **B 679**, 167-176] or peripheral (plasma too; dosage with piperacillin) [Nolting, A. et al. (1996) J Pharm Sci **85**, 369-376].

Vitreous carbon particles (graphite coated with a resin and pyrolyzed) in a pre-column selectively removed interferants from TCA homogenates of rat brain prior to choline and AcCh assay (on a polymer column; pH 8.4); enzymic reactor; EC detection.- Ikarashi, Y. et al. (1995) J Chr'y **A 718**, 267-272. **AcCh** was similarly assayed in microdialysates, with **optimized Pt electrodes**.- Carter, A.J. & Kehr, J. (1997) J Chr'y **B 692**, 207-212.

Sample handling, various - incl. membrane/disc/fibre use, SPE, liquid-liquid, protein binding [cf. end of #D], GC loading, SFE

Supported liquid membrane approach (pre-LC-EC) studied with urinary propofol (the active ingredient in diprivan): design and optimized use of the undecane-impregnated PTFE membrane system is described.- Trocewicz, J. et al. (1996) J Chr'y **B 685**, 129-134.

On-line 'microdialysis'/LC also features in 2 assemblies for studying enzyme reactions [Zhou, J. et al. (1996) J Pharm Biomed Anal **14**, 1691-1698] and in a system, with back-flushing, evaluated using verapamil and norverapamil [Ceccato, A. et al. (1996) J Chr'y **A 750**, 351-360].

Membrane adsorption and protein-holding aspects of dialysis/ trace enrichment/HPLC(-CN) on-line systems were studied using clozapine and desmethyl clozapine. Addition to the plasma of HCl/ R_4N^+/glycerol gave notable improvement.- Johansen, K. et al. (1997) J Chr'y **B 690**, 223-232; cf. K. Rasmussen's art. (from same lab.) in Vol. 23 (this series) & art. #A-10n above.

A novel hydrophobic porous membrane was prepared by a photopolymerization process and introduction of phenyl and diol groups which minimized adsorptiveness in tests with BSA.- Kubota, N. et al. (1995) J Chr'y **A 718**, 27-34.

A system for **pre-GC on-line dialysis** with a trapping stage has been developed.- Herráez-Hernándes, R. et al. (1996) J Pharm Biomed Anal **14**, 1077-1087. [Cf. #A-9 and, in Vol. 23, a review from the same Amsterdam lab. of various on-line treatments.]

Membrane concentration prior to CE-MS(ESP), with sulfonated particles on PTFE, was applied to metabolites of 'PAP' (a propanediol derivative) in incubates.- Benson, L.M. et al. (1996) J High Resolu Chr'y **19**, 291-294. A **mixed-mode disc** (hydrophobic and cation-exchange groups) served for methadone enantiomers in serum prior to HPLC.- Rudaz, S. & Veuthey, J-L. (1996) J Pharm Biomed Anal **14**, 1271-1279.

An SPE packing of porous polystyrene type (Styrosorb) was compared with C-18 as the first step in analyzing propranolol enantiomers in serum (achiral and chiral columns).- Rumiantsev, D.O. & Ivanova, T.V. (1995) J Chr'y **B 674**, 301-305.

Interferants from SPE cartridge packings (Waters C18, preconditioned) emerged, in a nitroimidazole assay, using an eluant (2% acetic acid in MeOH) that minimized rat-plasma interferences and gave good recovery. Remedial procedures included chilling during SPE (endcapping loss reduced) and rigorous drying down of eluates.- Kagel, J.R. et al. (1997) J Chr'y **A 763**, 129-137. In an assay (C-18) for topotecan in plasma, **low-temp. SPE operation** was likewise one stratagem to overcome cartridge-derived interferants.- Rosing, H. et al.(1996) J Pharm Biomed Anal **15**, 279-286.

SPE tip from R. Whelpton, for whole-blood samples: freezing/ thawing and centrifugation may give an aqueous phase apt for SPE.

SFE performed on cartridges carrying dried-down serum was investigated with various spiked-in drugs including ketoprofen, fenbufen and benzodiazepines; the applicable HPLC mobile phase served as the solvent trap.- Simmons, B.R. & Stewart, J.D. (1997) J Chr'y **B 688**, 291-302.

On-fibre SPE prior to GC was performed on neutralized PCA supernatants, in vials, from blood containing local anaesthetics. After 250° treatment, the polydimethylsiloxane-coated fibre was immersed in the analyte solution, then put into the GC port and heat-desorbed.- Kimazawa, T. et al. (1996) Chr'ia **43**, 59-62. Such pre-GC **micro-extraction ('SPME')** served for anorectics in urine.- Chiarotti, M. et al. (1997), J Microcol Sep **9**, 249-252; likewise for antidepressants in plasma - Ulrich, S. & Martens, J. (1997) J Chr'y **B 696**, 217-234.

A sample extractor/injector.- A syringe-type device ('Extra-shot-ODS'), tested with anti-epileptics, passed the serum sample through a mini-column into the main column (C-18; UV detection). The results (LLOQ 1 µg/ml) were in accord with enzyme immuno-assay.- Kouno, Y. et al. (1997) J Chr'y **B 695**, 349-353.

For GC, octanol-'modified' fibres, polyacrylate-coated, allowed initial micro-extraction by immersion in TCA supernatants from plasma containing diazepam; the fibre was desorbed in the injection port. Therapeutic-monitoring or forensic application was envisaged.- Krogh, M. et al. (1997) J Chr'y **B 689**, 357-364.

Liquid extraction on-line with GC, for β-blockers extracted from serum or urine by SPE, was performed by passage through a coil and a cooled phase separator, with UV monitoring; analyte-containing fractions with added derivatizing agent were conveyed by helium carrier gas to the GC column.- Hyotylainen, T. et al. (1997) J Chr Sci **35**, 380-206.

Cold-trap coupling of SFE to GC-MS was effective for PAH's in marine sediments.- Fuoco, R. et al. (1997) Anal Chim Acta **346**, 81-86.

SPE on-line with CZE (sleeve insert used) was explored with sulfonylurea antidiabetics, aiming to improve detectability in urine.- Strausbauch, M.A. et al. (1995) J Chr'y **A 717**, 279-291.

A series of CE reviews (1997; Electrophoresis **18**, 2123-2501) includes **CE-based IA's**, some using microchips (Schmalzing, D. & Nashabeh, W., 2184-2193; cf. 2203-2213) and affinity CE ('ACE'; Rippel, G. et al.; 2175-2183).

MIP-based **sensors**: Kriz, D. et al. [Lund group] (1997) Anal Chem **69**, 345A-348A.- See summary below, p. 191. Survey of MIP's of **non-covalent type**: Sellergren, B. (1997) TRAC **16**, 310-320. **Aqueous and semi-aqueous mobile phases** were compared (frontal and zonal modes) with a theophylline MIP: Baggiani, C. et al. (1997) J Chr'y **A 786**, 23-29.

Section #B

ANALYTE CASE HISTORIES

The suffix 'n' as in #B-3n distinguishes a 'NOTE' from main articles. COMMENTS (appearing on p.112) comprise discussion remarks and Editor's 'annotations'.

#B-1

COMPARISON OF RIA AND LC-MS-MS METHODS FOR DETERMINING AN LHRH ANTAGONIST

[1]H.J. Skrzipczyk,[†] [2]J. Sandow, [1]G. Ringsdorf,
[1]F.J. Müller, [1]J. Lapp & [1]K. Weimer

[1]Dept. of Pharmacokinetics [2]Pharma Research, Dept. of Pharmacology

Hoechst Marion Roussel, D-65926 Frankfurt/Main, Germany

To assay serum/plasma for the decapeptide HOE 013 - a third-generation LHRH antagonist potentially useful in cancer therapy - both a competitive RIA and an LC-MS-MS assay have been developed. For the RIA (on 5-20 µl samples) the immunogen used to raise the Ab was prepared by coupling to BSA* ***via*** *a hexanoyl spacer, and the radioligand was synthesized by the chloramine-T method. For HPLC after SPE (100-250 µl samples), detection was by MS-MS(API; Turbo Ion Spray) using a nonapeptide as i.s. Both methods proved to be sensitive (LLOQ, ng/ml: RIA, 0.6; LC-MS-MS, 0.5). However, LC-MS-MS was preferred, being of unparalleled specificity and advantageous with regard to quality criteria such as precision and accuracy.*

LHRH antagonists are reckoned to be potential substitutes for the commercially available agonists in different advanced hormone-dependent tumours, e.g. in prostatic cancer or premenopausal breast cancer. The immediate onset of gonadal steroid suppression, without a stimulatory phase as with the agonists, is one important advantage of the antagonists. HOE 013, a third-generation LHRH antagonist, is a decapeptide with glycosylation at position 6 [D-Ser(rhamnose)]. This glycosylation improves the biological tolerance by reducing and/or eliminating side effects previously found with this class of compound (release of histamine and related anaphylactoid reactions).

HOE 013

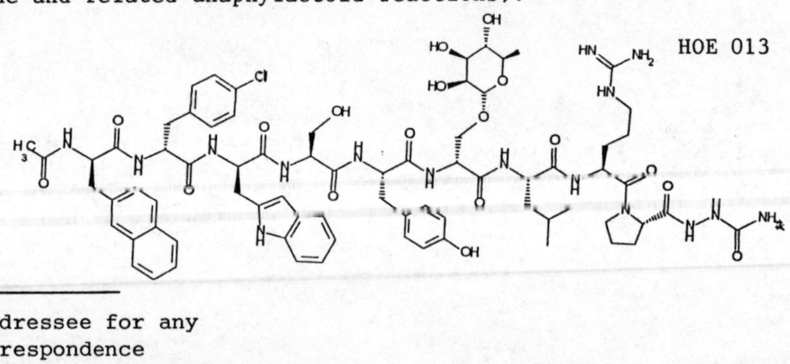

† addressee for any correspondence

*Abbreviations (see also Preface).- Ab, antibody; i.s., internal standard; LHRH, luteinizing hormone releasing hormone.

A sensitive and specific method was required to support the pre-clinical and clinical studies needed to develop this drug candidate. At the start of pharmacokinetic studies in aminals and in humans, only an RIA (research assay) was available. Due to the possible lack of specificity with RIA, however, the development of a highly specific and sensitive reference method had been considered necessary. LC-MS-MS was selected because of its potential to fulfil these requirements.

EXPERIMENTAL

RIA materials.- The sources were: phosphate, acetic acid, EDTA, Na azide & BSA (#05480), Fluka; goat-anti-rabbit gammaglobulin ('forte'), Behringwerke; rabbit gammaglobulin (HOE 013 antiserum, AS-K-1/3; 1:8000 working solution), Hoechst AG; ^{125}I-HOE 013 (1290 µCi/µg) & G-25 Sephadex, Pharmacia.

LC-MS-MS materials.- The sources were: acetic & formic acids, ammonium acetate, dichloromethane, acetonitrile (HPLC grade S) & methanol: Riedel de Haen; 1cc C-8 extraction column (100 mg), ICT; rechargeable column 125 × 4.0 i.d. mm packed with Nucleosil 300-5 (C4), MPN. The i.s. was prepared in-house.

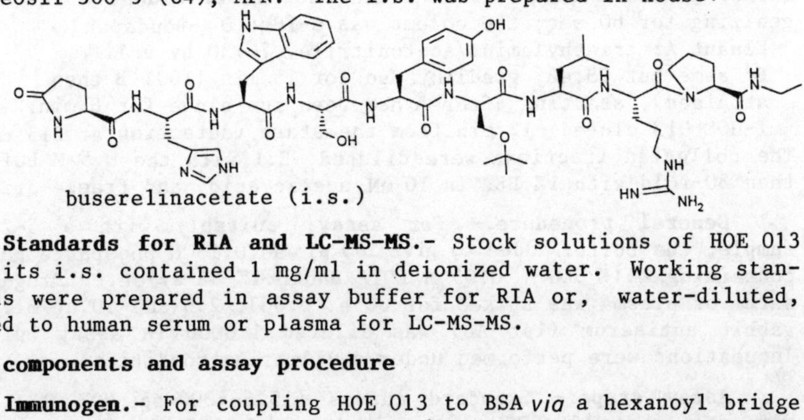

buserelinacetate (i.s.)

Standards for RIA and LC-MS-MS.- Stock solutions of HOE 013 and its i.s. contained 1 mg/ml in deionized water. Working standards were prepared in assay buffer for RIA or, water-diluted, added to human serum or plasma for LC-MS-MS.

RIA components and assay procedure

Immunogen.- For coupling HOE 013 to BSA *via* a hexanoyl bridge BSA (220 mg) was reacted with ANSA (8-anilino-1-naphthalenesulfonic acid; 110 mg) in 20 ml 0.1 M PBS buffer pH 7.4 (1 h, room temp.). Chromatography (G-25 Sephadex, 0.1 M pH 6.2 PBS), peptide addition (137 mg; under N_2) to collected eluate, stirring (17 h, room temp.), treating the clear layer with 10% acetic acid (3 × 100 ml; ultrasonic bath, 7 min), filtering the 'extract' and freeze-drying gave purified immunogen (274 mg).

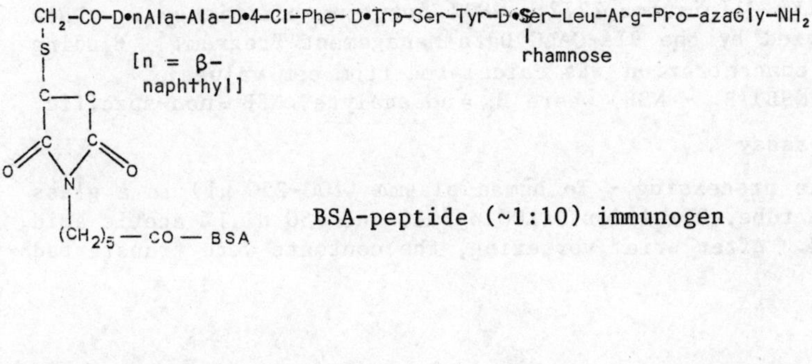

BSA-peptide (~1:10) immunogen

Immunization.- Three rabbits were immunized using the same procedure for all three, starting with 0.5 mg immunogen in 0.5 ml saline, supplemented by 0.5 ml complete Freund's adjuvant (cFA; Difco Labs.) injected s.c. at 4-6 abdominal sites. Two booster injections with 10-day intervals were accompanied by 0.5 ml of incomplete Freund's adjuvant (FA; Behringwerke) administered by the same route. After ~9 weeks and then 4 weeks later, 3rd and 4th boosters were given. Rabbits K1 and K3 received a 5th one ~50 weeks after the 4th. Blood was taken at adequate times after boosting, and antiserum aliquots were stored at -20°.

Radioligand.- HOE 013 was mono-iodinated in the 3-position of the molecule's tyrosine moiety. ^{125}I-Sodium iodide (0.13 kBq, 3.5 mCi) was mixed with 50 µl 0.5 M Na phosphate buffer pH 7.4, and added to 20 µg (12.6 nmol) HOE 013 acetate, dissolved in 100 µl 10 mM acetic acid. After adding 10 µl buffer (as above but only 0.05 M) which contained 20 µg of chloramin T, the reaction mixture was whirled (60 sec, room temp.). Excess oxidizing agent was neutralized by adding with whirl-mixing (30 sec, room temp.) 20 µg Na metabisulfite in 10 µl of 0.05 M buffer. HPLC was performed on the mixture after adding 300 µl of eluant **A** and homogenizing for 60 sec; the column was C-18 (10 µ-Bondapak).

Eluant **A**: triethylamine/acetonitrile, 70:30 by vol.
B: same but 58:42; gradient run for 15 min (100% **B** then attained), starting after **A** had been run alone for 8 min.
I-HOE 013 eluted ~12 min from the start (detection at 215 nm). The collected fractions were diluted ~2:1 with the 0.5 M buffer, then 30-fold with 1% BSA in 10 mM acetic acid, and freeze-dried.

General procedure.- For assay, suitably with a 5-20 µl sample, the buffer added to give 100 µl was 0.15 M phosphate pH 7.4 containing 0.1% BSA, 0.05 M EDTA and 0.1% Na azide. Drug-free serum or plasma was spiked for QC's, 0.667, 2.5 and 10 ng/ml. The rabbit antiserum (1st Ab) was diluted 1:8000 in assay buffer. Incubations were performed under equilibrium conditions.

Assay steps.- Standard curves (15.6-4000 pg HOE 013/tube) were set up in 12 × 75 mm glass tubes with 100 µl of tracer containing 8000-10000 cpm/tube (6.2-8 pg peptide) and 100 µl antiserum (AS K1/3) at 1:8000 working dilution. Incubation was at 4° for 48 h then, after adding 200 µl of 2nd Ab (1:90), 24 h. After centrifugal separation with 1 ml PBS (1800**g**, 15 min), the supernatant was decanted.

Radioactivity in the precipitate was determined in a gamma-counter (LKB-Clinigamma 1272; ~80% counting efficiency). Data were analyzed by the RIA-CALC Data Management Program. Binding % at each concentration was calculated from cpm values:
$100 \times (B - NSB)/B_0 - NSB)$ where B_0 = no analyte, NSB = non-specific.

LC-MS-MS assay

Sample processing.- To human plasma (100-250 µl) in a glass centrifuge tube, 25 µl i.s. (100 ng/ml) and 750 µl 1% acetic acid were added. After brief vortexing, the contents were transferred to a C-8 column (ICT, 1cc; pre-conditioned with 2 × 1 ml methanol and 4 × 1 ml re-dist. water) followed by elution under ~15 kPa vacuum. After washing with 1 ml re-dist. water and 0.5 ml water/methanol (1:1) under ~80 kPa vacuum, and sucking dry for ~15 min, HOE 013 was eluted with 1 ml dichloromethane/methanol/0.1% acetic acid under gravity. The eluant was re-applied to the column and allowed to run off into a tapered tube; collection of the elution mixture residue was done centrifugally (2300**g**, 10 sec). After drying down in a water bath (45°, N_2) the residue was dissolved in 100 µl mobile phase, and 90 µl (kept in a dark vial) injected.

Chromatography was performed with a Spectra-Physics pump (#P 4000) and a Gilson auto-injector (#231) with a 100 µl loop (90 µl injected). The 125 × 4.0 mm column was packed (replaceably) with Nucleosil 300-5 (C-4; MPN). The mobile phase (0.85 ml/min) was 60:40 ammonium acetate (2 mM)/acetonitrile.

Detection.- MS-MS was performed with a Sciex API-III Plus apparatus; the capillary comprised 1 m of 0.075/0.150 mm (i.d./o.d.) FS tubing. Post-column split, 1:20; nebulizer, 64 psi; curtain gas, 1.1 L/min; interface heater, 55°; aux. gas flow, 6 L/min; Turbo Ion Spray, 300°. Parameter values: DI 50; ISV 4000; IN 650; OR 40; RO 30; M1 1000; RE1 115.2; DM1 0.180; R1 27; L7 15; R2 15; M3 1000; RE3 114.5; DM3 0.180; RX -5; R3 5, L9 -100; FP -150; MU -4600; CC 1; CGT 250. The parent compound HOE 013 and the i.s. were double-ionized and fragmented to daughter ions. Product peak-area ratios in MRM (200.00 / 0.02 m-sec dwell/pause; 20.0 Q3 Park Mass), analyte/i.s. (775.2 → 693.2, 620 → 592.4), were computed with Sciex Mac Quan 1.4.

Calibration.- The $y = 1/x^2$ function was obtained by a linear least-squares regression forced through the origin ($a_0 = 0$) since there was no blank signal. The independent variable x was the set concentration and the dependent one y the peak-area ratio measured (Q); slope calculation: $a_1 = \sum_{i=1}^{n} x_i y_i / \sum_{i=1}^{n} x_i^2$

Analysis function for calculating results: $Q \times 1/a_1$ = concn. found Thus, $1/a_1$ corresponds to the response factor (RF) used for calculations in the data system of the LC-MS-MS.

RESULTS

The aim was to develop a sensitive (<1 ng/ml) and specific assay for HOE 013 in serum, meeting other quality criteria also (e.g. precision, accuracy). Validation results are now given.

RIA validation studies.- Fig. 1 shows a typical standard curve with a measuring range from 0.16 to 40 ng/ml. Using a 20% maximum for intra-assay C.V.'s the LLOQ was 0.6 ng/ml. Some intra- and inter-assay values lay outside the defined limits, 15% (20%) in the normal (lower) concentration range. Recovery and stability data also revealed that the available RIA could not fully meet accepted requirements. As an RIA re-development based on new Ab's would have taken months, LC-MS-MS was reckoned to be a suitable alternative.

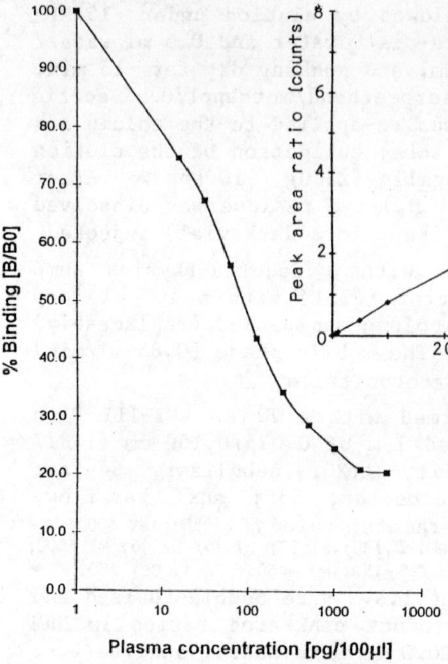

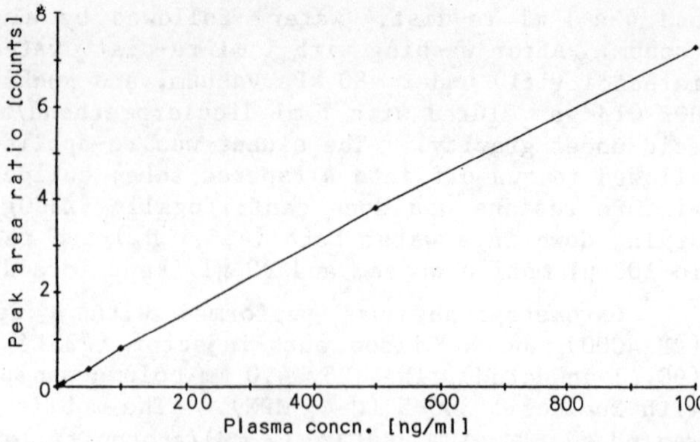

Fig. 2, *above*. Calibration curve for HOE 013 in human plasma by LC-MS-MS.

Fig. 1, *left*. Calibration curve for HOE 013 in plasma by RIA.

LC-MS-MS validation studies.- The calibration curve (Fig. 2) showed excellent linearity over a 0.5-1000 ng/ml measuring range and reproducibility using a weighted ($1/x^2$) regression. Peak-area ratios were blank for drug-free serum or plasma. The LLOQ was 0.5 ng/ml (both S.D. and difference between mean measured and known value <20%). Both precision and accuracy, within-day and day-to-day, for standards and QC's prepared independently were <15% over the measuring range. The extraction yield of HOE 013 and its i.s., ~30%, was relatively low, maybe due to the hydrophobicity of the molecule, but there was no detriment to assay quality. The stability data at different temperatures and after several freeze-thaws were also satisfactory. Assay specificity could be proven, e.g. with blank plasma from different subjects where HOE 013 ran free from endogenous interferants (Fig. 3).

Cross-validation of RIA and LC-MS-MS

Despite the RIA's quality problems a limited cross-validation between the two methods was conducted, to see whether RIA data from earlier studies could be correlated with LC-MS-MS values. For plasma samples from a Phase I study in volunteers, values were obtained by both methods. Quality problems with the RIA were again found with respect to precision and accuracy, manifest by QC samples. The kinetic profile of the investigated subjects,

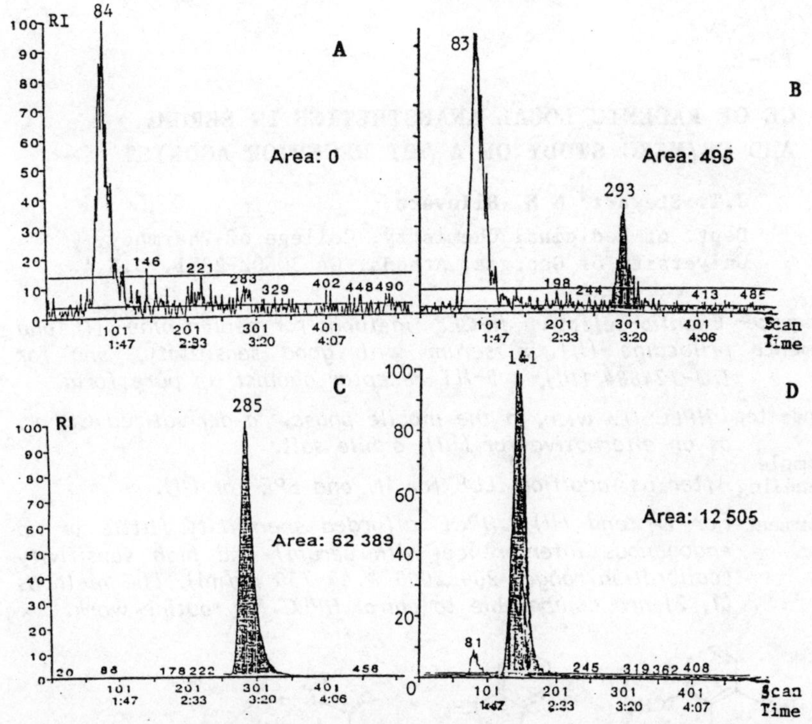

Fig. 3. LC-MS-MS of blank plasma - unspiked (**A**), spiked with HOE 013, 0.5 (**B**) or 100 (**C**) ng/ml, or spiked with i.s.(**D**). The values above peaks are MRM scan nos. Early non-hatched peaks (nature unclear) are irrelevant.

however, looked very similar with the two methods; but the RIA values were in general about twice the LC-MS-MS values. The explanation could be cross-reactivity of the RIA to metabolic degradation products of HOE 013 (hepta-, octa- and nona-peptide) as found in metabolism studies. This assumption has still to be confirmed with authentic metabolites.

CONCLUDING COMMENTS

The RIA available for determining HOE 013 (research assay) showed lack of specificity and quality problems. On the other hand, the LC-MS-MS method subsequently developed proved to be appropriate for analyzing HOE 013 in serum and plasma as demonstrated firstly in a validation study and later in different toxicological and human Phase I studies. It showed high specificity, sensitivity and robustness. Altogether the LC-MS-MS peptide assay using the ESI (Turbo Ion Spray) mode met all accepted criteria for quality, and its development and validation were completed within 2 months. As many as 80 test samples/day could be processed by one person and measured overnight.

#B-2

CE OF RACEMIC LOCAL ANAESTHETICS IN SERUM AND CE/MEKC STUDY OF A 5HT RECEPTOR AGONIST

J.T. Stewart† & M. Siluveru

Dept. of Medicinal Chemistry, College of Pharmacy, University of Georgia, Athens, GA 30602-2351, U.S.A.

Requirements *Enantio-selective HPCE* methods for mepivacaine (I) and prilocaine (II) in serum with good sensitivity, and for ±DU-124884 (III), a 5-HT receptor agonist, in pure form.*

End-step *HPLC-UV with, in the mobile phase, a derivatized CD or, as an alternative for (III), a bile salt.*

Sample handling *After i.s. addition, LLE for (I) and SPE for (II).*

Comment *For (I) and (II), HPCE afforded specificity (little or no endogenous interference with serum) and high sensitivity (calibration ranges 200-2000 & 45-750 ng/ml). The methods [1, 2] are comparable to chiral HPLC for routine work.*

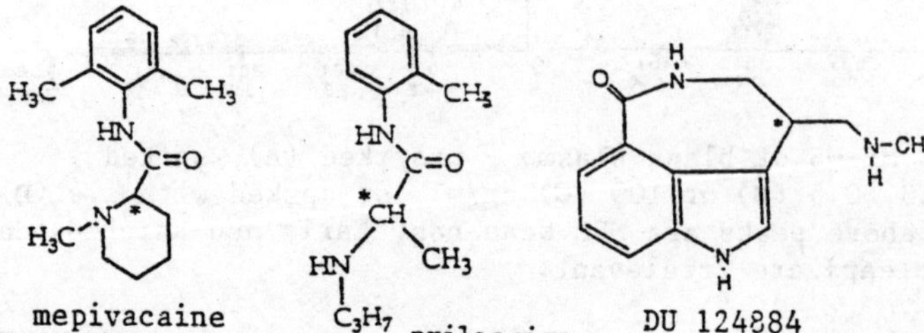

mepivacaine (I) prilocaine (II) DU 124884 (III) * denotes centre of asymmetry

HPCE* is currently a popular technique for resolving enantiomers. Principal advantages are simplicity, efficiency and small sample size. Chiral separations are achieved largely by adding native or derivatized CD's to the mobile phase; enantiomer recognition occurs due to a combination of inclusion phenomena and other interactions with functional groups on the CD rim.

As now outlined, HPCE assays on a fused silica capillary have been developed for enantiomers of mepivacaine [1] and prilocaine [2] in serum, using β-dimethyl-CD as selector. The background

---

†Addressee for any correspondence

**Abbreviations*.- (HP)CE, (high performance) capillary electrophoresis; MEKC, micellar electrokinetic capillary chromatography; i.s., internal standard; HTAB, hexadecyltrimethylammonium bromide; LLE, liquid-liquid extraction; SPE, solid-phase extraction; LLOQ, lower limit of quantitation; LOD, limit of detection. CD, cyclodextrin; 'β-dimethyl CD' = heptakis-(2,6-di-*O*-methyl)β-CD.

electrolyte contained a pH 2.5 phosphate buffer, the CD (at mM levels) and HTAB (to reduce serum interferant interactions with capillary-wall silanols). A variant procedure, MEKC with taurocholate, was explored in trial chiral separations on a new 5-HT receptor agonist, ±DU-124884 (a tetrahydro-oxo-azepino indole).

Materials.- The racemic drugs came from Sigma (St. Louis, MO; also sodium taurocholate) and Solvay Pharmls. (Norcross, GA). Astra (Westborough, MA) gifted mepivacaine and prilocaine enantiomers. Aldrich (Milwaukee, WI), Advanced Technologies (Whippany, NJ) and American Maize Products (Hammond, IN) supplied derivatized CD's. Other chemicals were the highest grade purchasable.

CE.- An ABI model 270A system was used, its UV detector being placed at the cathode end of the 72 cm × 50 μm i.d. capillary. The sample was introduced with a 24 sec vacuum injection. **SPE** was done with 1 cc C-18 BondElut cartridges attached to a vacuum manifold; acidified methanol served to elute the drug.

Assay conditions for particular analytes

For mepivacaine spiked into serum, with prilocaine as i.s., LLE was performed with diethyl ether/ethyl acetate, 1:1 by vol. For prilocaine (procainamide as i.s.) SPE was used; resolution in the capillary was performed at 25 kV, 30°. Detection was at 215 nm for both, but 254 nm for ±DU-124884 (pure; not in serum). Extraction recoveries were ~70% and ~85% respectively.

BEHAVIOUR OF THE SERUM ASSAYS, AND COMPARISON WITH HPLC

Assessment details, intra- and inter-day, have been published along with illustrative electropherograms [1, 2]. In brief, for mepivacaine with 200-2000 ng/ml calibration range, the LLOQ was 200 ng/ml; precision and accuracy were in the 2-8% range for both enantiomers. For prilocaine (range 45-750 ng/ml) the LLOQ was 45 (LOD 38) ng/ml for each isomer; precision and accuracy 2-9%. For both drugs the CE assays compared well with HPLC (Table 1).

OPTIMIZATION OF ENANTIOMER SEPARATIONS

For the anaesthetics, various native and neutral or charged derivatized CD's were studied as chiral selectors in the mobile phase [1, 2]. Only the neutral β-dimethyl CD gave baseline resolution of the enantiomers. With optimized conditions (buffer and CD concentrations, buffer pH, applied voltage) and the detergent HTAB present in the mobile phase, the migration times (min) were:

mepivacaine: (+), 14.9; (-), 15.4; (-)-prilocaine, 17.9
prilocaine: (-), 16.0; (+), 16.4; procainamide, 8.9
β-Dimethyl-CD worked for ±DU-124884 but not charged (+ or -) CD's.

Method development for ±DU-124884 enantiomers

Optimized conditions were developed with, as chiral selector, β-dimethyl CD (Table 2; electropherogram, Fig. 1A) or taurocholate (Table 2; Fig. 1B). Migration times for both enantiomers were <25 min. In general, peak shape and resolution were better with β-dimethyl CD than with taurocholate.

Table 1. Comparison of HPCE and HPLC assays for *R*(+) and *S*(-) mepivacaine (I) and prilocaine (II) in human serum: α, selectivity; Rs, resolution; N, theoretical plates; t, migration/retention time, min. Range (of linearity), LOD and LLOQ are ng/ml. *Data as in refs. [1] & [2], courtesy of Elsevier Science Publishers.*

	(I), HPCE		(I), HPLC		(II), HPCE		(II), HPLC	
	R	*S*	*R*	*S*	*R*	*S*	*R*	*S*
α	1.03		1.23		1.03		1.13	
Rs	1.75		2.80		1.82		1.95	
N	42238	52056	4245	2164	84624	88912	9101	8666
t	14.9	15.4	9.3	11.4	16.0	16.4	9.3	10.1
Range	200-2000		150-2400		45-750		10-1000	
r^2 (n=3)	>0.999		>0.999		>0.999 (n=4)		>0.999	
LOD	150	150	100	100	38	38	4	5
LLOQ	200	200	150	150	45	45	10	10

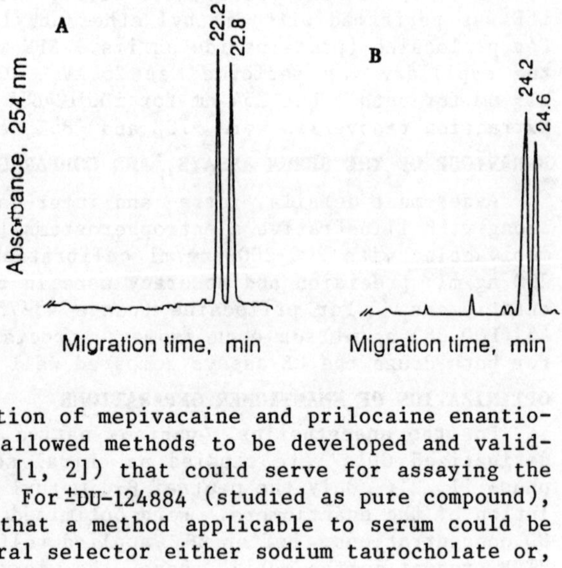

Fig. 1. Electropherograms of racemic DU-124884 with (**A**) β-dimethyl CD, (**B**; MEKC) sodium taurocholate as chiral selector. Parameters were set as marked in Table 2.

CONCLUDING COMMENTS

In summary, resolution of mepivacaine and prilocaine enantiomers achieved by HPCE allowed methods to be developed and validated (for details see [1, 2]) that could serve for assaying the two analytes in serum. For ±DU-124884 (studied as pure compound), HPCE trials indicated that a method applicable to serum could be developed using as chiral selector either sodium taurocholate or, preferably, β-dimethyl CD.

Table 2. Influence of various parameters (organic modifier as % v/v) on the resolution and migration times (min) for ±DU-124884 with β-dimethyl-CD or sodium taurocholate as the chiral selector.

Parameter		β-dimethyl-CD			taurocholate		
		Rs	Pk.1	Pk.2	Rs	Pk.1	Pk.2
Chiral selector concn., mM	10	-	12.6	-	-	12.8	-
	20	0.6	14.1	14.2	0.7	15.2	15.4
	30	0.9	15.8	16.0	0.8	19.0	19.2
	40	1.0	17.2	17.4	0.9	20.9	21.3
	†50	1.1	18.5	18.6	1.2	24.2	24.6
	60	1.3	19.7	20.1	1.1	26.6	27.0
	*70	1.9	22.2	22.8	0.8	35.0	35.3
	80	1.8	23.9	24.5			
	90	1.7	25.3	25.9			
Buffer concn., mM	25	1.7	17.5	17.9			
	50	1.7	19.1	20.0			
	75	1.8	19.6	20.0			
	100	1.9	22.1	22.7			
pH	*2.5	1.7	22.5	23.1	1.2	24.1	24.6
	4.5	1.3	21.1	21.0	0.6	21.3	24.6
	6.5	0.9	14.3	14.6	-	12.0	-
	8.5	-	6.9	-	-	7.5	-
Voltage, kV	10	1.7	49.5	50.6	-	-	-
	15	1.8	31.6	32.1	1.0	37.1	37.7
	20	1.9	22.1	22.5	1.2	24.1	24.6
	*25	1.5	16.2	16.5	1.0	16.2	16.5
	30	1.2	12.2	12.4	0.6	10.9	11.1
Temp., °C	*30	1.8	22.6	23.1	1.2	24.0	24.5
	40	1.5	18.9	19.3	1.1	20.4	20.9
	50	1.3	16.2	16.5	1.0	16.9	17.3
Methanol *or acetonitrile* %v/v	5	1.4, *1.3*	21.7, *21.3*	22.2, *21.8*			
	10	1.1, *1.1*	22.9, *20.2*	23,5, *20.5*			
	20	0.8, *0.6*	26.7, *19.0*	26.9, *19.2*			

For each variation studied, generally the other parameters were set at the * value († in taurocholate trial if different)

Values adopted in the final method were those marked * (cf. Fig. 1, **A**).

References

1. Siluveru, M. & Stewart, J.T. (1997) **J. Pharm. Biomed. Anal. 15,** 1751-1756.
2. Siluveru, M. & Stewart, J.T. (1997) **J. Chromatog.** B **693**, 205-210.

ADDED by Editor.- HPLC methodology is described in other 1997 arts. by these authors - as for 2., **690,** 359-362, and **Anal. Lett. 30,** 1167-1178; see also category #5a in the Assay Compendium, #ABC.

#B-3 n

A Note on

OVERCOMING VERY LATE ELUTING INTERFERENCES IN A TRACE BIOASSAY FOR A NEW CHEMICAL ENTITY IN EARLY PHASE I DEVELOPMENT

Alan M. Edgington & Richard F. Venn

Pfizer Central Research, Sandwich, Kent CT13 9NJ

To support Phase I clinical trials, a method was needed to assay plasma for a basic heterocyclic compound (Pfizer; 'UK-X'). Initial experiments using SPE gave low recoveries and variable results. Ultimately a method was developed that involved 'liquid-liquid' extraction with ethyl acetate using Chem-Elut cartridges (Phenomenex), the analyte being adsorbed onto the inert packing; this mode had worked well for other analytes and was readily automated. Pre-HPLC trace enrichment was done by an SCX cartridge (as used successfully for other such compounds). Initial results with this method were encouraging; peaks for the compound and internal standard (i.s.) were seen at 9.5 and 12.5 min in clear areas of the baseline (Fig. 1A).

When the first validation was performed a problem was obvious with all but the very earliest samples: the baseline, previously clear, began to show significant large and wide peaks, precluding accurate quantitation of the compound (Fig. 1B). Extending the run time (Fig. 1C) disclosed a series of repeating peaks, the time between which was identical with the time between injections.

The interfering peaks were late eluters; they did not appear in the first three chromatograms and they had very long retention times, ~80 min for one injected sample. Several other extraction solvents were tried; but each was found to reduce assay recovery. An effective remedy that removed the late-eluting material was switching of the outflow from a cartridge (HiRPB†) that served as a guard column just before the analytical column.

Switching procedure.- The set-up that gave improved results (example: Fig. 2) is shown in Fig. 3. Back-flushed UK-X and i.s. from the SCX cartridge are fed rapidly through the 1 cm HiRPB cartridge onto the analytical column, from which the analytes are eluted using a second pump while the later eluting interferants pass slowly (<13 min) through the HiRPB cartridge to waste.

Extraction procedure.- The plasma (human-based; 1 ml), with 2 ml pH 10 borate buffer added, was loaded onto a 12 ml Chem-Elut cartridge, followed by 7 ml ethyl acetate. The eluate was dried down (37°, N_2), reconstituted in 0.5 ml of loading mobile phase and centrifuged (2000 **g**, 30 min); 0.3 ml was injected.

---

†HiRPB is C-8 + C-18. Points from Hichrom manual: residual acidic sites eliminated... monolayer coverage of high purity silica... combines the robustness of C-18 with the high coverage of a C-8.

Fig 1. Typical chromatograms with original method.
(A) Early run in a series.
(B) Later run.
(C) Extended run time, showing regular pattern of interfering peaks.

Plasma (human) containing 'UK-X' (cationic) and i.s. was extracted with ethyl acetate; the residue from drying down was taken up in mobile phase ('analytical'; cf. final method - similar column). For some details see foot of next p.

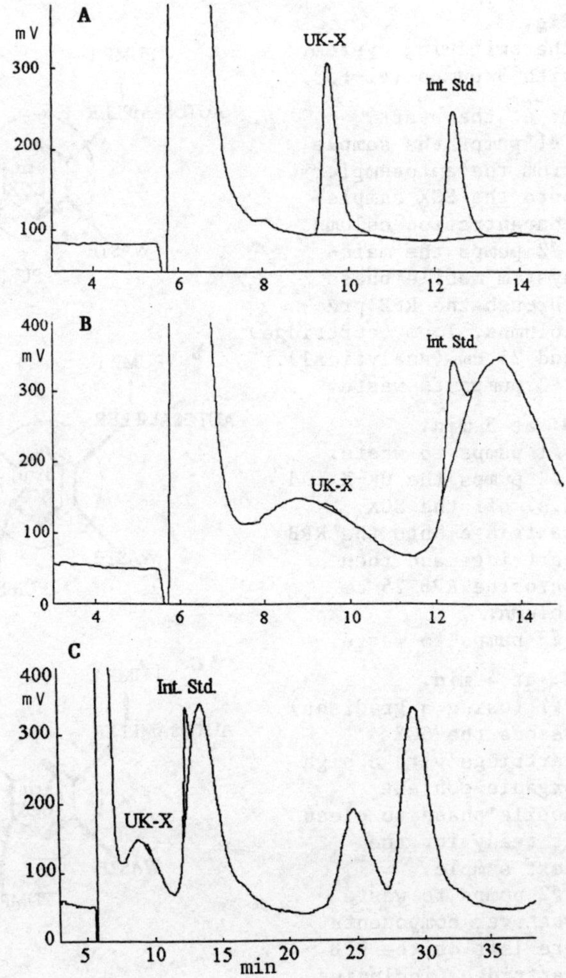

Fig. 2. Typical late chromatogram (cf. Fig. 1B) with the improved method described in the text, entailing pre-HPLC switching as in Fig. 3.

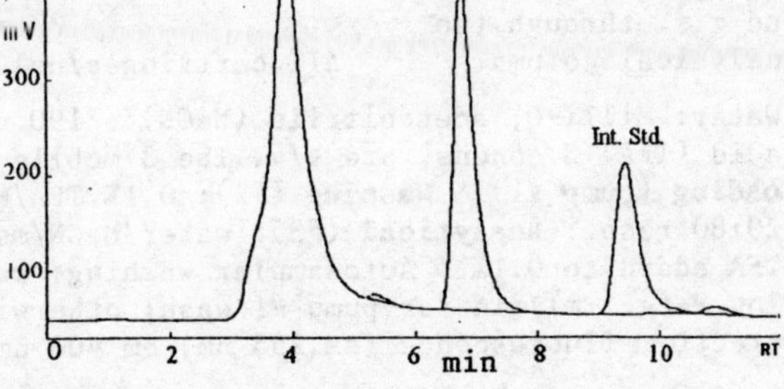

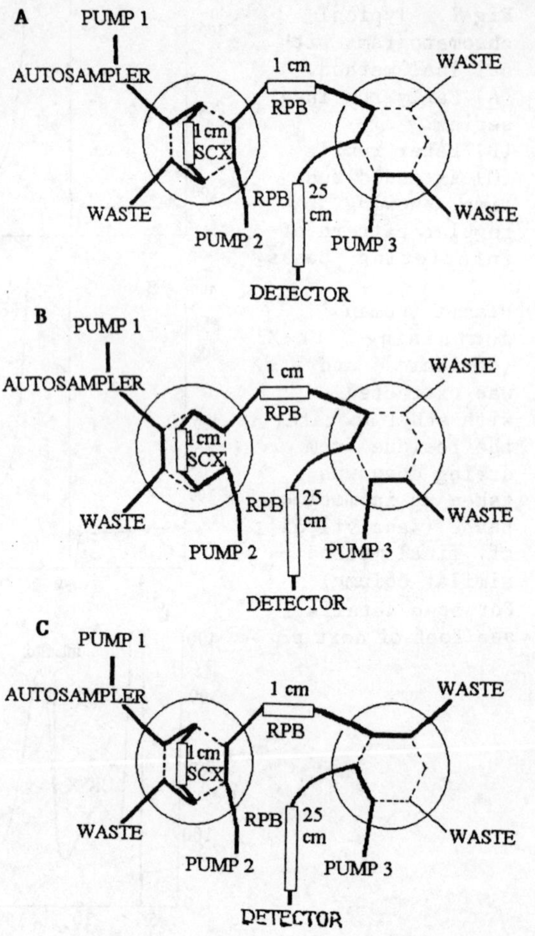

Fig. 3.
The switching system, with 3 pumps (#1-#3).

A: at the start.
#1 pumps the sample from the autosampler onto the SCX sample-concentration column
#2 pumps the main-system mobile phase through the RPB pre-columns, 1 cm (cartridge) and 25 cm (analytical).
#3 pumps to waste.

B: at 3 min.
#1 pumps to waste.
#2 pumps the UK-X and i.s. off the SCX cartridge onto the RPB cartridge and then onto the RPB 25 cm column.
#3 pumps to waste.

C: at 4 min.
#1 (using a gradient) washes the SCX cartridge with a high organic content mobile phase to clean it ready for the next sample.
#2 pumps to waste whatever components are left on the RPB cartridge, including the late interferants.
#3 takes over from #2 for pushing the UK-X and i.s. through the analytical column.

All cartridges/columns were 3.2 mm, 5 μm.

Water: Milli-Q; acetonitrile (MeCN): '190 nm'. Trifluoroacetic acid (TFA) % concns. are v/v. The 3 mobile phases comprised (%): **Loading** (pump #1) & **Washing** (#2): 0.1% TFA/MeCN, 90:10 and 20:80 resp.; **Analytical** (#3): water/MeCN/methanol, 47:30:23; TFA added to 0.1%. Autosampler washing: water/MeCN, 50:50. Flow rate 1 ml/min for pump #1 wash; otherwise 0.5 ml/min. Detection: fluorescence (ex 255 nm, em 400 nm).

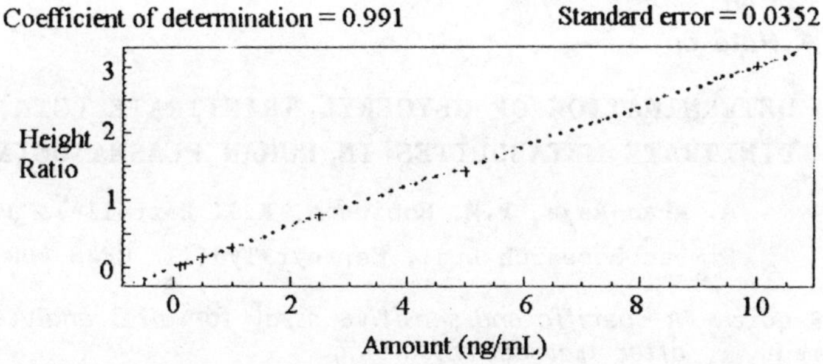

Fig. 4. Improved method: a calibration curve.

Table 1. Validations (3) of the improved method.

	Conc. Prepared (ng/mL)	Conc. Measured (ng/mL) Mean	±	SD.	n	% Inacc.	% Imprec.
Batch 1	0.1	0.106	±	0.003	7	6.4	2.8
	1.0	1.00	±	0.036	7	0.0	3.6
	10.0	10.6	±	0.370	7	6.1	3.5
Batch 2	0.1	0.093	±	0.008	7	7.4	8.6
	1.0	1.04	±	0.061	7	3.8	5.9
	10.0	11.0	±	0.601	7	9.9	5.5
Batch 3	0.1	0.106	±	0.015	7	6.7	13.6
	1.0	1.02	±	0.074	7	1.7	7.3
	10.0	10.7	±	0.776	7	7.4	7.2

OUTCOME

The addition of the extra column-switching step dramatically improved the cleanliness of all later samples. Even after 90 samples there was no indication of the changing baseline which was found after 4 samples in the previous system. Fig. 4 shows a typical calibration curve. The assay was successfully validated on three occasions (Table 1) and found to be rugged and robust in practice. Assay performance was unaffected by lipaemic or haemolyzed plasma or by repeatedly frozen and thawed plasma, nor by the presence of various co-administered compounds. The method served to determine UK-X concentrations in ~2000 samples from clinical trials.

In summary, a combination of a column switching valve and a 1 cm SCX guard column was used successfully, without greatly affecting sample throughput, to remove late-eluting interferants which had troubled the assay.

The developed assay was rugged and robust and found to work with a wide range of plasmas. Evidently guard columns can have wider usefulness than merely protecting analytical columns.

Such an HPLC system that eliminates late-eluting interferants may be used in conjunction with extraction techniques that have previously been regarded as too unselective, even if optimized.

#B-4n
A Note on

DETERMINATION OF GLYCERYL TRINITRATE (GTN) AND ITS TWO DINITRATE METABOLITES IN HUMAN PLASMA USING GC-MS

A. Khan-Raja, P.R. Robinson, K.J. Barrell* & J. Maddock
Simbec Research Ltd., Merthyr Tydfil CF48 4DR, U.K.

Requirement *A specific and sensitive assay for all 3 analytes in plasma after transdermal dosing.*

End-step *GC-MS(nCI)*[†] *with an appropriate i.s.; splitless injection.*

Sample handling *Liquid-liquid extractions; pooled extracts dried down. $AgNO_3$ in the sample pre-storage (prevents GTN degradation).*

GTN is a potent vasodilator widely used in the treatment of angina pectoris. It is rapidly metabolized in man to the active 1,2- and 1,3-dinitrates. There are numerous GC methods for measuring GTN [1-4]; however, few incorporate MS detection [5][▫]. The sensitive and specific GC-MS method described here, with *p*-nitrobenzyl alcohol (in methanol) as i.s., determines GTN and the two metabolites simultaneously. The extraction procedure provides equivalent recoveries for each component, through re-extraction with a solvent apt for recovering the metabolites.

To the plasma sample (2 ml) were added 80 µl 20 mM $AgNO_3$ and i.s. Liquid-liquid extraction was performed with pentane and then with methyl *t*-butyl ether; the pooled extracts were taken to dryness (N_2 stream). The residue was reconstituted in anhydrous toluene and transferred to an autosampler vial. A 1.2 µl aliquot was then introduced into the GC column by splitless injection.

The MS(nCI) analysis was performed with a Finnigan Incos XL quadrupole mass spectrometer coupled to a Varian 3400 gas chromatograph with CTC A200S autosampler. The reagent gas was ammonia. The fused silica capillary column (30 m × 0.32 mm i.d., 1 µm film thickness) was run with helium as carrier gas (1 ml/min). The oven was temperature-programmed: 1 min at 100°, then a 50°/min rise to 280°, where it was held for 2.9 min. The injection port and transfer line were both maintained at 150°.

The selected ions monitored were the established base peaks, m/z 62 both for GTN (which had shown a subsidiary peak at m/z 46) and for both GDN's, and m/z 153 for i.s. (minor peaks manifest also, notably at 92, 119, 135, 190, 213, 229). The GDN reference compounds were obtained from Promochem (Welwyn Garden City).

---

*addressee for any correspondence
†*Abbreviations*.- nCI, negative ion chemical ionization; i.s., internal standard; LLOQ, lower limit of quantitation.
▫Analyte Index listings in this series: **Ia**, "Glyceryl...".-*Ed*.

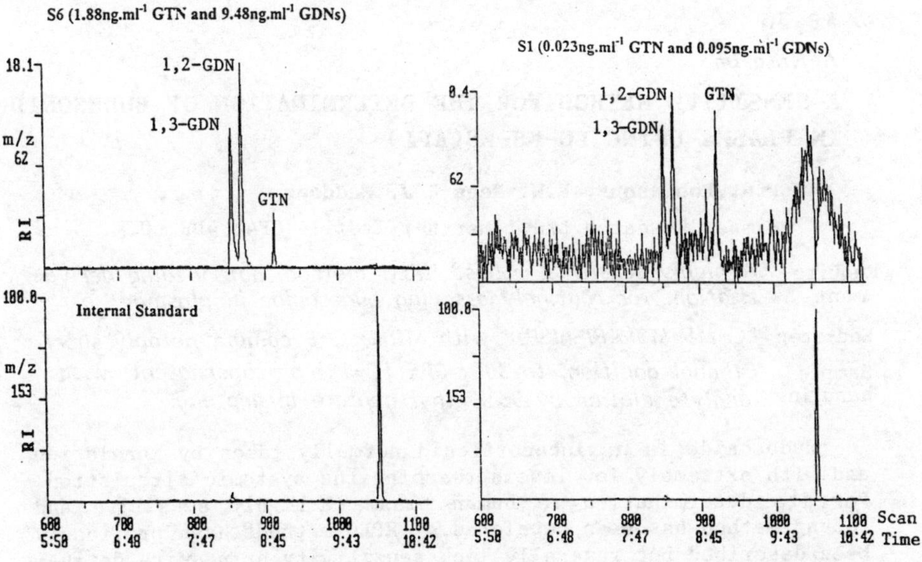

Fig. 1. GC-MS(nCI) peak pattern (m/z 62 and 153) for plasma spiked with GTN and GDN's at the levels shown [note the RI axis difference], the RI being based on the peak for i.s. (50 ng/ml). Peak areas rather than heights were used for quantitation, being more accurate.

Fig. 1 shows typical chromatograms for spiked plasma. The LLOQ's were 0.025 ng/ml (GTN) and 0.1 ng/ml (GDN's). Consistent extraction recovery (>75%) was obtained for the three analytes. A linear GC-MS response was obtained over the concentration ranges 0.025-2 and 0.1010 ng/ml for GTN nd GDN's respectively (mean correlation coefficients of 0.9999). Acceptable inter-batch and intra-batch imprecision (maximum 5.82%) and inaccuracy (range -3.06% to +2.63%) were established, and it was demonstrated that the three components were stable in plasma through three freeze-thaw cycles.

This method has been successfully used for the determination of GTN and GDN's in human plasma following dosing by application of several transdermal patches.

References

1. Jorgensen, M. & Anderson, M.P. (1992) J. Chromatog. **B 577**, 167-170.
2. Carlin, A.S., Simmons, G.K., Shiu, A.O., Sager, V.K., Prasad, J.P. & Skelly, J.P. (1988) Pharm. Res. **5**, 99-102.
3. Sioufi, A., Pommier, F. & Dubois, J.P. (1987) J. Chromatog. **413**, 101-108.
4. Wu, C.C., Sokoloski, T.D., Burkman, A.M., Blanford, M.F. & Wu, L.S. (1982) J. Chromatog. **228**, 333-339.
5. Settlage, J.A., Gielsdorf, W. & Jaeger, H. (1983) J. High Resolu. Chromatog. - Chromatog. Comm. (HRC-CC) **6**, 68-71.

#B-5n

A Note on

A SENSITIVE METHOD FOR THE DETERMINATION OF BUDESONIDE IN PLASMA USING LC-MS-MS(API)

P.R. Robinson*, L.W. Rees & J. Maddock

Simbec Research Ltd., Merthyr Tydfil CF48 4DR, U.K.

Requirement *A highly sensitive, robust HPLC method, not needing derivatization, for routinely assaying budesonide in plasma.*

End-step *LC-MS-MS(API-nISP)[a] with MRM; C-8 column, notably short.*

Sample handling *Ethanol addition, to 30%; SPE (C-18) on supernatant, with analyte elution by 35% ethyl acetate in heptane.*

Budesonide is a glucocorticoid normally taken by inhalation and with extremely low levels reaching the systemic circulation. For its determination in human plasma a highly sensitive and robust method has been developed. HPLC methods have previously been described but generally lack sensitivity or require derivatization [1-3]. Our method provides the increased sensitivity necessary for reliable pharmacokinetic analysis and improves sample throughput with reduced chromatographic run times. The plasma sample (1.0 ml), after deproteinization, is transferred to C-18 SPE cartridges for isolation of budesonide and, as i.s., spiked-in beclomethasone dipropionate.

Assay of SPE eluates is done by automated injection onto a Pecosphere 3 C-8 (3 µm; 33 × 4.6 mm) analytical column with a Spherisorb C-8 guard column (5 µm; 30 × 2.1 mm), with as mobile phase acetonitrile/10 mM ammonium acetate (70:30 by vol.). C-8 was chosen for the analytical column to achieve the necessary short run times and sharp peak shape.

LC-MS-MS was conducted using a Perkin Elmer Sciex API III Plus with nISP. MRM was performed on the negatively charged parent ions, using argon to promote collisionally activated dissociation. The resulting parent/daughter ion transitions were m/z 489 to 357 (drug) and 579 to 519 (i.s.). Representative chromatograms are shown in Fig. 1 (R_T's <5 min).

*addressee for any correspondence

[a]*Abbreviations.*- API(nISP), atmospheric pressure ionization, negative-ion mode with Ionspray; MRM, multiple reaction monitoring; i.s., internal standard.

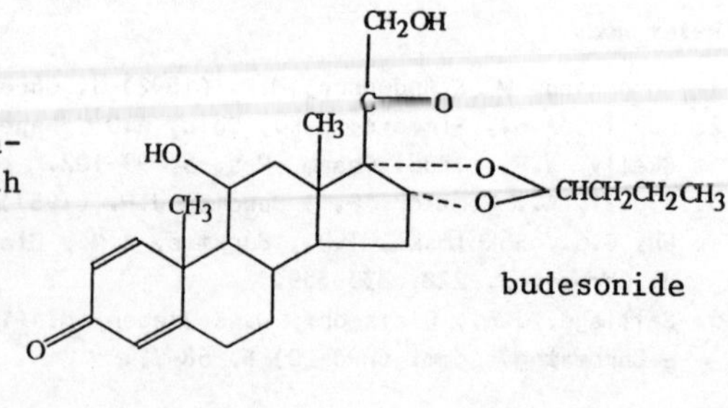

budesonide

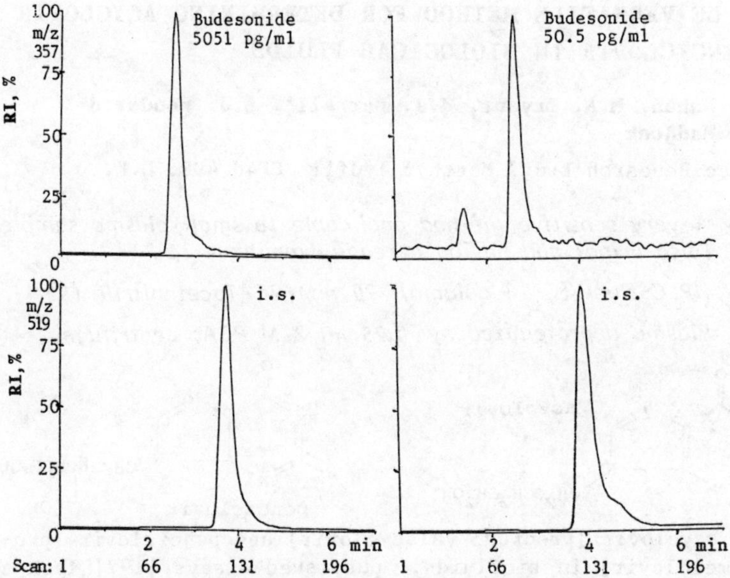

Fig. 1. HPLC-MS-MS of extracts of plasma spiked as shown, with MRM: m/z 357 for budesonide, 519 for i.s.

Clarification furnished to Editor.- "The 100% y-axis relates only to the intensities of peaks on the MRM chromatogram for the stated parent-to-daughter transition. Several such transitions can be monitored during each HPLC run. Integration of the relevant peak area on individual chromatograms provides analyte-to-i.s. peak area ratios for calculation of plasma concentrations from calibration curves."

Analysis of spiked calibration standards has shown that the assay is linear over the concentration range 50 to 5000 pg/ml in plasma, with a mean correlation coefficient of 0.999 and >80% recovery. Inaccuracy and imprecision of the back-calculated standard data did not exceed -2.73% to +3.98% and ±5.08% respectively. Replicate analysis of quality control samples showed an inaccuracy ranging from -2.22% to +2.09% and imprecision ±5.98%. The method has been routinely used in several clinical studies.

References

1. Lindberg, C., Blomqvist, A. & Paulson, J. (1992) Biol. Mass Spectrom. **21**, 525-533.
2. Ryrfeldt, A., Tönnesson, M., Nilsson, E. & Wikby, A. (1978) J. Steroid Biochem. **10**, 317-324.
3. Faouzi, M.A., Dine, T., Luyckx, M., Brunet, C., Gressier, B., Cazin, M., Wallaert, B. & Cazin, J.C. (1985) J. Chromatog. **B 664**, 463-467.

#B-6n

A Note on

A SIMPLE VERSATILE METHOD FOR DETERMINING ACYCLOVIR AND PENCICLOVIR IN BIOLOGICAL FLUIDS

G.A. Laban, M.A. Bryant, K.J. Barrell*, A.J. Woodward & J. Maddock

Simbec Research Ltd., Merthyr Tydfil CF48 4DR, U.K.

Requirement — *A very sensitive method applicable to small plasma samples (also urine) and having a readily available i.s.*[a]

End-step — *HPLC-fluor (C-18 column); 20 mM PCA/acetonitrile (99:1).*

Sample handling — *Plasma deproteinized by 0.25 ml 2 M PCA; centrifuge.*

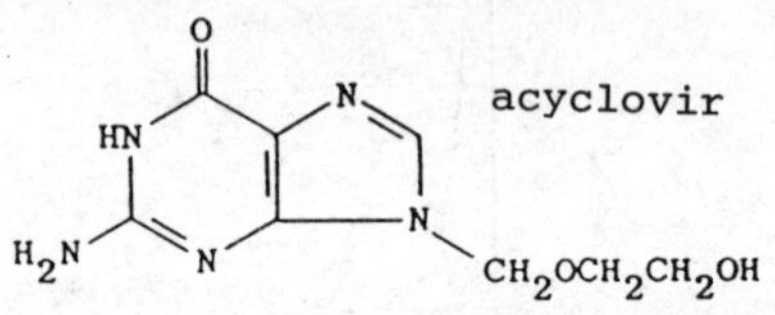

acyclovir

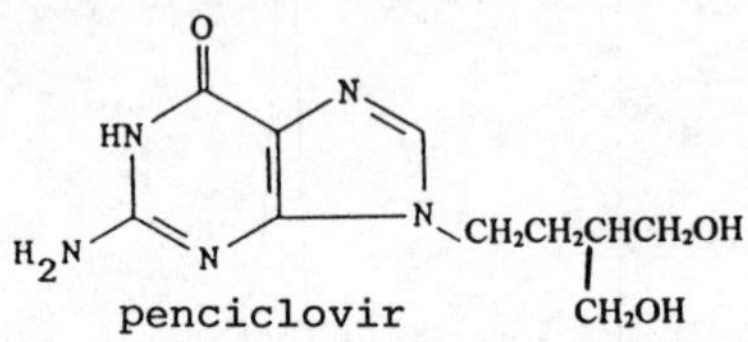

penciclovir

For acyclovir [prodrug: valacyclovir] and penciclovir [prodrug: famciclovir] in biofluids, published assays [1-7] lack a readily available i.s. and/or are too insensitive for low-level work. One [5] has been adapted to assay either drug, the other serving as i.s., down to 10 ng/ml in human plasma, or even in <100 µl animal samples if diluted to 500 µl with human plasma. Human or animal urine is diluted to 1000 µl with mobile phase.

With plasma samples (500 µl), spiked to 1 µg/ml with i.s., 50 µl of PCA supernatant was injected onto a 250 × 4.6mm column (1.5 ml/min; fluor: ex 260, em 360 nm; R_T (min): acyclovir, 5.6; penciclovir, 7.6). There was linearity up to 5 µg/ml (r: acyclovir, 0.9996; penciclovir, 0.9990). The respective inaccuracies, intra- or inter-batch, did not exceed +4.5%/+2.6%, and precisions were within 3.0%/5.4%. Each drug gave 100% recovery and survived 3 freeze-thaws. Neither prodrug interfered.

*addressee for any correspondence | [a]*Abbreviations:* i.s., internal standard; PCA, perchloric acid; r, mean correlation coefficient.

References [Chromatography abbreviated to Chr'y]

1. Svensson, J.O., Barkholt, L. & Sawe, J. (1997) J. Chr'y B **690,** 363-366.
2. Swart, R. & Hundt, H.K. (1994) J. Chr'y **A 663,** 65-69.
3. Nebinger, P. & Koel, M. (1993) J. Chr'y **B 619,** 342-344.
4. Macka, M., Borak, J., Semenkova, L., Popl, M. & Mikes, V. (1993) J. Liq. Chr'y **16,** 2359-2386.
5. Mascher, H., Kikuta, C., Metz, R. & Vergin, H. (1992) J. Chr'y **B 583,** 122-127.
6. McMeekin, J.R., Fowles, S.E., Winton, C.F. & Pierce, D.M. (1992) Anal. Proc. **29,** 178-180.
7. Winton, C.F., Fowles, S.E., Pierce, D.M. & Hodge, A.V. (1990) Anal. Proc. **27,** 181-182.

#B-7

BIOANALYSIS OF ANTIBIOTICS AND ANTIVIRALS IN UNUSUAL MATRICES BY LC-MS-MS(API)

S.D. Clarke†, H.M. Hill, T.N. Liddicoat & T.A.G. Noctor
Covance Laboratories Ltd., Dept. of Biopharmaceutical Analysis, Otley Road, Harrogate, N. Yorks. HG3 1PY, U.K.

The analysis of drugs in uncommon biological matrices such as semen, breast milk, saliva, CSF and MEE is sometimes necessary. The use of LC-MS-MS(API) has greatly eased the development of methods involving unusual matrices, although different matrices may call for different sample pre-treatment approaches. Methods are described for the analysis of a guanosine-based antiviral and a cephalosporin-based antibiotic in unusual matrices.*

Bioanalysis in unusual matrices may be required for various reasons. It may be a matter of proximity to the site of action of the drug (e.g. for compounds acting on the CNS in either an efficacious or a toxic manner, analysis in CSF may be required), or demonstrating absence of a drug from a particular site (e.g. breast milk is analyzed to establish whether drugs may be transferred from mother to offspring; fluid from the infection site may give a better indicator of a compound's therapeutic efficacy than plasma), or of non-invasive sampling (e.g. saliva may be more suitable for sampling from infants, and may be representative of plasma levels for non-ionizable drugs)[1].

Biological fluids vary greatly in complexity. The nature and relative levels of endogenous constituents vary greatly, significantly affecting analytical methods (Table 1) [2, 3].

Table 1. Matrix comparisons.

	Plasma	Semen	Breast Milk	Saliva	CSF
pH	7.3 - 7.4	6.9 - 7.4	6.4 - 7.6	6.4 - 7.6	7.3
Dry Mass (g/l)	80	176	129	6.0	10.8
Water Content (%)	93 - 95.5	92	90	99.4	>99
Protein (g/l)	60 - 75	33 - 69	7.3 20	1 4 - 6.4	0.14 - 0.41
Carbohydrate (g/l)	4.0	8.3	49 - 95	0.15 - 0.33	0.6
Lipids (g/l)	5.5 - 6.0	2.6 - 3.1	13.4 - 82.9	0.03	0.01

---

†addressee for any correspondence
**Abbreviations (see also Preface)*.- CSF, cerebrospinal fluid; LLOQ, lower limit of quantitation; MEE, middle ear effusate. In tandem MS (MS-MS) context: API, atmospheric pressure ionization; MRM, multiple reaction monitoring.

Application of LC-MS-MS(API) greatly eases the development of methods in unusual matrices [4]. The developmental starting point is usually an established method for plasma or serum. There may, however, be difficulties, such as: obtaining control matrix (very limited supplies may be available; cf. the following art.); steps in a plasma method that may be superfluous and reduce the assay's robustness and reliability; high levels of fats, carbohydrates or proteins or vastly different ionic strengths; high viscosity; different interference profiles; variability in pH. See Table 2.

Table 2. Suitable sample preparation techniques.

	Semen	Breast Milk	Saliva	CSF	MEE
Ease of transfer of plasma method	Difficult	Moderate	Simple	Simple	Moderate
Dilution/ direct Injection	No	No	Yes	Yes	No
Protein Precipitation	No	Yes	No	No	Yes
Liquid-Liquid Extraction	Yes	Yes	Yes	Yes	Yes
Solid Phase Extraction	Yes (see text)	Yes	Yes	Yes	Yes

Problems encountered with unusual matrices

Semen.- High viscosity of semen causes pipetting difficulties (resolved by using wide-orifice pipet tips). Protein precipitation is unsuitable as a single-step sample preparation procedure, due to precipitation of co-extractives on the mass spectrometer interface plate causing loss of sensitivity, but is useful as a first step prior to liquid-liquid extraction or SPE. However, cation-exchange SPE is unsuitable due to the high levels of polyamines, which competitively reduce extraction of basic drugs leading to poor and variable recoveries. The pH of semen can increase on prolonged standing to 7.6-8.0 due to CO_2 release.

Breast milk.- High levels of lipids may cause problems with the extraction of basic drugs. Removal of fats is achievable by washing the milk with hexane; ionized drug remains in the aqueous matrix at appropriate pH. Control of pH may be required due to the natural variability of breast milk (6.4-7.6).

Saliva. High viscosity causes pipetting difficulties, resolvable as for semen. Control of pH is generally required due to variability of the pH of saliva (5.8-7.1).

CSF.- If the intended assay is for an anti-infective agent, bacterial or viral load may significantly increase the levels of proteins and carbohydrates. Control of pH is generally not needed due to the small natural variability (unless needed to facilitate extraction).

Analysis of a guanosine-based antiviral in plasma, semen and CSF

Analysis was performed using a SCIEX API III+ in positive-ion MRM mode using either IonSpray or Turbo IonSpray. HPLC details appear in the legend to Fig. 1.

Plasma and CSF method (LLOQ 100 ng/ml).- To a 50 µl sample 250 µl methanol was added to precipitate proteins. After vortex-mixing and centrifugation, the supernatant was reduced to dryness (N_2). The residue was dissolved in 200 µl mobile phase, and 10 µl analyzed by LC-MS-MS (IonSpray).

Semen method (LLOQ 500 pg/ml).- To a 200 µl sample, 600 µl methanol was added to precipitate proteins. After mixing, centrifuging and drying down as for plasma, the residue was dissolved in 200 µl pH 11 NH_4OH and submitted to ASPEC-automated SPE with Isolute ENV+ cartridges (Internat'l Sorbent Technology, Hengoed, U.K.; 100 mg/1 ml reservoir). The eluate was reduced to dryness and dissolved in mobile phase (100 µl); 50 µl was analyzed by LC-MS-MS(API; Turbo IonSpray). Fig. 1 shows a chromatogram.

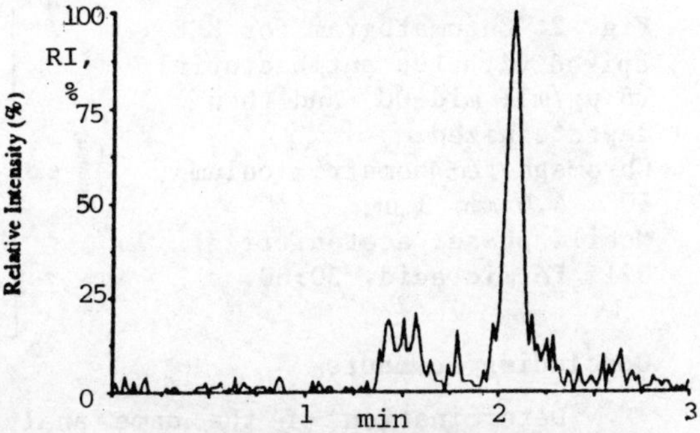

Fig. 1. Chromatogram for an extract of semen spiked with the antiviral at the LLOQ, 500 pg/ml. Cyano column (Spherisorb; 125 × 4.6 mm, 3 µm). Mobile phase: methanol/ 10 mM ammonium acetate, 90:10 by vol.

Initial investigations using protein precipitation only were unsuccessful, primarily due to a large amount of co-extractants deposited in the mass spectrometer ion source, leading to loss of sensitivity. Due to the basic nature of the drug, extraction using cation-exchange SPE was investigated. This proved ineffective because of the high levels of polyamines in semen. Effective extraction required pH >10.5, the pKa of the analyte being high. Polymer-based non-polar SPE cartridges* are stable at high pH, leading to increased reproducibility over conventional silica-based sorbents.

CSF method (LLOQ 500 pg/ml).- To a 200 µl sample 1200 µl pH 11 NH_4OH was added. After vortexing and centrifugation, the supernatant was subjected to automated SPE as for semen. The resulting eluate was dried down. The residue was dissolved in 100 µl mobile phase and 50 µl analyzed as for semen.

*ENV+ is of polystyrene-divinylbenzene type and may suit well for polar analytes which are poorly retained on C-18; cf. art. by M.R. Taylor & co-authors in Vol. 24, pp. 288-298.-*Ed.*

Analysis of an antibacterial in serum and MEE using LC-MS-MS

Analysis of a cephalosporin-based antibacterial was done with HPLC (Fig. 2 legend), then MS-MS (Turbo Ionspray, +ve ion). In the **serum method** (LLOQ 20 ng/ml), 400 µl acetonitrile was added as precipitant to 30 µl. After vortexing, centrifuging, drying down and redissolution in 100 µl 0.1% formic acid, 70 µl was loaded.

Middle ear effusate method (LLOQ 20 ng/ml).- Samples of MEE (30 mg) from glue-ear patients were analyzed (Fig. 2) exactly as for serum. Supplies of control matrix being extremely limited, the compound was spiked into samples at the mid-QC level as part of the validation and compared to extracted serum. The signal reflecting detector response/extraction efficiency for the MEE samples was 30% of the serum value, although consistent between different individual MEE samples (3.0% precision). Concentrations in MEE samples can therefore only be evaluated in relative rather than absolute terms.

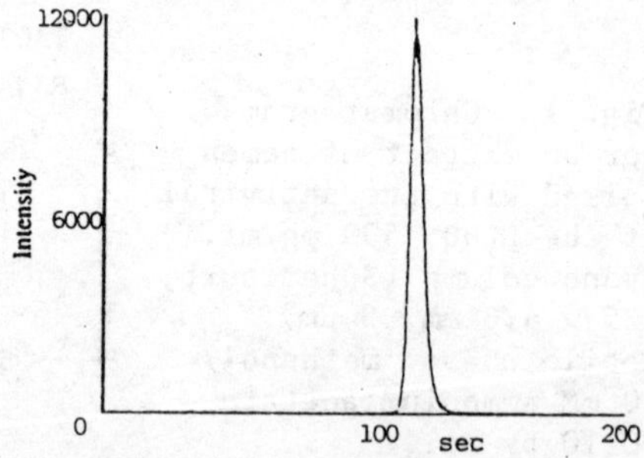

Fig. 2. Chromatogram for MEE spiked with the antibacterial (8 µg/ml; mid-QC) and then deproteinized.
Chromsphere 5 Biomatrix column, 50 × 4.6 mm; 3 µm.
Mobile phase: acetonitrile/ 0.1% formic acid. 20:80.

Concluding comments

Determination of the same analyte in different biological matrices can require very different approaches in sample preparation, even when using LC-MS-MS(API) which effectively eliminates specificity problems. Knowledge of the physicochemical characteristics of the relevant matrices greatly assists systematic development of effective methods.

References

1. Chamberlain, J. (1995) 'Analysis of Drugs in Biological Fluids' (2nd edn.), CRC Press, Boca Raton - see pp. 35-48 (properties of different matrices).
2. Letner, C., ed. (1981) Geigy Scientific Tables, Vol. 1 (8th edn.), Ciba-Geigy, Basel.
3. as for 2. (1984), Vol. 3.
4. Thomas, D., Clarke, S.D., Hill, H.M., Noctor, T.A.G. & Robson, J.N. (1996) in 'Biofluid Assay for Peptide-Related and Other Drugs' [Vol. 24, this series] (Reid, E., Hill, H.M. & Wilson, I.D., eds.), Royal Society of Chemistry, Cambridge, pp.230-238.

#B-8n

A Note on

HPLC DETERMINATION OF DICLOFENAC AS ITS FLUORESCENT CARBAZOLE ACETIC ACID IN HUMAN AQUEOUS HUMOUR WITH A BOVINE SUBSTITUTE MATRIX

K.J. Barrell*, G.A. Laban, A.H. Jenkins, A.J. Woodward & J. Maddock

Simbec Research Ltd., Merthyr Tydfil CF48 4DR, U.K.

Requirement: *A simplified HPLC method for diclofenac (DC), sensitive enough to be used for aqueous humour.*

End-step: *Photoderivatization/fluorescence following HPLC [cf. 1].*

Sample handling: *Sample (50 µl) loaded after adding acetonitrile (30 µl).*

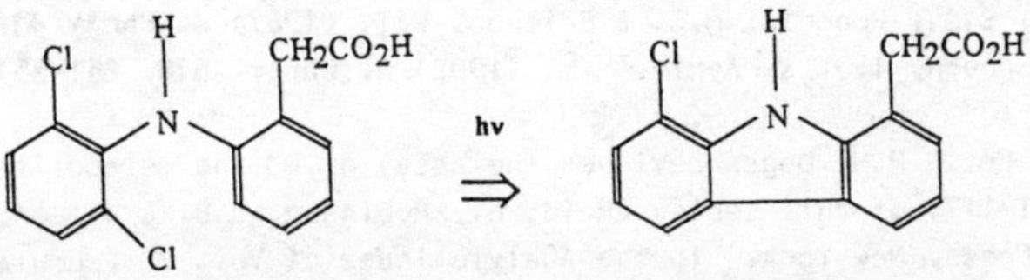

DICLOFENAC CARBAZOLE ACETIC ACID

DC, an NSAID used in rheumatic diseases, is becoming applied to reducing inflammation in cataract surgery. Only one [2] of the numerous HPLC assays for DC in biofluids [1-5] concerns aqueous humour; it calls for a large sample (100 µl or more) to achieve adequate sensitivity and entails complicated sample preparation. A relatively simple and sensitive HPLC method, based on a post-column technique reported [1] for DC in plasma, was therefore developed and rigorously validated for use on aqueous humour.

Method.- After dilution of the sample in an autoinjector vial, 55 µl was injected onto a 150 × 5 mm RP8 column. DC was eluted isocratically, at 4.9 min, with 42 mM 6.5 sodium phosphate pH 6.5 / acetonitrile (67:33; 1.0 ml/min). Fluorimetry (ex 288, em 360 nm) was performed on the carbazole derivative formed in the photochemical reactor (Aura Industries) [1].

Assay validation.- The method was linear up to 200 ng/ml (r = 0.9947). Inaccuracy and imprecision of the back-calculated standard data did not exceed -3.8% and 10.6% respectively. Assay sensitivity down to 10 ng/ml was demonstrated. Replicate analyses of intra- and inter-batch QC samples showed inaccuracy ranging from +10.3% to +11.7% and +5.5% to +9.8% respectively. For the

---

*addressee for any correspondence

same QC samples within a batch, imprecision did not exceed 6.8%. Recovery of DC was consistently high (mean ±S.D. 99.8±3.2%. DC was found to be stable through at least 3 freeze-thaw cycles. Validation data were processed using the peak height of DC only.

Bovine and human aqueous humour samples were spiked with DC over the nominal concentration range, 5-200 ng/ml, and subjected to the assay procedure. Peak-height data demonstrated comparable recovery of DC from the two matrices, indicating that bovine aqueous humour was a suitable substitute matrix for human aqueous humour.

Application.- The procedure has been successfully applied to measure DC in human aqueous humour samples following dosing with eye drops containing 1% (w/v) DC.

References [J. Chr'y = J. Chromatography]

1. Wiese, B. & Hermansson, J. (1991) **J. Chr'y 567**, 175-183.
2. Riegel, M. & Ellis, P.P. (1994) **J. Chr'y B 654**, 140-145.
3. Godbillon, J., Gauron, S. & Matayer, J. (1985) **J. Chr'y 338**, 151-159.
4. Owen, S.G., Roberts, M.S. & Friesen, W.T. (1987) **J. Chr'y 416**, 293-302.
5. Blagbrough, I.S. & Daykin, M.M. (1992) **J. Chr'y 578**, 251-257.

Note by Ed..- P.H. Degen reviewed the assay of DC and metabolites in Vol. 18 (pp. 101-117) of this series (Reid, E., Robinson, J.D. & Wilson, I.D., eds.), Plenum Press, New York. In the Analyte Index of Vol. 22 (cumulative), 23 & 24, DC entries will be found in the chemical category 'IIa'.

#B-9n

A Note on

A FAST, SPECIFIC LC-MS-MS METHOD FOR THE DETERMINATION OF THE ANTI-DIABETIC AGENT GLIBENCLAMIDE IN HUMAN PLASMA

L.W. Rees, P.R. Robinson, H.C. Morgan† & J. Maddock

Simbec Research Ltd., Merthyr Tydfil CF48 4DR, U.K.

Requirement *A specific and sensitive assay for glibenclamide in human plasma samples (1 ml), with short HPLC run times.*

End-step *LC-MS-MS(MRM*) using a short C-18 column.*

Sample handling *Liquid-liquid extraction with dichloromethane (10 ml); centrifuge, and dry down the organic phase (N_2).*

Comment *LC-MS parent-ion m/z distinguished i.s. (glibornuride) and drug, indistinguishable by retention or daughter-ion m/z.*

Glibenclamide is one of the most widely used orally active sulfonylureas in the treatment of maturity-onset diabetes. After administration it is absorbed fairly rapidly from the GI tract. It acts by stimulating insulin release from the pancreas.

Although several HPLC methods have been described for its determination in plasma, most have LLOQ 5 ng/ml and HPLC run times >15 min [1-4]. The present LC-MS-MS method benefits from R_T's <2 min and can measure the compound accurately and reproducibly down to 0.5 ng/ml in plasma. This sensitive and specific method, utilizing glibornuride as i.s., has been used for pharmacokinetic and also bioavailability studies in man with instrument run times of 25 samples per hour.

The samples were spiked with i.s., buffered with pH 5 0.1 M sodium acetate, and solvent-extracted. The residue from drying down the organic phase was reconstituted in acetonitrile/water (90:10; mobile phase), and put into an autosampler vial; a 20 µl aliquot was injected. The C-18 4 µm guard and analytical columns (Genesis) were 10 × 4 and 30 × 3 mm, with 500 µl/min flow rate. A Perkin-Elmer API III instrument was used, in negative ionspray (at 3600 V) and MRM mode with a nebulizer gas pressure of 30 psi. The curtain gas (0.6 L/min) was N_2 and the collision gas argon. The transitions, monitored throughout each analytical run, were from m/z 492 (glibenclamide) or 365 (i.s.) to 170 in each case.

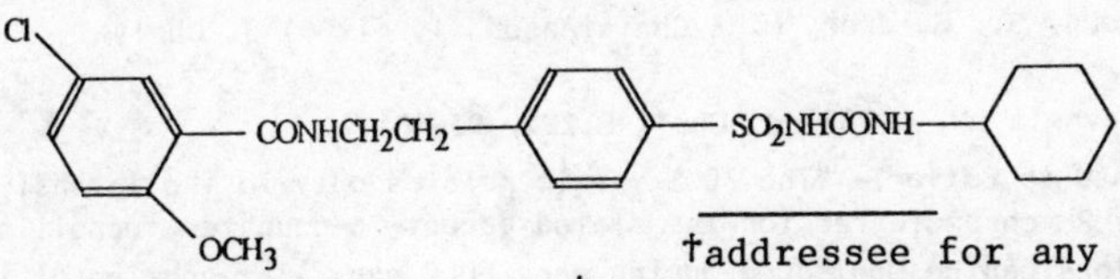

glibenclamide

†addressee for any correspondence

*Abbreviations.- MRM, multiple reaction monitoring; i.s., internal standard

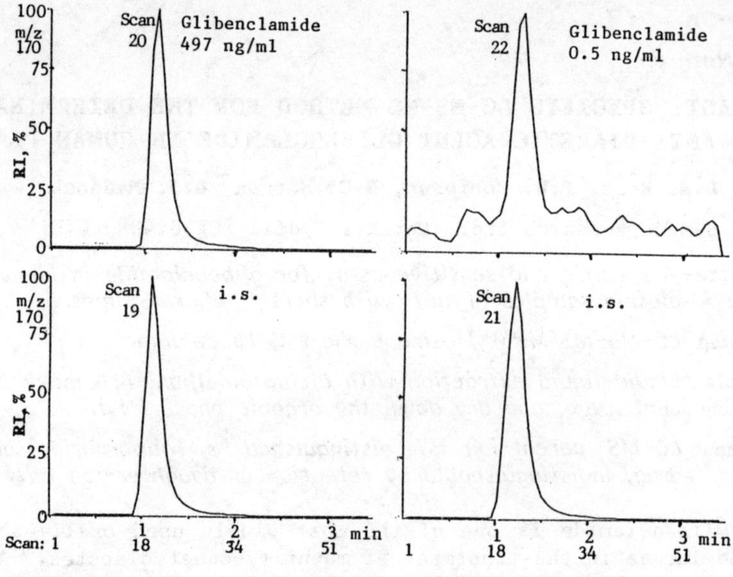

Fig. 1. LC-MS-MS applied to extracts of plasma spiked with glibenclamide at two levels as indicated (i.s. level constant).*

Fig. 1 illustrates LC-MS-MS runs. R_T's for the two analytes were <2 min, near-identical. Extraction of spiked calibration standards demonstrated linearity over the range 0.5-500 ng/ml plasma with a mean correlation coefficient of 0.999. The back-calculated standard data showed acceptable inaccuracy (-1.33 to 2.71%) and imprecision (4.32%); likewise for replicate analyses of QC's: inter-batch, -1.53 to 1.35%; intra-, -3.30 to 2.04%; the assay imprecision didn't exceed 5.58%. Drug recovery was consistently >80%, and 3 freeze-thaw cycles did not affect stability.

The procedure has been successfully employed in the determination of glibenclamide in several thousand plasma samples from clinical studies.

References [**Chr'y** denotes **Chromatography**]

1. Al-Dhawailie, A., Abdulaziz, M., Tekle, A. & Matar, K. (1995) **J. Liq. Chr'y 18,** 3981-3990.
2. Zecca, I., Trivulzio, S., Pinelli, A., Colombo, R. & Tofanetti, O. (1985) **J. Chr'y B 339,** 203-209.
3. Emilsson, H., Sjöberg, S., Svedner, M. & Christenson, I. (1986) **J. Chr'y B 383,** 93-102.
4. Uihlein, M. & Sistovaris, N. (1982) **J. Chr'y B 227,** 93-101.

***Clarification furnished to Editor.**- "The 100% y-axis relates only to the intensities of peaks on the MRM chromatogram for the stated parent-to-daughter transition. Several such transitions can be monitored during each HPLC run. Integration of the relevant peak area on individual chromatograms provides analyte-to-i.s. peak area ratios for calculation of plasma concentrations from calibration curves."

#B-10

QUANTITATION OF TROGLITAZONE AND ITS MAJOR METABOLITES IN PLASMA USING LC-MS-MS(API)

S.D. Clarke[†], L. Laugher, T.A.G. Noctor, [□]D. Thomas & S. Verity
Covance Laboratories Ltd., Dept. of Biopharmaceutical Analysis, Otley Road, Harrogate, N. Yorks. HG3 1PY, U.K.
and [□]Perkin-Elmer, Post Office Lane, Beaconsfield HP9 1QA, U.K.

Troglitazone (Tgz) is an orally active antidiabetic drug, the first of a new class of thioazolidinedione compounds. Analysis of Tgz and its sulfate and quinone metabolites in human plasma was required. The existing HPLC-UV method was problematic and slow (run time >30 min). Development of an LC-MS-MS(API) assay was therefore undertaken to dramatically reduce sample analysis time and increase sensitivity. The very different physicochemical properties of these three compounds led to a variety of problems, including: variable adduct formation in positive ion mode, poor precision with gradients and poor precision with MeCN-based mobile phases. These problems were overcome, and analysis was performed through two separate assays with a combined run time of 5 min.*

Tgz* (GR92132) is an antidiabetic drug, the first of a new chemical class for treatment of non-insulin-dependent diabetes. A variety of metabolites are formed, the sulfate and the quinone being of greatest interest.[□] The analysis of Tgz and these metabolites in human plasma was needed to support a phase III clinical trial in elderly non-insulin-dependent diabetics. The analysis of >3000 samples was required with LLOQ 10 ng/ml for Tgz and the quinone metabolite and 200 ng/ml for the sulfate metabolite.

The existing method of analysis was based on liquid-liquid extraction followed by HPLC-UV analysis. The analytical column used was a 300 × 6 mm YMC ODS at 35° giving a run time of 30 min per sample (typical chromatogram: Fig. 1). The plasma volume required was 500 µl, and the LLOQ for all three components was 100 ng/ml. This method was beset by a number of problems: late-eluting peaks from the previous sample interfering with the Tgz peak, needing readjustment of run time; occasional adjustment of phosphoric acid content in the mobile phase was required to move the sulfate peak independently of the other analytes to resolve the sulfate from endogenous interferants; adsorption of Tgz was observed at low concentrations on older columns. The method was

---

[†]addressee for any correspondence
**Abbreviations* (*one by Ed.*: Tgz, troglitazone).- In tandem MS context: API, atmospheric pressure ionization; pI, nI: positive/negative ionization; inj. marker: injection (load) standard.- Cf. i.s., internal standard (added at start). LLOQ, lower limit of quantification; MeCN, acetonitrile; QC, quality control.
[□]Formulae on p. 103.

judged unsuitable for the current study, particularly as the required LLOQ's were unattainable. Hence we sought to develop a simple, robust and reliable method based on LC-MS-MS(API).

EXPERIMENTAL AND METHOD DEVELOPMENT

MS characterization.- The LC-MS-MS(API) system comprised a Perkin-Elmer Series 200 autosampler and LC pump coupled to SCIEX API 300 MS-MS system. The relative sensitivities of the three compounds were assessed in both ionization modes, pI and nI:

	Positive Ion (pI)			Negative Ion (nI)		
Compound	Precursor Ion $[M+H]^+$ *(m/z)*	Product Ion *(m/z)*	Relative Sensitivity	Precursor Ion $[M-H]^-$ *(m/z)*	Product Ion *(m/z)*	Relative Sensitivity
Troglitazone	442	165	Good	440	397	OK
Quinone Metabolite	458	217	Good	456	145	OK
Sulphate Metabolite	522	-	Very Poor	520	397	Good

Ideally the mass spectrometer should have been run in nI mode for the sulfate metabolite and pI mode for Tgz and quinone; but due to polarity-switching limitations initial analyses had to be done by nI. Hence the Turbo Ionspray interface was exploited to attain adequate sensitivity for Tgz and the quinone. Use of chemical ionization (APcI) was precluded due to variable breakdown of the sulfate in the high-temperature region of the interface.

Chemicals for pre-MS processing were from Romil (MeCN, methanol - ultra gradient grade) or Merck-BDH.

Chromatography (gradient chromatogram: Fig. 2)

A rapid gradient of MeCN and ammonium acetate buffer was developed using a Hypersil BDS C-8 3 µm column, 50 × 4.6 mm. Flow

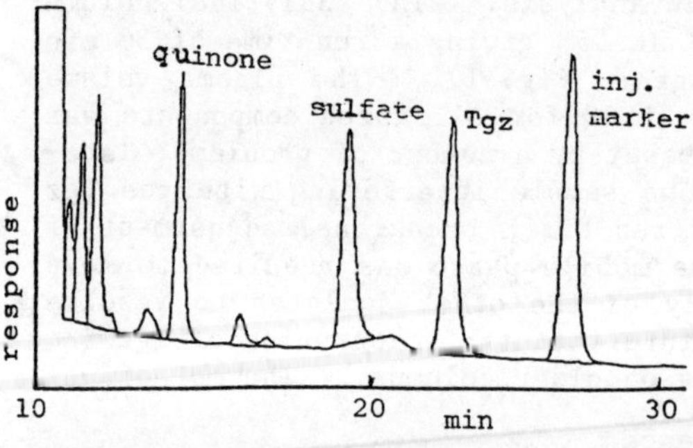

Fig. 1. HPLC-UV chromatogram of human plasma spiked at 1600 ng/ml with Tgz and the two metabolites. (Formulae on p. 103.)

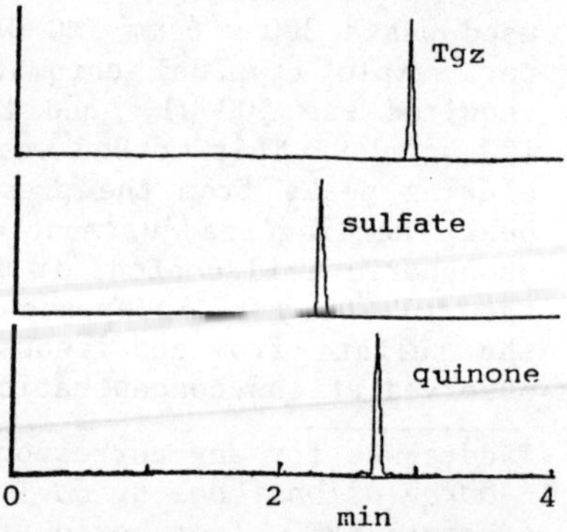

Fig. 2. An LC-MS-MS(APnI) rapid gradient run: Tgz and metabolite standards (low calibration levels).

rate was 1 ml/min split ~200 µl/min to the mass spectrometer. Run time was 4 min, with 6 min gradient cycle time (Fig. 2). Poor precision was observed in pure and extracted samples. Precision of pure standards (C.V. 12%) improved on changing mobile-phase organic component from MeCN to methanol (C.V. 8%), although the precision was still not within acceptable limits. Further adaptations to the gradient system failed to improve the precision.

The system was split into two isocratic methods, one method for Tgz and the quinone metabolite in pI mode (monitoring the $[M+H]^+$ ion) and the other for the sulfate metabolite in nI mode (monitoring the $[M-H]^-$ ion). This improved the precision of the pure standards to within acceptable limits (C.V. 3%). Due to relatively high sensitivity for the sulfate and the high LLOQ (100 ng/ml), this method was switched to ionspray. Both methods used the same column as for the gradient method. The mobile phase for Tgz and quinone was 75:25 (v/v) methanol/10 mM ammonium acetate pH 4.0, giving R_T 1.7 min for Tgz and 1.1 min for the quinone. The mobile phase for the sulfate metabolite assay was 70:30 but otherwise the same. Due to time constraints it was not possible to combine these two methods into a single isocratic run with polarity switching. The i.s. problem is considered later.

There was benefit to **pre-chromatographic steps** through the increased specificity and sensitivity of MS-MS(API) compared with UV for HPLC detection. Sample volume could be shrunk, from 500 to 200 µl, and complexity reduced by using protein precipitation rather than liquid-liquid extraction.

Sample preparation procedure

MeCN (600 µl) was added to 200 µl of plasma to precipitate proteins. After vortex-mixing and centrifugation, 200 µl was transferred from the supernatant into a clean tube labelled for Tgz and quinone analysis and reduced to dryness (N_2); a further 200 µl was put into a tube labelled for sulfate analysis and similarly dried down. Each residue was redissolved for LC-MS-MS in 200 µl of the appropriate mobile phase.

Performance and assay design aspects

Precision of 6% was achieved with extracted mid-level QC's for the sulfate analysis. Poorer precision (>12%) was observed with corresponding Tgz and quinone samples, not improved through investigation of alternative protein precipitation methods.

Possible adduct formation was investigated. In extracted samples Tgz and the quinone showed both the protonated pseudo-molecular ion $[M+H]^+$ (Tgz, 442; quinone, 458) and the ammoniated equivalent $[M+NH_4]^+$ (Tgz, 459; quinone, 475]. The latter type of ion, when fragmented, gave a characteristic daughter ion (Tgz, 165; quinone, 217). By varying the temperature of the Turbo Ion-spray heater, the ionspray voltage (ISV) and the orifice voltage (OR), the ratio of $[M+H]^+$ to $[M+NH_4]^+$ could be modified.

Monitoring of either adduct obtained with Turbo Ionspray failed to produce a stable response, primarily because lack of temperature control on the interface plate leads to a gradual increase in temperature of the ion source. This in turn affects the ratio of protonated:ammoniated adducts present. Switching the analysis to IonSpray gave predominantly the $[M+NH_4]^+$ form, monitoring of which gave good precision with extracted samples.

Injection markers (load standards)

To further improve the robustness of these methods, screening for possible injection markers was performed - ~20 compounds for each method. Only two proved suitable for the Tgz method, namely naproxen and memantine. The patients included in this study were elderly diabetics (on uncontrolled medication); so naproxen (a commonly prescribed anti-inflammatory) could not be used due to the possibility of its presence in patient samples. Memantine was therefore used as an injection marker. No suitable injection markers were found for the sulfate analysis.

THE DEVELOPED METHODS: RESULTS AND DISCUSSION

The use of rapid gradient systems in LC-MS-MS(API) has, in our experience, proved unsuccessful. They have often resulted in very poor precision. This is likely to be due to micro-changes in the eluent composition during ionization of the compound of interest, and also to small changes in the split ratio, between HPLC runs. Possible solutions to the imprecision problem are to divide the assay into two separate isocratic methods, to run a slower gradient system or to use a smaller bore column and not split the effluent flow. Separating the assay into two isocratic methods resolved the problem of poor precision.

The adduct formation observed with Turbo Ionspray caused problems due to the variation in the ratio of the $[M+H]^+$ and $[M+NH_4]^+$ adducts. The main factor governing variation was that the temperature of the interface and surrounding chamber varied, so much as to cause the high imprecisions observed. Because IonSpray is performed at ambient temperature, it caused no such instability. After running the instrument in Turbo Ionspray or APcI mode for other assays, a 2-h 'cool down' period was required prior to performing this analysis in IonSpray.

A typical chromatogram is shown in Fig. 3 for an extracted human plasma sample spiked at 150 ng/ml for the Tgz and quinone metabolite assay, and similarly in Fig. 4 for the sulfate metabolite assay with 1500 ng/ml spiking.

The methods were validated to current international regulatory requirements. The SCIEX API 300 was run continuously 24 h a day for 7 days, to analyze >800 samples and re-assays for all three compounds. To date we have successfully analyzed >3800 plasma samples using these methods.

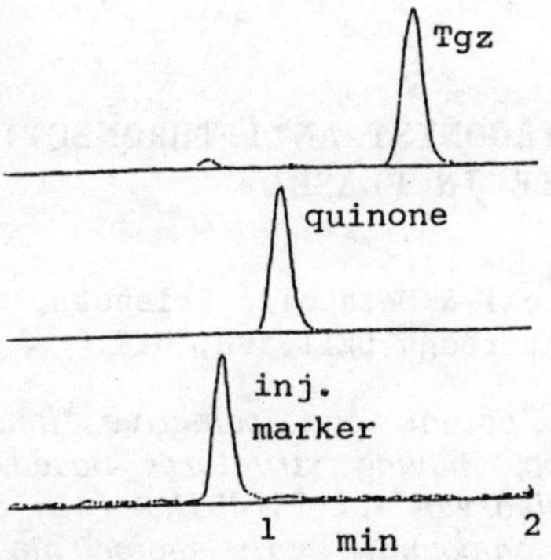

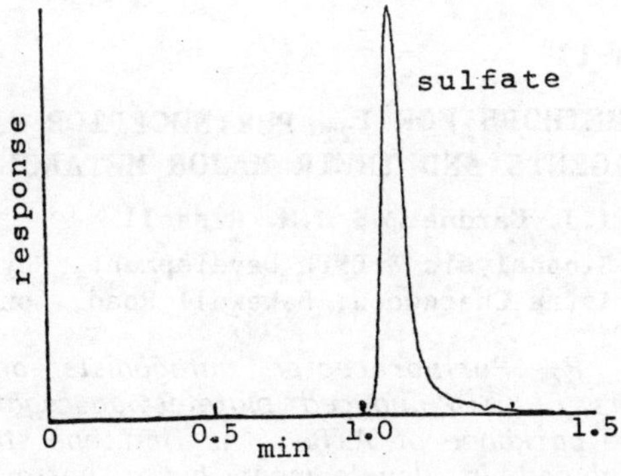

Fig. 3. Isocratic LC-MS-MS(API) chromatogram of human plasma spiked at 150 ng/ml with Tgz and the quinone metabolite.

Fig. 4. As for Fig. 3, but sulfate metabolite, spiked at 1500 ng/ml.

CONCLUDING COMMENTS

The different physico-chemical properties of Tgz and its major metabolites caused a number of problems when attempting to analyze all the compounds simultaneously. The formation of $[M+NH_4]^+$ adducts in extracted samples may easily escape notice and lead to an increased imprecision in the method. Although some problems may be encountered in the development of an LC-MS-MS(API) method, these are easily resolved and do not impact on the selectivity and sensitivity that LC-MS-MS(API) offers over conventional techniques.

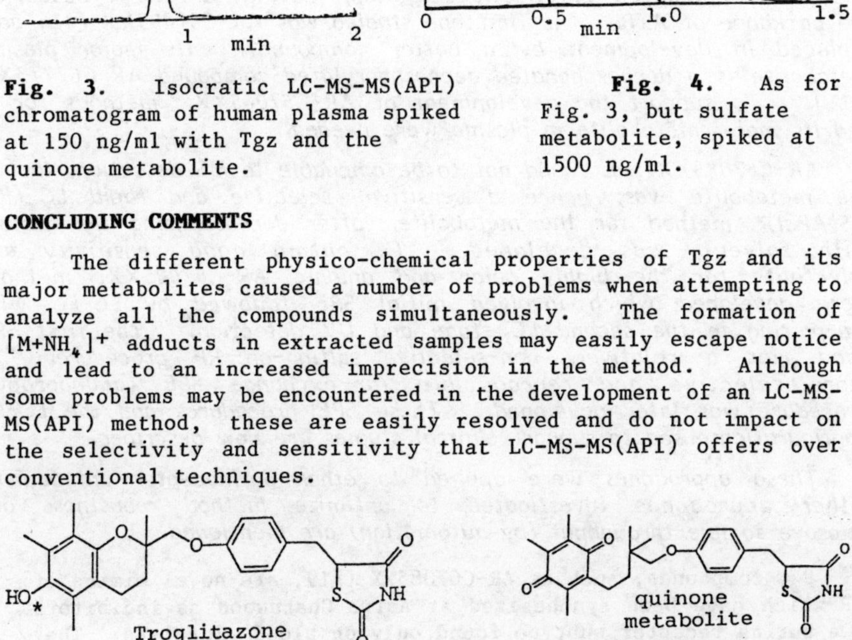

HO₃S-O- at * in sulfate metabolite

Even for Tgz itself it was difficult to find a compound with suitable properties to serve as a conventional i.s. throughout the assay; hence only at the column load stage was an 'injection standard' added.

#B-11

METHODS FOR P_{2T} PURINOCEPTOR ANTAGONIST ANTI-THROMBOTIC AGENTS AND THEIR MAJOR METABOLITE IN PLASMA

J.J. Gardner† & J.M. Russell

Bioanalysis & DMPK Development, Physical & Metabolic Sciences, Astra Charnwood, Bakewell Road, Loughborough LE11 5RH, U.K.

P_{2T} Purinoreceptor antagonists are potent and selective inhibitors of ADP-induced platelet aggregation, having structures based on the backbone of ATP. The first one studied was AR-C67085XX (#I; now replaced in development by a better compound). Its major plasma metabolite is a dephosphonated dephosphorylated compound AR-C67775XX (#II). To support the development of AR-C67085XX, methods for it and its major metabolite in plasma were needed.

AR-C67085XX was found not to be amenable to MS detection, though the metabolite was; hence a sensitive, selective and rapid LC-MS-MS(APcI) method for the metabolite, after deproteinizing the plasma with solvent, was developed. To obtain good sensitivity and selectivity for the highly polar and anionic AR-C67085XX, methods were developed which involved initial SPE followed by LC-LC with ion-pairing in the second LC stage and UV detection. The first SPE used was a relatively non-selective 'salting-on' RP procedure. A more selective and reproducible ion-exchange SPE (aminopropyl) procedure was later developed. These SPE procedures and the LC-LC configurations used to support clinical studies are now described.*

These approaches were applied to other purinoceptor antagonists. Other approaches investigated to optimize method robustness and improve sample throughput (by automation) are mentioned.

P_{2T} compounds, such as AR-C67085XX (#I), are novel mimics of ATP which have been synthesized at Astra Charnwood as inhibitors of a purine receptor subtype found only on blood platelets. They are potent and selective inhibitors of platelet aggregation induced by ADP. One platelet-dependent process, haemostasis, is important in health to prevent bleeding, and another, arterial thrombosis, is important in various cardiovascular diseases.

---

†Addressee for any correspondence. [**Formulae** on facing page.

**Abbreviations (see also Preface).*- In tandem MS (MS-MS): AP(c)I, atmospheric pressure (chemical) ionization. LLOQ, lower limit of quantification; i.s., internal standard; MeCN, acetonitrile; QC, quality control.

The P_{2T} compounds were designed to be rapidly cleared from the plasma to allow rapid reversibility of their anti-platelet action, and are being developed as anti-thrombotic agents for intravenous infusion. To support this development, bioanalytical methods now described were developed for #I using various SPE-LC-LC approaches, and for the metabolite #II using LC-MS-MS, in plasma samples.

Background to assay development

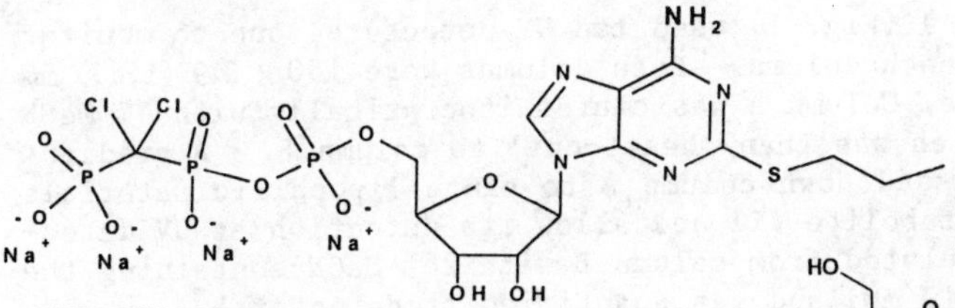

AR-C67085XX, #I

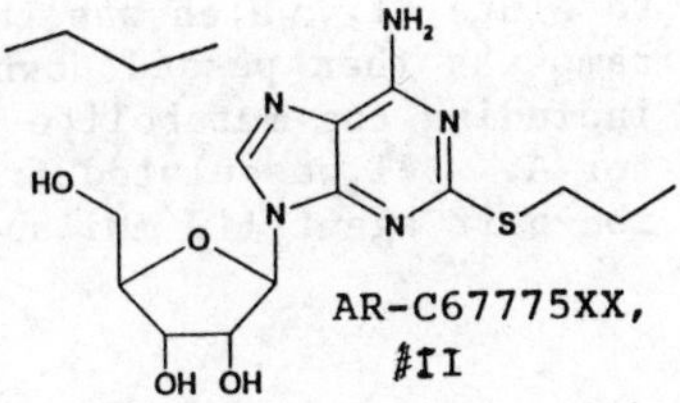

AR-C67775XX, #II

The structure of #I includes a dichlorocarbon bridge between the terminal and adjacent phosphorus atoms which renders the compound resistant to hydrolysis, compared to ATP with an oxygen bridge. #I is a polyvalent anion with three highly acidic sites, is very polar, and has a UV absorbance maximum at 277 nm. It is initially metabolized by loss of the phosphonate group giving a putative monophosphate product, which is further hydrolyzed to a dephosphorylated product (AR-C67775XX; #II) that is the main plasma metabolite. Unusually, this metabolite is much more lipophilic, due to the loss of the polar negative charges, than the parent compound.

Various approaches were considered for determining #I in plasma, including the formation [1] of a fluorescent chloracetaldehyde derivative. Initial work showed that the fluorescence of this derivative was less useful than the UV absorbance of the underivatized compound. Although RIA methods have been developed for other trinucleotides [2], this approach was not actively considered for #I. Evaluation of LC-MS-MS showed poor detection for #I, though the metabolite #II was sensitively detected. LC approaches for #I were therefore evaluated.

SPE-LC-LC System 1 for #I and #II

It was found that #I could be 'salted-on' to C-18 SPE supports in the presence of 2 M NaCl, providing a non-selective means to concentrate the polar compound from plasma, and that methanol eluted it from the support. #I showed some retention on Waters C-18 Novapak LC columns with low MeCN concentrations (4%, v/v) in pH 6 phosphate buffer, but was highly retained on the same C-18 columns in the presence of the ion-pairing agent tetrabutylammonium bromide, needing ~25% v/v MeCN for elution. This difference between the MeCN concentrations required to elute #I from the two HPLC systems, which had different mechanisms and selectivities, suggested a column-switching LC system, with the 'salting-on' SPE as the initial work-up, as a possible analytical approach for #I in plasma. The 'salting-on' SPE was tedious, involving large volumes to be applied, but it had the potential advantage that the lipophilic metabolite #II might also be extracted and then eluted in methanol.

LC-LC System **1** (Fig. 1) used two UV detectors, one to monitor the efflux from each column. Both columns were 150 × 3.9 (i.d.) mm cartridge columns. Column A was eluted isocratically with 4% MeCN to elute #I, which was then 'heart-cut' to column B. A gradient ramp was then passed down column A to elute lipophilic materials including the metabolite #II and allow its detection at UV detector A. #I was eluted from column B with 26% MeCN containing the ion-pair agent (10 mM) and was quantified at detector B.

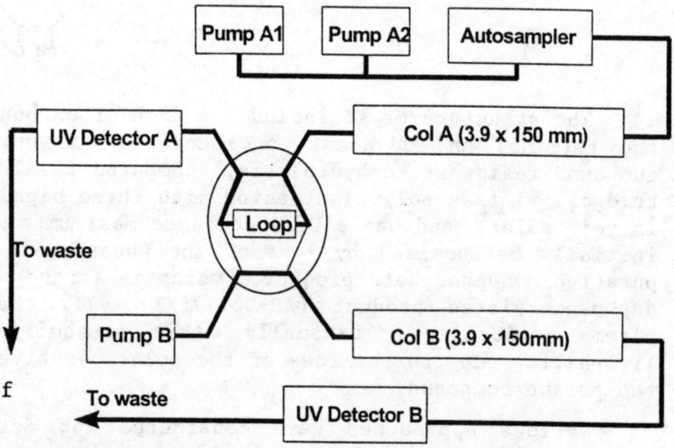

Fig. 1. Configuration of LC-LC System **1**.

Fig. 2 shows a chromatogram on column B of a plasma sample from a volunteer infused at 4000 ng/kg/min, which was found to contain #I at 140 ng/ml, confirming that the approach provides a selective system for #I in human plasma with LLOQ ~1 ng/ml. The method was not formally validated for human plasma; whilst back-calculated calibrators indicated satisfactory mean accuracy, the precision was poor, ~10-20%. Although the method was used to generate useful data [3] for the first clinical study on #I, a more precise method was needed for further studies.

Chromatograms (not illustrated) from column A showed that, although metabolite #II could be detected with an LLOQ in the region of 20-50 ng/ml, there were many surrounding peaks due to endogeous plasma constituents which eluted around #II on the gradient ramp. Only for subjects infused at the highest rates

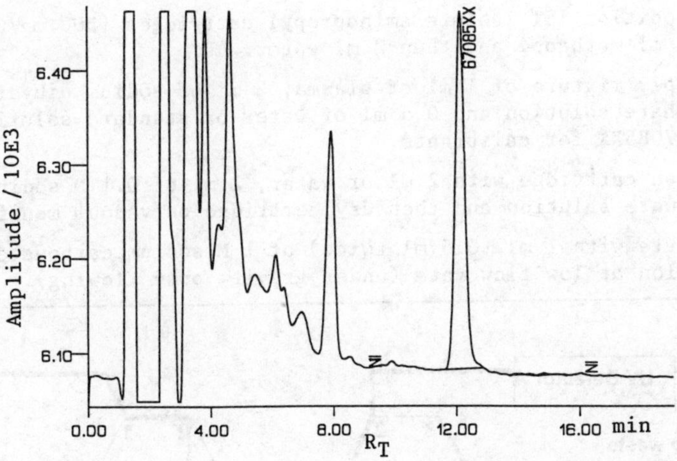

Fig. 2. Chromatogram on column **B**, in LC-LC system **1**, of a clinical sample containing AR-C67085XX (#I) at 140 ng/ml.

(2000 and 4000 ng/kg/min) were measurable concentrations of #II found. A more selective and sensitive method for #II in plasma was therefore needed (LC-MS being the favoured approach).

SPE-LC-LC System 2 for #I only

Since only the parent drug #I would be determined using the improved System 2 (metabolite #II being determined by a different technique), a selective and specific new procedure was sought since it was believed that the poor precision was caused by the variability in the 'salting-on' SPE procedure. The high anionic charge on #I suggested that binding to cationic supports might provide a selective SPE procedure. Both strong (SAX) and weak (aminopropyl) cationic supports were found to bind #I strongly. In fact, #I could not be eluted from the SAX support. To elute #I from the less basic aminopropyl support (pKa = 9.7), the support had to be deprotonized with strong sodium carbonate solution, which had a pH of ~11 and strong buffering capacity.

In the SPE (Scheme 1) developed using the aminopropyl cartridges, plasma was loaded in slightly acidic phosphate solution, to decrease the binding of weak acids. The wash with dilute carbonate solution (which did not elute #I) decreased the volume of strong carbonate solution, added in divided portions, needed for good recovery of #1.

The configuration used for System **2** (Fig. 3) had a short 3.9 mm i.d. Novapak C-18 column for column A and a narrow Novapak C-18 column, 2.1 mm i.d., for column B. After flushing the strong

Scheme 1. Specific aminopropyl SPE procedure for AR-C67085XX (#I) in plasma.

1. Condition IST Isolute aminopropyl cartridges (500 mg/6 ml) wih 2 ml methanol and then 2 ml water.
2. Apply mixture of 1 ml of plasma, 2 ml of sodium dihydrogen phosphate solution and 0.5 ml of water or standard solution of AR-C67085XX for calibrants.
3. Wash cartridge with 2 ml of water, 2 ml of 0.1 M sodium carbonate solution and then dry cartridge on vacuum manifold.
4. Elute with 1 ml (0.5 ml twice) of 1 M sodium carbonate solution at low flow rate (under gravity once flowing).

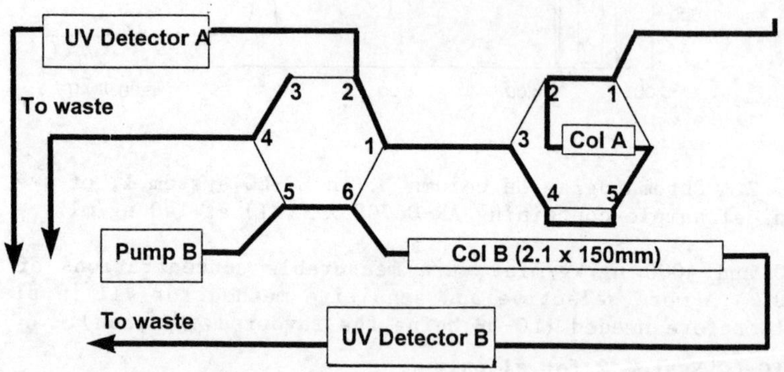

Fig. 3. Configuration of LC-LC System **2**.

alkaline carbonate solution to waste, column A was back-flushed *via* a switching valve using a step gradient and was then directly coupled to column B *via* the second valve for transfer of the #I to column B.

The resulting method using the aminopropyl procedure and LC-LC System **2** validated well; the accuracy and precision data (Table 1) showed improvement compared to System **1**, with inter-assay C.V. values between 5% and 10% (except at the LLOQ of 1 ng/ml), compared to 10-20% with System **1**.

SPE-LC-LC System 3 (for cognate compounds)

Subsequently the systems described above were applied to the determination of other P_{2T}-antagonists, which had similar UV absorbances and the same highly anionic nature. To analyze these compounds a third system (System **3**) was developed. The selective aminopropyl SPE was retained but was now automated using a Gilson ASPEC XL system to process plasma samples overnight. The plasma

Table 1. Intra- and inter-assay data for the aminopropyl SPE-LC-LC System **2** method for AR-C67085XX (#I) in plasma.

AR-C67085XX	Intra-assay		Inter-assay	
concn., ng/ml	Accuracy, %	Precision, %	Accuracy, %	Precison, %
1	78	12.5	88	23
2	86	4.3	95	9.4
5	102	3.6	99	4.6
10	104	2.2	99	4.6
20	109	1.7	101	6.6
50	113	2.2	104	5.9
100	110	3.4	104	5.2
200	113	2.2	104	5.4

extracts in strong carbonate solution were partially neutralized with dilute phosphoric acid before injection onto the LC-LC system. Simple heart-cutting with a loop was restored (to avoid the high pressures seen with System **2** when columns A and B were in line when transferring #I), and a more lipophilic decyltrimethylammonium ion pair was used, this ion-pair having a linear structure compared to the spherical tetrabutyl ion-pair used previously. Both columns in the system, A and B, were now 2.1 mm i.d., with the hope of maximizing sensitivity. Column A was now a short Shandon Hypersil C-18 Elite column. Column B was changed to a Waters Symmetry C-18 column (initially with 5 µm and later with 3.5 µm packing), which gave increased retention with narrower peaks. The new ion pair also gave increased retention.

Although this System **3** was used successfully by two analysts on two separate HPLC systems, various limitations became apparent. The major one was progressive deterioration of the short narrow column A. Similar columns from two other manufacturers showed the same instability. Pragmatically, a fresh column for every batch and medium-sized batches allowed satisfactory use of the method, with minimal batch rejection due to QC drift. Design of the method to allow two sample injections from each plasma extract was useful when the first LC-LC analysis failed. The slow speed of sample processing overnight by the ASPEC XL was a minor limitation compared to the instability of column A.

Other approaches

The use of a Varian Prospekt* system to insert a new small C-18 cartridge for every plasma extract to replace column A of System **3** has shown promise, which is now being defined. Faster SPE instrumentation (e.g. Zymark Rapidtrace) to automate the aminopropyl SPE is also being evaluated.

LC-MS-MS(APcI) method for #II (typical chromatogram: Fig. 4)

It had already been shown that the metabolite #II was sensitively detected by atmospheric pressure chemical ionization MS.

* Manufacturer of 'Prospekt': Spark Holland, Emmen, The Netherlands

Implementation of this approach using a Perkin-Elmer Sciex API+ triple quadrupole instrument showed initially variable responses, which were not corrected for using analogues of #II as an i.s. Use of a hepta-deuterated isomer as the i.s. stabilized the responses. It was then realized that leaks in the instrument caused 'icing' on the quadrupoles, with progressive changes in mass detection and, accordingly, varying response. Stopping the leak stabilized the responses; so the i.s. is now less essential.

The analysis procedure (Scheme 2) involved plasma protein precipitation with MeCN (added in divided portions) and extract dilution with water to improve chromatographic peak shape (by decreasing solvent strength). The chromatographic analysis was rapid, with a run-time, including the autosampler operation, of only 3.5 min which allowed 90 samples to be processed and analyzed per day. The method validated well for human and, later, pre-clinical samples. The inter-assay precision data (Table 2) plasma gave precision values of <5% at all concentrations other than the LLOQ of 1 ng/ml. This method has performed well in the hands of several analysts.

Scheme 2. LC-MS-MS(APcI) method for AR-C67775XX (#II) in plasma.

1. To 50 µl plasma add 50 µl [$^2H_7$]-AR-67775XX in MeCN.
2. Mix, add further 50 µl of MeCN, re-mix and centrifuge. Dilute 100 µl of supernatant with 200 µl of water.
3. Inject 80 µl of mixture onto a Kromasil-100 C-18 column; elute at 1 ml/min with 27% v/v MeCN in 25 mM ammonium acetate (pH 6).
4. MS-MS(APcI) detection with a heated nebulizer interface.

Table 2. Inter-assay data for LC-MS-MS(APcI) method for AR-C67775XX (#II) in plasma.

AR-C67775XX	Human plasma		Pre-clinical plasma	
concn., ng/ml	Accuracy, %	Precision, %	Accuracy, %	Precison, %
1	96	16	96	10
2	94	11	102	5
5	100	7	103	3
10	100	9	104	1
20	102	7	104	4
50	101	6	102	3
100	99	5	99	2

The Scheme **2** approach has also been successfully applied to the determination of the major plasma metabolite of another P_{2T} antagonist.

Acknowledgements

We wish to thank our many colleagues at Astra Charnwood who have contributed, or continue to contribute, bioanalytical support to the P_{2T} project.

References

1. Perrett, D. (1987) J. Chromatog. **386**, 289-296.
2. Piall, E.M., Curtin, N.J., Aherne, G.W., Harris, A.L. & Marks, V. (1989) Anal. Biochem. **177**, 347-352.
3. Nassim, M.A., Gardner, J.J., Wilkinson, D., Corfield, J.E., Rudol, L. & Wyld, P.J. (1995) Br. J. Pharmacol. **39**, 98P.

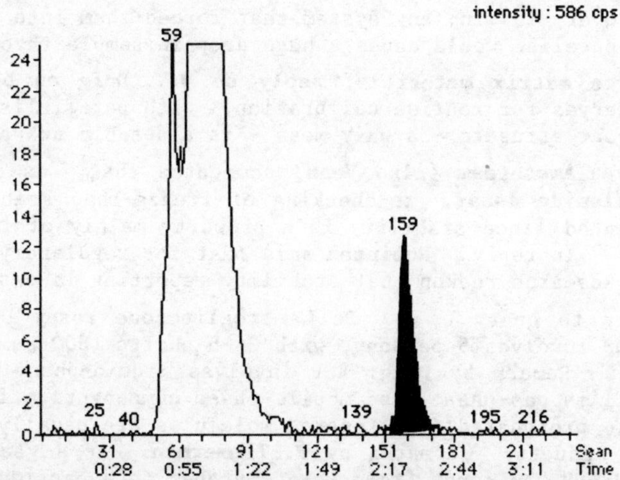

Fig. 4. LC-MS-MS chromatogram of a plasma sample from a subject infused with AR-C67085XX (concentration found to be 1.1 ng/ml; R_T 2.24 min).

Comments on #B-1, H-J. Skrzipczyk: RIA & LC-MS-MS: A COMPARISON
#B-3n, A.M. Edgington: LATE-ELUTING INTERFERENCES
#B-7, T.A.G. Noctor: ASSAYS WITH MATRIX PROBLEMS
#B-9n, P.R. Robinson: GLIBENCLAMIDE BY LC-MS-MS
#B-10, T.A.G. Noctor: TROGLITAZONE BY LC-MS-MS
#B-11, J.J. Gardner: AN ANTI-THROMBOTIC AGENT

Comment by **P. End to Skrzipczyk**: the very long incubation times (48 h), entailing (e.g.) precipitate formation, may account for the low precision. Reply: no time to re-develop the assay!

Reply to **K. Barrell: Edgington** said the late-eluting peaks (elution with the solvent front not achievable) are shapeless and spread over ~15 min; any system that forced them into quiet areas of the baseline would cause a huge drop in sample throughput.

Scarce matrix materials: reply to M.V. Doig on blanks etc.- Serum serves for routine calibration, with parallelism checking. Middle-ear effusate - a waxy mess - is a notably awkward matrix.

P. van Amsterdam (also Venn) commented that, as done in the **glibenclamide assay, re**-checking of freeze-thaw stability seems unwarranted since stability is a property mainly of the compound itself. In reply, **Robinson** said that for regulatory validation Simbec Research reckon that stability reporting is essential.

Reply to query by **D. Dell: troglitazone assay** by automated LC-MS-MS involves 5 persons, with 24-h shifts (800 samples/week). output. Remark by **Doig**: ESP involves liquid-phase ionization, but APcI is gas-phase, so mobile-phase components affect quantitation; protonated/deprotonated molecules are usually preferable to ion adducts. Remarks by **P. Timmerman**.- At a gradient start in LC-MS-MS the spray (temp. etc. matched to a particular solvent mixture) goes dry, and noise impairs sensitivity. Conjoint use of nCI and pCI could best be sequential [as remarked in #B-10].

To **Gardner, J. Schmid** put the thought that oxidative metabolism **might** occur at the S atom. **Reply**: for the metabolite, yes!

Citations contributed by Senior Editor (not merely 'model' analytes), complementing #ABC & ('endogenous' emphasis) end of #E

Extracting propranolol and furosemide from plasma for mixed-mode HPLC (C-18/SCX) can be done with solvents or, as a shortcut, with a C-18 pre-column and switching.- Walshe, M. et al. (1966) J Pharm Biomed Anal **14**, 478-481.

Metabolite studies on **phosphothioate oligonucleotides** (ISIS 11061, 11637; 2'-deoxy) in rat bio-samples entailed MS(nCI, ESP) on RP-LC eluates - revealing composition even with equal-length products - and capillary gel electrophoresis, whereby products having a one-nucleotide length difference were distinguishable.- Gaus, H,J. et al. (1997) Anal Chem **69**, 313-319.

1-*O*-Acyl glucuronides (furosemide, tolmetin) solvent-extracted from urine were purified (for study): LC (not *HP*LC) on micronized Amberlite XAD-2.- Mizuma, T. et al.(1998) Prep. Biochem. **28**,37-47.

#ABC

Compendium by E. REID (Guildford)

RECENT PAPERS ON DRUG ASSAY OR METABOLITE DELINEATION

mostly HPLC/GC; therapeutic-group listing as in Vols. 22-24 but transfers from #6b (#6c new) & #7 ('immuno')

1. **MUSCULO-SKELETAL AGENTS** incl. analgesic [anti-pain/-pyrexia in #5a], anti-inflammatory [cf. #2], skeletal-muscle relaxant
2. **RESPIRATION-EASING/IMMUNO-AFFECTING AGENTS**: incl. bronchodilator, mucolytic [expectorant: #7]; anti-histamine/-inflammatory
3. **(ANTI)ADRENERGICS/CHANNEL MODULATORS/ACE INHIBITORS** ['straddles' #2, #4 & #5; (anti-)dopaminics are in #5b] SEE ALSO #4 heading
4. **CARDIOVASCULAR AGENTS** incl. anti-thrombotic/-cholesterol, diuretic, inotropic [PAF-/eicosanoid-/IL-related: SEE #2; renin: #3]
5. **NEUROACTIVE AGENTS** [& see #1 for some analgesics, #3 for (anti)adrenergics, #7 for smooth-muscle relaxants]:-
   5a: 'damping-down' action, esp. anaesthetic (incl. local), narcotic/analgesic, hypnotic/sedative; anxiolytic; **anti-** -convulsant/-emetic/-insomnia/-migraine [#5b for anti-Parkinson]
   5b: stimulatory/various: AcCh-/dopamine-/5HT†-related; **anti-** -depression/-hyperkinesis/-schizoid/-MAO; nootropic; psychotic
6. **ANTI-INFECTIVE/ANTI-CANCER (or cytostatic) AGENTS**:-
   6a: some structural groups [A-to-Z within each]: #cef/ceph/cillin/penem/β-lactamase-inhibitory; #oxacin; #nucleoside-type
   6b: anti-infective, not a #6a type [Housed #6c also in past vols.]
   6c: anti-cancer (or cytostatic) [nucleoside type in #6a]
7. **VARIOUS** [Housed immunosuppressants; now #2]: #anti-diabetic; #retinoid; #UNCLASSIFIED: e.g. anti-CCK, anti-ulcer [anti-emetic: #5a], hormone-related [some in #1 or #6c], smooth-muscle relaxant

† #5b is not the sole 'home' for drugs that affect 5-HT (serotonin)

Classes featured in **earlier vols.** [address for info.: p. vi]: V.14: #6; V.16: #5 etc.; V.18, #1, #3 & #4; V.20: #6; V.22: #2; V.23: #4, hypolipidaemics. V.24 touches on #6 (peptide-related).

At end of each A-Z group: agents designated by a **Code; generic names** may later appear in Merck (has company Code Letters list)/ Martindale; both books have therapeutic-category listings.

#ABC entries (mainly 1996/97), not 'quality-controlled', bear on automation, on drug efficacy/metabolism/safety and on therapeutic monitoring, but not on screening (e.g. forensic or for residues). Generally the sample is plasma/serum (or, if specified, whole blood) and - not clear in some cited arts.! - from humans.

UP-DATING SEARCH recommendation.- For a listed paper, ascertain in *Science Citation Index* (by first author) whether any later paper cites it; also seek the analyte, helped by any Cumulative Index, in *Anal. Abs.* and in post- **B 698** vols. of *J. Chromatog.* **[676** had Index to **652-675]**.

LAYOUT AND ABBREVIATIONS KEY (& see p. vi)

Mostly human plasma samples

* (at margin) signifies **noteworthy**, e.g. methods exploration or discussion (maybe with a 'model' compound), of general interest.
Enantiomer studies/agents are implied by BOLD-letter **chiral.**
For metabolites, **mtb(s).**, there may be an investigative slant (**invstg**; maybe with radioisotope use). Most are Phase I (II if 'glucuronide', studied intact); their nature may be indicated - e.g. oxid. (or -OH) if including oxidation products; **des(-N).**
[Hitherto used for **mtbs.** but now *abandoned* - superscripts ¹ ¹ ².]
For conjugates (Phase II), cleavage is implied by **hydrol.**

LE, liquid-liquid extraction; **pptn.** = protein precipitation. (Commonly - in SPE too - dry down and redissolve; not stated.)
Automation implied by (e.g.) **swtch** (= switching) or **on-line.**
Types of column (col.)/cartridge packing include: C-18, CN, AGP (α_1-acid glycoprotein); SCX/SAX = cation/anion-exch(ange).
For HPLC etc. (aq., aqueous; grad. = gradient): acet., acetate; phos., phosphoric/phosphate (ac., acid); MeCN, acetonitrile; MeOH, methanol; iPrOH, isopropanol; TEA, Et_3N; DEA, Et_2NH
See Preface for PCA, TCA; also EC, ECD, nCI, etc.

Derivatization/conversion: **deriv.** (post-column maybe implicit).
Where **nm** value given, peak detection was by UV (or visible).

Approx. sensitivity denoted **sens.**; if '(Q)', LLOQ rather than LOD.
PK, pharmacokinetics; TM, therapeutic monitoring.

References: J = *Journal*
Chr = *Chromatog.*
Bm = *Biomedical*
HR = *High Resolution*

So: Chr'ia = *Chromatographia*
J Lq Chr = *J. Liq. Chromatog. Relat. Tech.*
J PhBmAn = *J. Pharm. Biomed. Anal.*
AtMc Ag Ch = *Antimicrob. Agents Chemother.*

---

1. MUSCULO-SKELETAL AGENTS incl. analgesic [anti-pain/-pyrexia in #5a], anti-inflammatory [cf. #2], skeletal-muscle relaxant

Various **NSAID's (12; diclofenac, felbinac, sulindac, etc.**, measurable simultaneously - urine).- SPE (silica); C-18: MeCN/phos. pH 5; 230 or 320 nm; sens. 5-50 ng/ml. Hirai, T. et al. (1997) J Chr **692**, 375-388.

* **Baclofen.**- SPE(SAX); deriv.; C-18: MeOH/phos. pH 7; EC; sens. 2 ng/ml. Millerioux, I. et al. (1996) J Chr **A, 729**, 309-314.

Budesonide.- LE; SPE; acetylate (i.s.: acetate, ³H), C-18: aq. MeOH; MS(CI); sens. 0.2 ng/ml(Q). Optional follow-on: enantiomer separation on **a chiral col.** Li, Y.N. et al. (1996) J Chr **B 683**, 259-268.

Chlorzoxazone & 6-hydroxychlorzoxazone.- (1; hydrol.; incl. urine): LE; C-18: MeCN/THF/acet.; 283 nm; sens. 0.1(urine 4) µg/ml(Q). (2; microsomal incubates): LE; C-8: MeCN/phos. ac.; 287 nm. (1) Frye, R.F. & Stiff, D.D. (1996) J Chr **B 686**, 291-296. (2) Chittur, S.V. & Tracy, T.S. (1997) J Chr **B 693**, 479-483.

Cisatracurium & mtbs. incl. laudanosine (urine).- SPE; SCX (anion-exch.) col.: MeCN/SO_4^{2-}, grad.; fluor; sens. 10 ng/ml(Q). Bryant, B.J. et al. (1997) J Lq Chr **20**, 2041-2051.

* **Diclofenac**.- (1; oxid. **mtbs**.; urine, ±hydrol.; invstg.): dry-state SFE; GC-MS(nCI). (2): pptn.(acetone); LE; GC-MS; sens. 0.2 ng/ml. (1996) J Chr **B 685**: (1) Blum, W. et al., 251-263; (2) Borestein, M.R. et al., 59-66.

Felbinac/Fenbufen/Rufloxacin.- LE; SAX col.: MeCN/phos. pH 7; 280 nm; sens. 0.1 μg/ml. Carlucci, G. et al. (1996) J Chr **B 682**, 315-319.

* **Fenbufen & mtbs**..- (1): SPE; CN col.: MeOH/MeCN/phos. ac.; photo-irrad.; fluor; sens. 50 ng/ml(Q). PK use. Siluveru, M. et al. (1996) J Chr Sci **34**, 182-184 (also J Chr **B 682**, 89-94).

Flunixin (horse, incl. urine, ±hydrol.).- SPE; CE with SDS; 286 nm; sens. 3(urine:17) ng/ml. Gu, X. et al. (1997) J Chr **B 692**, 187-198.

Flurbiprofen (rat).- Pptn.(MeCN); C-18: MeCN/phos. ac.; fluor; sens. 0.05 μg/ml(Q). Park, K-M. et al. (1997) J Lq Chr **20**, 1849-1855.

Furprofen/Rufloxacin.- LE; SAX: MeCN/phos. pH 7; 280 nm; sens. 50 ng/ml. Carlucci, G. & Mazzeo, P. (1996) J Chr Sci **34**, 182-184.

Ibuprofen.- (1): pptn.(MeCN); C-18: MeCN/phos. ac.; 220 nm; sens. 1 μg/ml. PK use. (2; **mtbs**. too): LE; TLC; UV. (3): pptn.(MeCN); CE: borate pH 8.5; 214 nm; sens. 8 μg/ml. (4; incl. urine): LE; deriv.; SPE; C-18: MeCN/phos. pH 3.5. (1) Rifai, N. et al. (1996) Clin Chem **42**, 1812-1816. (2) Save, T.K. et al. (1997) J Chr **B 690**, 315-319. (3) Shihabi, Z.K. & Hinsdale, M.E. (1996) J Chr **B 683**, 115-118. (4) Tan, S.C. et al. (1997) Chr'ia **46**, 23-32.

* **Ketorolac**.- (a) LE; **chiral** deriv.; NP: solvents; OR (b; good method-agreement) SPE; AGP (**chiral**): iPrOH/phos. pH 5.5. (a) & (b) 317 nm; sens. 20 ng/ml. Tsina, I. et al. (1996) J PhBmAn **15**, 403-417.

Ketoprofen.- (1): pptn.($AmSO_4$); SPE; **Chirex** col.: MeOH/acet.; 254 nm; sens. 0.1 μg/ml(Q). (2; incl. rat, 0.1 ml only): **Chiralpak** AD, then NP: solvents/TFA; 254 nm; sens. 0.02(rat:0.2) ng/ml. (3): pptn.(MeCN); **chiral** CE, pH 8.9 with β-cyclodextrin; 254 nm; sens. 1 ng/ml(Q) [a CN-col. method surpasses, but more costly]. (1) Boisvert, J. et al. (1997) J Chr **B 690**, 189-193. (2) Lovlin, R. et al. (1996) J Chr **B 679**, 196-198. (3) Friedberg, M. & Shihabi, Z.K. (1997) J Chr **B 695**, 193-198.

Nimesulide mtb., reduced (invstg.; urine).- SPE, batch mode (XAD-2); GC-MS or FTIR. Sarkar, P. et al. (1997) J Anal Tox **21**, 197-202.

Piroxicam & 5'-hydroxypiroxicam (urine).- Pptn.(TCA); C-18 (40°): MeCN/acet.; 330 nm; sens. 50 ng/ml(Q). Avgerinos, A. et al. (1995) J Chr **B 663**, 142-146.

Sulindac mtb. (FGN-1); oxid., active (incl. hydrol. urine & faeces).- LE; C-8: MeCN/acetic ac.; 329 nm; sens. 25 ng/ml(Q). Ray, G.F. et al. (1995) J PhBmAn **14**, 213-220.

Tepoxalin & mtb.- LE; HPLC-fluor [no details]; sens. 50 ng/ml. Depré, M. et al. (1996) Int J Clin Pharmacol Res **16**, 1-8.

* **Tiaprofenic acid** (exclude light!).- LE; **chiral** col. (AGP): iPrOH/phos. pH 6.5; 317 nm; sens. 25 ng/ml(Q). Geisslinger, G. et al. (1996) J Chr **B 675**, 77-81.

Tramadol.- SPE; GC-MS; sens. 1 ng/ml. Merslavic, M. & Zupancic-Kralj, L. (1997) J Chr **B 693**, 222-227.

Triamcinolone & Hydrocortisone.- LE; C-18: aq. MeOH/THF; 252 nm; sens. 2 ng/ml(Q). Doeppenschmidt, D.A. et al. (1996) J Chr **B 682**, 79-87.

CI 1004 ['dual-inhibitor' NSAID] (incl. rat etc.).- SPE; C-8(45°/human, 40°): MeCN/acet. pH 4; 360 nm; sens. ≤8 ng/ml. Andress, L.D. & Guttendorf, R.J. (1997) J Chr **B 698**, 171-179.

CNI 1493 [an aromatic guanylhydrazone] (rat).- C-8 [C-18 poorer]: MeCN/phos.ac./ion-pair agents, grad.; 265 nm; sens. 1 μg/ml. Cerami, C. et al. (1996) J Chr **B 675**, 71-75.

DA 5018 [non-narcotic analgesic] (rat; incl. urine & bile).- LE; C-18: MeCN/phos./methanesulfonate pH 2.5; fluor; sens. 20 ng/ml. Shim, H.J. et al. (1997) J Chr **B 689**, 422-426.

A collagenase inhibitor & mtb. [hydroxamic & valinamide moieties] (incl. urine).- SPE; C-18: MeOH/formate pH 3.0; MS-MS(ionspray); sens. 1 ng/ml. Knebel, N.G. et al. (1995) J Chr **B 673**, 213-222.

2. RESPIRATION-EASING AGENTS/ANTI-ALLERGICS: bronchodilator, mucolytic [expectorant: #7]; anti-histamine/-inflammatory [cf. #1]; ALSO (even if arguably apt for #4) PAF-/eicosanoid-/IL-related

***N*-Acetylcysteine** (rat, incl. urine & tissue homogenates).- Deriv. (under pptn. conditions; filter); C-18: MeCN/acet./phos. pH 3.75; fluor; sens. 2 ng/ml(Q). Ercal, N. et al. (1996) J Chr **B 685**, 329-334.

Albuterol: SEE Salbutamol

Amiprilose.- Deriv., 2-stage (2nd for fluor); C-18: MeOH/amm.acet./diMe-octylamine; fluor; sens. 0.1 µg/ml(Q). PK use. Wu, S.T. et al. (1997) J Chr **B 692**, 149-156.

Azacyclonol (Terfenadine mtb.; aids its absorption).- LE; GC-MS(EI); sens. 2 ng/ml. No need to make deriv. Martens, J. (1996) J Chr **B 678**, 349-353.

Azathioprine.- LE; C-18: MeCN/phos. pH 2.3; 285 nm; sens. <3 ng/ml(Q). PK use. Binscheck, T. et al. (1996) J Chr **B 675**, 287-294.

* **Bambuterol**- Add physostigmine to avoid hydrol.; SPE (memb.); CZE with double stacking (allows large vol. load) and, if **chiral** mode, a cyclodextrin; 205 nm; sens. 1.5 ng/ml(Q). Palmasdottir, S. et al. (1997) J Chr **B 688**, 127-134.

Brequinar (rat).- Pptn.(MeCN); C-8(40°): MeCN/phos. ac.; 254 nm; sens. 0.05 µg/ml(Q). PK use. Dasgupta, S. et al. (1995) J Chr **B 673**, 107-112.

Clenbuterol (incl. urine).- LE; **Chirex** 3022 col.: solvents/TFA; 254 nm; sens. 2 ng/ml(Q). Abou-Basha, L.I. & Aboul-Enein, H.Y. (1996) Bm Chr **10**, 69-72.

Cromoglycate, Sodium (inhaled; urine).- SPE; C-8: MeOH/phos./R_4N^+ pH 2.3; 238 nm; sens. 0.2 µg/ml(Q). Aswania, O.A. et al. (1997) J Chr **B 690**, 373-378. [SEE end of #7, **SDZ...**, for another **cyclosporin**.]

Cyclosporin A (whole blood).- (1): LE (1-step); C-8 (65°): MeCN/MeOH/iPrOH; 208 nm; sens. 20 ng/ml(Q). (2; oxid. **mtbs**. also): LE; CN col.(50°): 212 nm; sens. 10 ng./ml. (1) Li, K. et al. (1996) J Lq Chr **19**, 2179-2188. (2) Khoschorur, G. et al. (1997) J Chr **B 690**, 367-372. [154.

Dextromethorphan.- SPE; LC-MS-MS; 50 pg/ml. Eichhold, T.H. et al. (1997) J Chr **B 698**, 147-

Dimethindene & mtbs., invstg. (incl. rat; urine).- (<u>a</u>; simple and rapid): CE, phos. pH 3.2; 200 nm. (<u>b</u>) C-8: MeCN/ammon. acet.; MS(CI); sens. 1 ng/ml. (<u>c</u>, for enantiomers): **Chiracel OD** col. Prien, D. & Blaschke, G. (1997) J Chr **B 688**, 309-318.

Epinastine.- LE; C-18: MeOH/phos. pH 4.5; 220 nm; sens. 4 ng/ml. Ohtani, H. et al. (1996) J Chr **B 683**, 281-284.

Levodropropizine.- LE; polymer-RP col.: MeOH/THF/phos. pH 3; fluor; sens. 1 ng/ml. Tagliaro, F. et al. (1996) J Chr **B 685**, 167-170.

Loraditine & Pheniramine.- LE; GC-MS; sens. 0.5 & 2 ng/ml. Martens, J. (1995) J Chr **B 673**, 183-188.

Mebeverine mtbs., alcohols incl. 'des').- Add their ^{3}H-glucuronides; hydrol. (gives ^{3}H-i.s.); deriv.; GC-MS(EI); sens. 0.5 ng/ml. Tulich, L.J. et al. (1996) J Chr **B 682**, 273-281.

Mizolastine (to test micro-HPLC systems): see Diltiazem in #3.

Montelukast sodium [MK 0476; anti-LT_1].- (1): pptn.(MeCN); SPE (pH 3.6, 'Biomatrix'; swtch); **chiral** (AGP) col.: MeCN/acet. pH 5.8; fluor; sens. 10 ng/ml. (2; **mtbs**. invstg., radio): pptn.(MeCN); C-18: MeCN/phos. pH 3.5; fluor. (1) Liu, L. et al. (1997) J PhBmAn **15**, 631-638. (2) Balani, S. et al. (1997) Drug Met Disp **25**, 1282-1287.

Mycophenolate mofetil.- Pptn. (HCl); SPE; C-18: MeCN/citric & phos. ac./heptanesulfonate pH 3.6; 254 nm; sens. 0.4 µg/ml(Q). Degradation studied. Tsina, I. et al. (1996) J Chr **B 681**, 347-353.

Peptide T (rabbit).- Dilute; C-18: MeCN/phos. ac.; fluor; sens. 5 ng/ml. Bioavailability studied. Bechgard, E. et al. (1997) J Chr **B 693**, 237-240.

Salbutamol (Albuterol) (inhaled).- SPE; deriv.; GC-MS(EI); sens. 50 pg/ml(Q). Logsdon, T.W. et al. (1997) J Chr **B 692**, 472-477. | Cf. art. #D-4

Sirolimus (Rapamycin; mtbs. also; whole blood).- Pptn. (MeOH & Zn^{2+}); SPE: C-18: MeCN/formic acid.; MS(ESP; on Na^+ adducts); sens. 0.25 ng/ml. Streit, F. et al. (1996) Clin Chem **42**, 1417-1425.

Tacrolimus [FK 506; a macrolide, lactonic].- (1): SURVEY of TM, esp. by HPLC (incl. ELISA and MS detection), also by FPIA. (2; whole blood): pptn.(MeCN); SPE; C-4 Brownlee col.: MeOH/acet. pH 5; MS-MS(ionspray, MRM); sens. 0.2 ng/ml(Q). (1) Dasgupta, A. et al. (1995) J Lq Chr **18**, 4029- 4038. (2) Taylor, P.J. et al. (1996) Clin Chem **42**, 279-285.

Terfenadine [& SEE ABOVE, Azacyc...].- (1): LE; C-18: MeOH/acet. pH 4; 273 nm; sens. 0.2 ng/ml(Q). (2): LE; C-18: MeCN/MeOH/ammon. acet., grad.; MS-MS(API-CI); sens. 1 ng/ml(Q). (3): SPE; CN col.: MeCN/MeOH/ammon. acet.; MS(ESP); sens. 0.1 ng/ml. (1) Lau, Y.Y. et al. (1996) J Lq Chr **19**, 2669-2679. (2) Xu, A.S. et al. (1996) J PhBmAn **14**, 1675-1680. (3) Adam, D.A. et al. (1997) J Chr **B 693**, 345-351.

Theophylline & mtbs. (incl. urine).- LE; C-18: MeOH/acet. pH 4; 273 nm; sens. 0.02(urine:0.2) µg/ml. Rasmussen, B.B. & Brosen, K. (1996) J Chr **B 676**, 169-174.

Tipredane mtbs. (urine, incl. rat).- silica col.; C-18 cols. (swtch): MeCN/MeOH; 242 nm; sens. 25 ng/ml(Q). Baker, P.R. et al. (1997) J Chr **B 694**, 193-198; also (human urine) 199-209.

Zafirlukast [LT antagonist].- Pptn.; CN col.: THF/hexane/acet. ac.; fluor; 0.7 ng/ml(Q). Bui, K.H. et al. (1997) J Chr **B 696**, 131-136.

MK 591 [inhibits LT biosyn.].- Pptn. (MeCN); SPE, swtch; NP col.: MeCN/TFA; 290 nm; sens. 10 ng/ml(Q). Lin, C. et al. (1996) J PhBmAn **14**, 1601-1608.

SCH 44643 [PAF antagonist].- (1; rat): LE, with a -70° step; SPE($-NH_2$); GC-NPD; sens. 5 ng/ml(Q). (2; monkey): LE; GC-NPD; sens. 20 ng/ml(Q). (1) Kim, H. et al. (1996) Chr'ia **43**, 533-536. (2) Lin, C.C. et al. (1996) J PhBmAn **15**, 343-347.

ZD 9583 [thromboxane receptor antagonist] (incl. urine).- Pptn.(MeCN); C-8: MeCN/MeOH/TFA; swtch for MS-MS(CI); sens. 1 ng/ml(Q). Gao, C.X. et al. (1997) J Chr B 695, **299-307**. [S 1033, at end of #7, is also of PG type]

3. (ANTI)ADRENERGICS/CHANNEL MODULATORS/ACE INHIBITORS/renin area ['Straddles' #2, #4 & #5; (anti-)dopaminics are in #5b]

β-Blockers often serve as test compounds. SEE General Index. HPLC for pre-GC clean-up (incl. hydrol. urine): Hyotylainen, T. et al. (1997) J Chr Sci **35**, 280-286.

Alprenolol & mtb. (-OH, active; incl. saliva).- SPE(C-2); chiral deriv.; GC-MS; sens. 2 ng/ml(Q). Hold, K.M. et al. (1996) J Chr Sci **34**, 13-19.

* **Amlodipine.**- (1; incl. gingival crevicular fluid): LE; GC-ECD; sens. 0.5 µg/ml. (2) SPE; 'mobile phase scavenger' step; chiral col. (AGP): PrOH/acet./TBA^+; then ('compressed') phenyl col.: MeOH/acet./EDTA/TBA'; EC; sens. 0.2 ng/ml. (1) Monkman, S.C. et al. (1996) J Chr **B 678**, 360-364. (2) Josefsson, M. & Norlander, B. (1996) J PhBmAn **15**, 267-277.

* **Captopril.**- (1): LE; SPE (silica) on benzene layer; C-18: MeCN/acetic ac.; 263 nm; sens. 15 ng/ml. (2; 150 µl sample): maleimide deriv.; C-18: MeCN/acetic ac.; fluor; sens. 10 ng/ml. (3): C-18-based methods.- (3a; plasma): SPE; deriv. (for UV) by halopyridinium salt. (3b), redox study (blood, urine): for -S-S-, reduce by triphenylphosphine. (4; plasma): SPE; deriv. (for fluor), pre-col. (monobromobinane) OR post-col. (oPA);

JOURNAL TITLES: Chromatog... - Chr; Liquid - Lq; Pharm. Biomed. Anal. - PhBmAn	Prefix * if notable, e.g. for a method study. Metabolites, mtb(s).; invstg. if investigated. For (e.g.) conjugates: hydrolysis, hydrol. Pptn., protein precipitation. LE, liquid-liquid extraction. (Commonly - in SPE too - dry down and redissolve; unstated.) Packings: e.g. C-18, CN (cyano), AGP (α_1-acid glycoprot.) MeCN, acetonitrile; iPrOH, isopropanol; DEA, diethylamine. If gradient: grad. Enantiomer studies/agents: chir.... Automation implicit if stated on-line or switching (swtch). Derivatization/conversion: deriv, (sometimes post-column). If nm stated, absorptiometric detection; EC, electrochemical. Approx. LOD or (Q) LLOQ: sens. Pharmacokinetic studies, PK.	*KEY (abridged. Generally human plasma/serum*

Captopril, *continued*
C-18: MeCN/TFA/borate pH 12 or 11; fluor; sens. 25 ng/ml(Q). (4): LE; C-18: MeCN/acetic ac.; 258 nm; sens. 2 ng/ml. PK use. (1) Bahmaei, M. et al. (1997) J PhBmAn **15**, 1181-1186. (2) Arroyo, C. et al. (1997) J Chr **B 688**, 339-344. (3a) Bald, E. et al. (1996) J Chr **B 681**, 283-289; (3b, same lab.) Sypniewski, S. et al. (1996) J Chr **A 729**, 335-340. (4) Kok, R.J. et al. (1997) J Chr **B 693**, 183-189. (5) Li, K. et al. (1996) Biomed Chr **10**, 237-239.

Carteolol.- Deriv.; LE; SPE(silica); GC-MS; sens. 30 pg/ml(Q). Nagasawa, M. et al. (1995) J Chr **673**, 294-298.

Celiprolol.- (1): LE; C-18 (silanol-deactivated): MeOH/TEA/phos. pH 3; fluor; sens. 5 ng/ml. (2): LE; EITHER C-6: MeCN/TEA/phos. pH 3.5, 238 nm; OR (**chiral**) similar but deriv., C-18 & fluor. (1) Chiu, F.C.K. & Raymond, K.(1996) J Chr **B 687**, 462-465. (2) Verbesselt, R. et al. (1997) J Chr **B 683**, 231-236.

Clotrimazole & mtbs. (incl. whole blood).- SPE; ultrafilter if value for **free** drug wanted; C-8: MeOH/phos. pH 6.3; 210 nm; sens. 0.03 µg/ml. Rifai, N. et al. (1995) Clin Chem **41**, 387-391.

* **Diltiazem**.- (1; rat): LE; 2 NP cols.: solvents/aq. NH_3; 240 nm; sens. 20 ng/ml(Q). PK use. (2; heart, homogenized frozen): pptn.(acetone) on trypsin incubate; SPE(CN); C-6: MeCN/acet. pH 4; 237 nm; sens. 0.1 ng/mg(Q). (3; desacetyl **mtb**. also): LE; CZE [accord with HPLC]: phos. pH 2.5; 238 nm; sens. 10 ng/ml(Q). (1) Scully, P. et al. (1996) J Chr **A 729**, 297-300. (2) Laeer, S. et al. (1997) J Chr Sci **35**, 93-96. (3) Coors, C. et al. (1996) J Chr **A 717**, 235-243.

Diltiazem (& Mizolastine, as test cpds. for micro-HPLC systems): Malavasi, B. & Ascalone, V. (1996) J HR Chr **19**, 503-510.

Dofetilide & analogs (dog).- LE; NP col.: MeCN/phos. pH 6.5; 230 nm; sens. 5 ng/ml(Q). Walker, D.K. et al. (1996) Xenobiotica **26**, 1101-1111.

Eliprodil (±hydrol.).- LE; **chiral** deriv.; C-8 (swtch after pre-col.): MeOH/MeCN/phos. pH 2.6; fluor; sens. 0.15 ng/ml. PK use. Malavasi, B. et al. (1996) J Chr **A 729**, 323-333.

* **Enalaprilat, Lisinopril**.- IA by time-resolved fluor (FIA); sens. 0.5 ng/ml(Q); cf. double-Ab RIA (wider range). Yuan, A. & Gilbert, J.D. (1996) J PhBmAn **14**, 773-781.

Isradipine & mtbs. (incubates).- LE/[cells] pptn.(MeOH); C-8: MeOH/THF/ phos./TBA^+ pH 3; 240 nm. Bidouil, S. et al. (1997) J Chr **B 693**, 359-366.

Lisinopril (urine).- SPE; C-18 (45°): MeCN/MeOH/THF/phos. pH 2.9; 206 nm; sens. 0.5 µg/ml(Q). Wong, Y-C. & Charles, B.G. (1996) J Chr **B 673**, 306-310. SEE ALSO Enalprilat, above.

Losartan & mtb. [active; EXP 3174].- (1; rat): Pptn.(MeCN); SPE (on-line; swtch/back-flush); C8: MeCN/acet. pH 3.5; 254 nm; sens. 50 ng/ml(Q). incl. urine): LE; CN col.(35°): phos.pH 2.3/MeCN; fluor; sens. 1 ng/ml. (1) Lee, H. et al. (1996) Chr'ia **42**, 39-42. (2) Ritter, M.A. et al. (1997) J PhBmAn **15**, 1021-1029.

Manidipine.- SPE(C-8); SPE(silica); C-8: MeCN/phos. pH 4.5; EC; sens. 0.3 ng/ml. PK use. Ohkubo, T. et al. (1996) J Chr **B 687**, 413-418.

Monatepil & mtbs. (S-oxid.; rat).- SPE; C-18: MeCN/phos. pH 6; EC; sens. 2 ng/ml(Q). Kurono, M. et al. (1997) J Chr **B 689**, 427-432.

Nifedipine (incl. urine).- SPE; **chiral** deriv.; C-18: MeOH/acet. pH 4; EC; sens. 0.5 ng/ml. Horvath, V. et al. (1996) J Chr **B 686**, 211-219.

Propranolol & 4-hydroxy-.- LE; **chiral** deriv.; C-18: MeCN/phos./TEA; fluor; sens. 1 ng/ml(Q). Wu, S.T. et al. (1997) J Chr **B 692**, 133-140.

Sotalol.- (1; incl. urine): SPE; **chiral** deriv.; C-18: MeCN/phos. pH 4.6; 225 nm; sens. 0.02(urine 0.2) μg/ml. (2; incl. heart homogenates): SPE; C-6: MeCN/phos. pH 3; 230 nm; sens. 30 ng/ml, 0.3 ng/mg. (1) Shimizu, T. et al. (1995) J Chr **B 667**, 77-83. (2) Laer, S. et al. (1996) J Chr **B 681**, 291-298.

CP 80,794 & mtb. - a fragment with morpholine moiety [renin inhibitor] (incl. dog).- LE; incub. with chymotrypsin; deriv.; C-18: aq. MeCN; fluor; sens. 10 ng/ml. Variant procedure for **mtb.**, with initial SPE (anion-exch.). Allen, M.C. et al. (1997) J Chr **B 696**, 243-251.

CP 191,166 [ANG II antagonist; imidazo, pyridine & thiophen moieties] (rat & dog).- LE; C-8: aq.MeCN/TEA/phos. ac.; 289 nm; sens. 0.01 μg/ml(Q). Allen, M.C. & Day, W.W. (1996) J Chr **B 675**, 265-271.

E 4177 [ANG II receptor antagonist] (incl. urine).- LE; C-18: MeCN/phos. ac.; fluor; sens. 2 ng/ml(Q). Seki, T. et al. (1997) J Chr **B 688**, 101-106.

Ro 31-7837 & mtb. [K^+ channel opener] (rat).- Pptn.(EtOH); trapping col. ('ABZ'; back-flush); C-18: formate pH 3.6/MeOH; MS(API); sens. 0.2 ng/ml(Q). Zell, M. et al. (1997) J Chr **B 694**, 135-143.

SIM 6080 [Ca^{2+} antagonist] & 'des' **mtbs.** (incl. hydrol. urine, & tissues, aq. sonicate).- LE; NP: MeCN/ClO_4^-/phos. pH 3; 210 nm; sens. 5 ng/ml. Castelnovo, P. (1996) et al. (1996) J Chr **B 676**, 77-85.

TCV 116 & mtbs. [ANG II antagonist; tetrazole/biphenyl/benzimidazole moieties] (incl. urine).- LE; 2 C-18 cols.(swtch): MeCN/phos. pH 4; fluor; sens. 0.5(urine:10) ng/ml. Variant conditions for mtbs. (e.g. 210 nm) or urine. Miyabayashi, T. et al. (1996) J Chr **B 677**, 123-132.

XR 510 [ANG II receptor antagonist] (dog & rat).- LE; SPE(amino-propyl); C-18: MeCN/phos. pH 6.5; 272 nm; sens. 5 ng/ml(Q). Emm, T.A. et al. (1996) J Chr **B 675**, 273-278.

4. CARDIOVASCULAR AGENTS incl. anti-thrombotic/-cholesterol, diuretic, inotropic [PAF-/eicosanoid-/IL-related: SEE #2; renin: #3]

Almokalant.- SPE (C-2, by 'Prospekt'); C-18 to concentrate (swtch); C-18: iPrOH/MeCN/diMe-octylamine/phos. pH 3.1; fluor; sens. 1 ng/ml(Q). Svennberg, H. & Lagerstroem, P-O. (1997) J Chr **B 689**, 371-377.

Amiloride & analogues (mouse, incl. kidney/liver/tumour).- SPE; C-18: MeCN/PCA; 365 nm; sens. 0.02(tissues:0.2) μg/ml. Lee, C. & Tannock, I. (1996) J Chr **B 685**, 151-157.

Bezafibrate, ciprofibrate & fenofibric acid.- LE; C-18: MeCN/phos. ac. 287 or (fenofibric) 232 nm; sens. 0.1 μg/ml. Masnatta, L.D. et al. (1996) J Chr **B 687**, 437-442.

Chlorthalidone (urine).- SPE(CN); C-18: MeCN/phos. pH 7; 214 nm; sens. 20 ng/ml. Salado, C. & Vera-Avila, L.E. (1997) J Chr **B 690**, 195-202; cf. **698**, 187-194 (Giachetti, C. et al.): plasma, incl. **atenolol**; LE at start.

Clofibric acid (rat).- Pptn.(MeCN); C-18: MeCN/acetic acid; 230 nm; sens. 1 μg/ml(Q). Lau-Cam, C. et al. (1995) J Lq Chr **18**, 3945-3954.

Coumarin & (its glucuronide too) **7-hydroxycoumarin**.- Pptn.(TCA); C-18: MeOH/acetic ac., grad.; 320 nm; sens. 0.2 μg/ml. Bogan, D.P. & Kennedy, R. (1996) J Chr **B 686**, 267-273.

Dextrazoxane (incl. urine).- Pptn.(MeCN; omit for urine); LE; phenyl col.: MeOH/phos. pH 4.7; 208 nm; sens. 0.1(urine:10) μg/ml(Q). PK use. Rosing, H. et al. (1997) J Lq Chr **20**, 583-601.

* **Digoxin** & other **cardenolides** [cardiac glycosides].- SPE; C-18: aq. MeCN, grad.; deriv. (reactor; conditions discussed) to allow: fluor; sens. 0.1 ng/ml. Belsner, K. & Buechele, B. (1996) J Chr **B 682**, 95-107.

Dilazep.- Pptn. (MeOH); SPE; C-18: MeCN/phos. pH 3.2; 267 nm; sens. 2 μg/ml(Q). Gumieniczek, A. & Pryzborowski, L. (1997) J PhBmAn **15**, 997-1000.

Fluvastatin.- SPE, automated; EITHER C-8 (45°): MeOH/TBA^+/phos. pH 6; OR **Chiracel** OD col. (10°), pH 2.3; fluor; sens. 0.4 ng/ml. Toreson, H. & Eriksson, B-M. (1997) Chr'ia **45**, 29-34.

Frusemide: SEE #7

* **Guanethidine sulfate**.- Trap on C-18 pre-col.; swtch; C-18: MeCN/phos./octanesulfonate pH 3; deriv.(ninhydrin); fluor; sens. 1 ng/ml. Kinoshita, T. et al. (1997) J Lq Chr **20**, 2099-2108.

Hydrochlorothiazide LE; NP: C-18: aq. MeCN/phos. pH 7.3; EC; sens. 5 ng/ml(Q). Richter, K. et al. (1996) J Chr **A 729**, 291-296.

Isosorbide ~~dinitrate~~, & mononitrate mtbs.- SPE; GC-ECD; sens. 2 ng/ml(Q). Pommier, P. et al. (1996) J Chr **B 678**, 354-359.

* **Isosorbide-5-mononitrate**.- LE; GC-ECD; sens. 10 ng/ml. Losses minimized [evaporative, and adsorptive onto plastic; start with 10 injections of std., & change liner after 100 injections]. Pennings, E.J.M. et al. (1996) J Chr **B 675**, 332-335.

Metirosine (Metyrosine) & mtb.- SPE (SCX); C-18: MeCN/MeOH/phos./EDTA/heptanesulfonate pH 3; fluor; sens. 1 μg/ml. Hefnawy, M.M. & Stewart, J.T. (1997) J Lq Chr **20**, 3009-3016.

Mexiletine.- LE. (a): deriv.; **Chiracel OJ** col.: solvents; fluor; sens. 50 ng/ml(Q). TM & multiple-dose PK use. ALTERNATIVE: (b): oPA & **chiral** deriv.; C-18; fluor. Single-dose PK use. Lanchote, V.I. et al. (1996) J Chr **B 685**, 281-289.

1-Octacosanol [cholesterol-lowering].- LE; GC-FID; sens. 50 ng/ml(Q). Gonzalez-Bravo, L. et al. (1996) J Chr **B 682**, 359-363.

Pentisomide & mtb. (N-des).- LE; C-18: MeCN/alkylamines/octanesulfonate, grad.; 260 nm; sens. 0.03 μg/ml. Mewis, C. et al. (1996) J Chr Sci **34**, 161-165.

* **Perhexiline** & -OH **mtb**.- LE (simple) & deriv. - iPr_2EtNH_2 added as anti-adsorptive & catalyst; phenyl col.: MeCN/acetic ac.; 340 nm; sens. 30 μg/ml. TM use. Grgurinovich, N. (1997) J Chr **B 696**, 75-80.

Selfotel [NMDA antagonist] (urine).- Deriv. (LE to wash); C-18, then **Chiracel OD** col. (swtch): phos. pH 2/MeCN; fluor; sens. 0.25 μg/ml. PK. Knoche, B. et al. (1997) J Chr **B 695**, 355-363.

Simendan & Levosimendan (incl. rat & dog).- LE; **chiral** col.(β-cyclodextrin): acet. pH 6/MeOH; 380 nm; sens. 5 ng/ml. Wikberg, T. et al. (1996) Chirality **8**, 511-517.

* **Simvastatin & mtb.** (acid; active).- SE(C-8); de-lactonize; deriv.; SPE (phenylboronic); CH col., swtch; C-18 (40°): aq.MeCN; fluor; sens. 20 pg/ml. Ochiai, H. et al. (1997) J Chr **B 694**, 211-217.

Ticlopidine.- SPE; C-18: MeCN/MeOH/phos. pH 4; 215 nm; sens. 10 ng/ml. Rona, K. et al. (1997) J Chr **B 693**, 393-398.

* **Tirofiban** (MK 0383).- Pptn.(PCA); C-18: MeCN/acet. pH 6; MS-MS(nCI); sens. 0.2 ng/ml(Q). Accord with RIA [as for another fibrinogen receptor antagonist: Vol. 24, p. 65]. Ellis, J.D. et al. (1997) J PhBmAn **15**, 561-569.

Torasemide (urine).- LE; deriv.; GC-MS(EI); sens. 5 µg/ml(Q). Barroso, M.B. et al. (1997) J Chr **B 690**, 105-113.

Verapamil.- (1): LE; GC-NPD; sens. 2 ng/ml. (2; **nor-** also; tissues too): LE; SPE; C-8: aq. MeCN; fluor; sens. 50 ng/ml. (1) Shin, H-S. et al. (1996) J Chr **B 677**, 369-373. (2) Negrusz, A. et al. (1997) Chr'ia **46**, 191-196.

Vinpocetine mtb., apovincaminic acid.- LE (ion-pair step); C-18: MeCN/ phos./octanesulfonate pH 2.8; 254 nm; sens. 5 ng/ml(Q). PK use. Maya, M.T. et al. (1996). J PhBmAn **14**, 617-622.

An **arylacetamide** [pyrollidine & thiophen moieties] (rat, incl. tissues).- Pptn.(acetone); LE; CN col.: MeOH/MeCN/acet. pH 7; 215 nm; sens. 0.1 µg/ml. Walker, M.L. et al. (1996) J Chr **B 675**, 257-263.

DX 9065a [a substd. propionic acid; anticoagulant] (urine & faeces).- SPE (cation exch.); 'Asahipak ODP' col.: phos. pH 8.5/MeCN; 242 nm; sens. 25 ng/ml. PK use. Murayama, N. & Sudo, K. (1997) J Chr **B 692**, 389-396.

KC 11346 [an aminopropane; pyrazole moiety in one **N**-substituent] & des-**N mtbs**. (dog).- Brownlee pre-col.(CN); CN col.: MeCN/acet.; fluor; sens. 3 ng/ml(Q). Lau, Y. et al. (1996) J Lq Chr **19**, 3279-3292.

* **TP 9201** [antithrombotic peptide] (cattle).- LE; C-18: PrOH/MeCN, grad.; react with Cu^{2+}; EC; sens. 20 ng/ml. Woltman, S.J. et al. (1995) J PhBmAn **14**, 155-164.

5. NEUROACTIVE AGENTS

5a: 'damping-down' action, esp. anaesthetic (incl. local), narcotic/analgesic, hypnotic/sedative; anxiolytic; **anti-**convulsant/ -emetic/-insomnia/-migraine [#5b for anti-Parkinson; pain: cf. #1, some analgesics; #3, (anti)adrenergics; #7, smooth musc. relaxants]

For **local anaesthetics**, entries under individual names may have wider applicability; see also art. # B-2.

Acetylsalicylic (& salicylic) acid.- Pptn.(MeCN); C-18: MeCN/phos. pH 2.5; 237 nm; sens. 0.3 µg/ml(Q). Kees, F. et al. (1996) J Chr **B 677**, 172-177; cf. **675**, 61-70 (Liu, J-H. & Smith, P.C.): several **aspirin mtbs**. (incl. urine).

Alfentanil & [maybe co-dosed] **Fentanyl**.- LE; CN col.: MeCN/phos. pH 2.8; 195 nm; sens. 2 ng/ml(Q). Kumar, K. et al. (1996) J PhBmAn **14**, 667-673.

Alprazolam/Clonazepam/Nitrazepam [low-dose **benzodiazepines**].- SPE (auto - 'ASP'); C-18: MeCN/MeOH/phos. pH 3.7; 245 nm; sens. 2 ng/ml. TM use. Akerman, K.K. et al. (1996) Clin Chem **42**, 1412-1416.

Alprazolam & mtb., α-OH.- SPE(CN); C-8: MeCN/phos.pH 4.5; 230 nm; sens. 1 ng/ml(Q). TM use. Nagasaki, T. et al. (1997) Anal Sci **13**, 245-249.

3-(Aminomethyl)-5-methylhexanoic acid enantiomer [new anticonvulsant] (rat, incl. milk; 100 µl sample).- Dilute & centrifuge; LE with deriv. step; C-18 (40°): MeCN/acet. pH 4; 350 nm; sens. 1 ng/ml(Q). Windsor, B.L. et al. (1995) J Chr **B 674**, 143-148.

Apomorphine.- (1; PK use): SPE; C-18: MeOH/phos./heptanesulfonate; pH 3.3; fluor; sens. 0.5 ng/ml(Q). (2): SPE; **Chiracel** OD-R col.: MeCN/ClO_4^- pH 2; 273 nm; sens. 10 ng/ml(Q). (1) Priston, M.J. & Sewell, G.J. (1996) J Chr **B 675**, 61-70. (2) Ameyibor, E. & Stewart, J.T. (1996) J Chr **B 686**, 297-300.

Bezitramide mtb. (acidic; urine).- LE; deriv.; GC-NPD; sens. 10 ng/ml(Q). De Baere, S.M. et al. (1996) J Anal Tox **20**, 159-164.

Bromocriptine (incl. eye fluids).- LE; C-18: MeCN/TEA; fluor; sens. 0.2 ng/ml. Choi, Y-J. et al. (1997) J Chr **B 694**, 415-420. (Arguably should be in #5b]

* **Bupivacaine/Etidocaine.**- SPE (on-line); C-18(50°): MeCN/phos., varied pH, grad.; 210 nm; sens. 0.02 µg/ml(Q). Drewe, J. et al. (1997) J Chr **B 691**, 105-110.

Buprenorphine & norbuprenorphine (urine, hydrol.).- (1): LE/deriv. (extractive alkylation); GC-MS; sens. 0.2 ng/ml. (2; serum too): LE; C-18: MeCN/formic ac.; MS(ESP); sens. 0.25 ng/ml(Q). (1) Lisi, A. et al. (1997) J Chr **B 692**, 67-77. (2) Hoja, H. et al. (1997) Analusis **24**, 104-107.

Buspirone & mtb.† (active moiety).- SPE; C-18: MeOH/phos. pH 4.5; EC; sens. 1 ng/ml. Dontiadis, J. & Franklin, M. (1996) J PhBmAn **14**, 347-351.

Butorphanol & mtbs., -OH & nor-.- SPE; deriv.; GC-MS(nCI); sens. 20 pg/ml(Q). Mulvana, D.E. et al. (1996) J Chr **B 682**, 289-300.

Butylaminobenzoate & Lidocaine.- Add AcCh esterase blocker; SPE; GC-NPD; sens. 5 ng/ml(Q). PK/TM use. Grouis, R.J.E. et al. (1995) J Chr **B 673**, 51-57.

Carbamazepine & epoxy **mtb.** (active).- (1; TM): pptn.(MeCN); C-14 (60°): MeCN/phos. pH 7; 210 nm; sens. 1 µg/ml(Q). (2; incl. saliva): LE; C-18: MeCN/phos. pH 4; 210 nm; sens. 50 ng/ml(Q). PK use. (3; **oxcarbazepine** too): LE; C-18: MeCN/phos. pH 6.3/TEA; 212 nm; sens. 0.1 µg/ml(Q). TM use. (1) Chollet, D. et al. (1996) J Chr **B 683**, 237-243. (2) Alexishvili, M.M. et al. (1997) Biomed Chr **11**, 36-41. (3) Pienimski, P. et al. (1995) J Chr **B 663**, 97-105.

Chloroprocaine & mtb. (acid; NaF stopped formation in sample).- C-18: MeCN/phos. pH 5; 210 nm [**mtb**: pH 4; 285 nm]; sens. 0.1[0.5] µg/ml. Janicki, P.K. et al. (1996) J Chr **B 675**, 336-341.

Citalopram & mtb. (des).- LE; C-18: MeCN/phos./TEA pH 2.5; 240 nm; sens. 2 ng/ml(Q). TM use. Olesen, O.V. & Linnet, K. (1996) J Chr **B 675**, 83-88.

Clonazepam.- LE; deriv. GC-MS(nCI); sens. 0.1 ng/ml. Song, D. et al. (1996) J Chr **B 686**, 199-204.

Clozapine & mtb., N-des.- (1) Ppptn.(MeCN); LE; C-8: MeCN/Me_4N^+/PCA pH 4.2; 245 nm; sens. 15 ng/ml(Q). (2; **N-oxide** also; incl. red cells): LE; C-18: MeCN/phos./R_4N^+ pH 7; 254 nm; sens. 10 ng/ml. (3) SPE; C-8: MeCN/MeOH/phos. pH 3.7; 220 nm; sens. 5 ng/ml. (1) Gupta, R.N. (1995) J Chr **673**, 311-315. (2) Guitton, C. et al. (1997) J Chr **B 690**, 211-222; risk that any **N-oxide** (little present) may revert to parent drug during analysis - J Anal Tox (1997) **21**, 73 & 74. (3) Akerman, K.K. (1997) J Chr **B 696**, 253-259.

Cocaine & mtbs. (rat).- LE; C-18: MeCN/MeOH/acet. pH 2.2; 235 nm; sens. 2.5 ng/ml. PK use. Ma, F. et al. (1997) J Chr **B 693**, 307-312.

* **Codeine & dihydrocodeine** (incl. urine).- SPE; GC-SID [surface ionization; more sens. (x10) than NPD]; sens. 3 ng/ml. **No** need for a deriv. step. TM use. Suzuki, O. et al. (1995) J Chr **B 673**, 189-195. SEE ALSO Morphine.

Codeine/norcodeine & glucuronides - of **morphine** also (incl. urine): SPE; C-18: MeCN/KH_2PO_4/SDS; 214 nm or EC; diverse sens. values (~0.1 µg/ml at best). Svensson, J.O. et al. (1995) J Chr **B 674**, 49-55.

'DALDA' [opioid peptide] (sheep, infused).- Pptn.(MeCN); SPE; C-18: MeCN/heptafluorobutyric, grad.; 200 nm & MS (MS-MS for sequence study). Grigoriants, O.O. et al. (1997) J Chr **B 695**, 287-298.

Diamorphine, morphine & mtbs. (glucuronides etc.).- SPE; C-18 (40°): MeCN/phos. ac./hexylamine, grad.; 210 nm; sens. 25 ng/ml(Q). TM use. Bourquin, D. et al. (1997) J Chr **B 694**, 233-238.

Re **Enkephalins** entry on next p.- **Bioactive peptides** featured in Vol. 24 (assay refs.: pp. 60 62). For bradykinin and met-enkephalin in rat brain: pptn.(PCA); on-line deriv.; C-8; LIF).- Kanazawa, H. et al. (1997) J Chr **A 763**, 23-29. For an ala-leu-enkephalin & mtbs. (given in ^3H form to rats) in plasma: pptn.(TCA); RP-HPLC-UV; count with a synchronized accumulating radioisotope detector.- Hasegawa, H. et al. (1997) J Chr **B 689**, 287-294. Cf. **697**, 149-162 (Chen et al.): an enkephalin in rat serum; SPE, CE & laser-fluor.

†'Profile' study (rat): Kerns, E.H. et al. (1997) J Chr **B 698**, 133-145 [LC-MS(-MS)].

Diazepam & mtbs.- SPE (disc); **Chiracel OD** col.: MeOH/ClO_4^- ; 210 nm; sens. 7 ng/ml. Liu, J. & Stewart, J.T. (1997) Anal Lett **30**, 1555-1566.

Dolasetron & mtb. (oxid.; MDL 74156).- (1; urine also): LE; C-18: MeCN/BuOH/phos. pH 2.5; fluor; sens. 3 ng/ml(Q; urine poorer). (2): LE; **chiral col.**(OVM): MeCN/acet. pH 7.2; fluor; sens. 2 ng/ml(Q). (1) Huebert, N.D. et al. (1996) J Chr **B 685**, 291-297. (2) McElvain, J.S. et al. (1997) J PhBmAn **15**, 513-521.

Droperidol.- LE; CN col.: MeCN/EtOH/phos. pH 2.4; 200 nm; sens. 0.5 ng/ml. Kumar, K. et al. (1997) J PhBmAn **14**, 1529-1533.

Eltanolone.- Pptn.(MeCN); deriv.; C-18: MeCN/TFA; 367 nm; sens. 0.02 µg/ml. Jones, D.J. et al. (1997) J Chr **B 694**, 467-470.

Etidocaine: SEE **Bupivacaine**

Enkephalins.- SPE; C-18: MeCN/ammon. acet. pH 4.4, grad.; MS(ESP); sens. 0.25 ng/ml. Marquez, C.D. et al. (1997) J Chr **B 694**, 21-30.

See also prev. p.

Fentanyl.- (1): LE; GC-MS(EI); sens. 0.05 ng/ml(Q). (2): similar; sens. 20 pg/ml, exceeding RIA (good accord). (1) Szeitz, A. et al. (1996) J Chr **B 675**, 33-42. (2) Fryirsa, B. et al. (1997) J Chr **B 688**, 79-85.[§]

Flunitrazepam & mtbs. incl. des.- LE (+ salt); CN col.: MeOH/phos. pH 4; 242 nm; sens. 3 ng/ml. Berthault, F. et al. (1996) J Chr **B 685**, 383-387.

Fluphenazine (partly decanoate prodrug; incl. dog).- LE; C-8: MeCN/MeOH/acet.; EC; sens. 0.1 ng/ml. Luo, J-P. et al. (1997) J Chr **B 688**, 303-308.

Gabapentin.- (1): pptn.(PCA); oPA deriv.; C-18: MeCN/MeOH/acet. pH 3.7; fluor; sens. 0.5 µg/ml(Q). TM use. (2): SPE; deriv.; GC-FID; sens. 0.2 µg/ml. (1) Forrest, G. et al. (1996) J Chr **B 681**, 421-425. (2) Wolf, C.E. et al. (1996) J Anal Tox **20**, 498-501.

* **Haloperidol.**- (1): LE [no drying-down]; GC-MS(EI/SI); sens. 1 ng/ml [SI = surface ionization]. (2; reduced **mtb.** also): LE; GC-MS(EI-ESP); sens. 0.1 ng/ml(Q). (3; **mtb.**, a phenylpiperidine; rat tissue).- LE (high salt); GC-ECD; sens. 5 ng/ml. (1) Fujii, T. et al. (1996) J Chr **B 687**, 395-403. (2) Hoja, H. et al. (1997) J Chr **B 688**, 275-280. (3) Fang, J. et al. (1996) J Chr **B 682**, 283-288.

4-Hydroxymephenytoin (phenotyping use; urine, hydrol.).- SPE; C-18: MeOH/phos. pH 4; EC; sens. 2 µg/ml(Q). Tanaka, M. et al. (1996) J Chr **B 676**, 87-94.

Imidazenil [a benzodiazepine deriv.] (rat).- LE; CN: hexane/EtOH/acetic ac.; 255 nm; sens. 30 ng/ml(Q). PK use. Mariot, R. & Zangirolami, L. (1996) J Chr **B 677**, 190-193.

* **Lamotrigine.**- (1; ±hydrol.): pptn.(MeCN); CE: acet./acet. pH 4.8; 214 nm; sens. 0.3 mg/ml. (2; pediatric samples): LE; C-18: MeCN/phos. pH 3.7; 270 nm; sens. 20 ng/ml. (3; 200 µl sample): LE; diphenyl col.: MeCN/phos. pH 3.7; 265 nm; sens. 0.1 µg/ml. (4): LE; deriv.; GC-MS(EI); sens. 0.25 µg/ml. (1) Shihabi, Z.K. & Oles, K.S. (1996) J Chr **B 683**, 119-123. (2) Forssblad, E. et al. (1996) J PhBmAn **14**, 755-758. (3) Londero, D. et al. (1997) J Chr **B 691**, 139-144. (4) Dasgupta, A. & Hart, A.P. (1997) J Chr **B 693**, 101-107.

Levodopa & mtbs.- SEE #5b

Lidocaine: SEE Butylamino.... entry

* **Mepicavaine**: SPE; brush-type **CSP** (Siluveru & Stewart): SEE art. #B-2.

Methadone & mtb., EDDP (incl. urine & hair).- LE; **chiral** CE (with a cyclodextrin deriv.); sens. 2 ng/ml. Frost, M. et al. (1997) Elect'sis **18**, 1026-1034.

Midazolam.- (1; incl. rat; 40 µl sample): LE; GC-MS(SIM); sens. 2 ng/ml. (2; rat brain - NaOH homogenate): LE; C-8: MeCN/MeOH/phos. pH 5.6; 240 nm; sens. 80 ng/g. (3; TM use): LE; CN col.: MeCN/acet. pH 3; 240 nm; sens. 7 ng/ml. (1) Podkowik, B.I. & Masur, S. (1996) J Chr **B 681**, 405-411. (2) Jiang, Q. et al. (1996) J Chr **B 683**, 276-280. (3) Odou, P. et al. (1997) Bm Chr **11**, 19-21. FURTHER ENTRIES on next p.

[§] SEE ALSO **Alfentanil** entry

[CTD.] **Midazolam & mtbs.**, esp. 1-OH.- (1): LE; CN col.: MeOH/iPrOH/ClO_3^-; 215 nm; sens. 2 ng/ml(Q). TM use. (2; rat, 50 µl sample : LE; C-18: MeOH/MeCN/acet. pH 3; 230 nm; sens. 0.1 µg/ml(Q). PK use. (3; neonates): LE; CN col.: MeCN/THF/phos. pH 6.7; 220 nm; sens. 10 ng/ml. PK use. (4) LE; deriv.; GC-MS(EI); sens. 2 ng/ml(Q). (1) Lehmann, B. & Boulieu, R. (1995) J Chr **B 674**, 138-142. (2) Ma, F. & Lau, C.E. (1996) J Chr **B 682**, 109-113. (3) Lee, T.C. & Charles, B. (1996) Biomed Chr **10**, 65-68. (4) Martens, J. et al. (1997) J Chr **B 692**, 95-100.

Mivacurium (*cis/trans* isomerism).- (1): LE; SCX [C-1 for total drug]: MeCN/acid sulfate, grad.; fluor; sens. 1 ng/ml(Q). (2; **mtbs.** too; add stabilizer at start): Pptn.(MeCN); SCX, similar to (1). TM use. (3; dog also): SPE; C-18: MeCN/phos. pH 5.4/octanesulfonate, grad.; fluor; sens. 5 ng/ml(Q). PK use. (1) Biederbick, W. et al. (1996) J Chr **B 685**, 315-322. (2) Weindlmayr-Goettel, M. et al.: as for (1), 123-127. (3) Lugo, S.I. & Eddington, N.D. (1996) J PhBmAn **14**, 675-683.

Is in #1 in Vol. 24 (mis-housed)

Morphine.- (1; **glucuronides** too): SPE (on-line); C-18: MeCN/phos. pH 3; fluor; sens. 5 ng/ml(Q). (2; **codeine** too; TM use, rapid): SPE; C-18: MeOH etc.; 240 nm; sens. 50 ng/ml(Q). (1) Howyler, J. et al. (1995) J Chr **B 674**, 57-63. (2) Freirmuth, M. & Plasse, J-C. (1997) J PhBmAn **15**, 759-764. SEE ALSO Codeine & Diamorphine entries.

Nalbuphine.- (1): SPE or LE; C-18: MeOH/KH_2PO_4; EC; sens. 0.02 ng/ml. (2): LE, then similar; sens. 0.1 ng/ml. Both: PK use. (1) Nicolle, E. et al. (1997) J Chr **B 690**, 89-97. (2) de Cazanove, F. ...SAME; 203-210.

Nerisopam & mtb., N-acetyl.- SPE; C-18: MeCN/MeOH/phos./heptanesulfonate/DEA pH 2.7; 385 nm; sens. 5 ng/ml. Rona, K. et al. (1996) J Chr **B 678**, 63-72.

Olanzapine (LY 170053) & **mtbs., N-des** etc. (rat).- SPE; YMC basic col.: MeCN/MeOH/phos. pH 7; EC; sens. 1 ng/ml(Q). PK use. Chiu J.A. & Franklin, R.B. (1996) J PhBmAn **14**, 609-615.

Ondansetron enantiomers.- (1): SPE(CN); **chiral** CE (with a cyclodextrin deriv.): phos./R_4N^+ pH 2.5; 254 nm; sens. 10 ng/ml. (2) similar, but **Chiracel OD** col.: MeCN/ClO_4^-; sens. 7 ng/ml. (1) Siluveru, M. & Stewart, J.T. (1997) J Chr **B 691**, 217-222 [cf. their art. in this book]. (2) Liu, J. & Stewart, J.T. (1997) J Chr **B 694**, 179-184. SEE also #7.

* **Perphenazine.**- SPE(NP); SGX-CN col.: MeOH/acet.; 257 nm; sens. 1.5 ng/ml. Hrdlick, A. et al. (1997) J Lq Chr **20**, 935-942.

* **Phencyclidine** (incl. urine).- 'Headspace' SPE ('SPME' fibre); GC-SI (surface ionization); 1 ng/ml. Ishii, A. et al. (1996) Chr'ia **43**, 331-333.

Phenytoin & Fosphenytoin [prodrug].- LE or (much better sens.) ultrafilter; C-18: MeCN/Bu_4N^+/phos. pH 2.2; 210 nm; sens. 0.4(prodrug:0.08) µg/ml. Cwik, M.J. et al. (1997) J Chr **B 693**, 407-414.

* **Prilocaine.**- SPE; 'brush-type' **chiral** col. (YMC): MeOH etc./TFA; 220 nm; sens. 5 ng/ml. Chromatography affected if EtOH or PrOH replaced MeOH. Siluveru, M. et al. (1996) J PhBmAn **15**, 389-392.

Pimozide.- LE; C-18: MeCN/phos.pH 3; fluor; sens. 1 ng/ml(Q). PK use. Kerbusch, T. et al. (1997) J Chr **B 694**, 163-168.

Pramipexole (U 95,528).- LE; CN col.: MeOH/ammn. acet.; MS-MS; sens. 50 pg/ml(Q). Lau, Y.Y. et al. (1996) J Chr **B 683**, 209-216; also 217-223: a variant method (incl. urine) using C-8 col. and EC detection; same sens.

Promethazine.- SPE (mixed mode disc); **Chiracel OJR** col. Liu, J. & Stewart, J.T. (1997) J PhBmAn **16**, 303-309

* **Propofol.**- (1): pptn. (HCl); C-18: MeCN/MeOH/acet. pH 3; ; fluor; sens. 2 ng/ml(Q). (2; urine): 'LE' ('supported liquid memb.'); C-18: MeOH/phos. pH 3.8; EC; sens. 1 µg/**100** ml. (3; rat-liver microsomes): LE; GC-MS or GC-atomic emission (AE; slightly less sens.); sens. 20 ng/ml. (1) El-Yazigi, A. & Hussein, R.F. (1996) J Lq Chr **25**, 99-104. (2) Trocewicz, J. et al. (1996) J Chr **B 685**, 129-134. (3) Elbast, W. et al. (1996) J Chr **B 686**, 97-102.

Remifentanil (rat).- LE; C-8: MeCN/TEA/phos. pH 3.5; 210 nm; sens. 2 ng/ml(Q). Haidar, S.H. et al. (1996) J PhBmAn **14**, 1727-1732.

* **Sumatriptan.**- (1): SPE(carboxymethyl); CN-C-18 col. (mixed mode): MeOH/phos. pH 6.6; EC; sens. 0.5 ng/ml. (2): Robotic SPE; C-18: MeOH/phos. pH 7; EC; sens. 1 ng/ml(Q). (1) Franklin, F. et al. (1996) J Chr **B 681**, 416-420. (2) Dunne, M. & Andrew, P. (1996) J PhBmAn **14**, 721-726.

* **Thiopental (Thiopentone).**- (1): LE; **chiral** col.(AGP): iPrOH/MeOH/phos. pH 5; 280 nm; sens. 10 ng/ml. (2): pptn. (MeOH, -40°); C-18: aq.MeCN; 280 nm; sens. 0.2 µg/ml. (1) Jones, D.J. et al. (1996) J Chr **B 675**, 174-179; cf. **673**, 245-250 (Huang, J.L. et al.): a similar **chiral** method (sheep). (2) Russo, H. et al. (1997) J Chr **B 694**, 239-245.

Timiperone (rat).- LE; C-18: MeCN/MeOH/phos. pH 3.5; EC; sens. 0.5 ng/ml(Q). Takayasu, T. et al. (1996) J Chr **B 679**, 161-165.

* **Tramadol.**- (1): LE; C-18 SCX [invstg.]: MeCN/phos. acid/hexanesulfonate; fluor; sens. 17 ng/ml. (2): SPE ('ASPEC', robotics); **Chiracel OD** col.: phos./ClO_4^- /MeCN pH 6; 220 nm; sens. 0.5 ng/ml. (1) Nobilis, M. et al. (1996) J Chr **B 681**, 177-183.- Cf. **686**, 221-227 (Paar, W.D. et al.): microsomes (**mtbs**). (2) Ceccato, A. et al. (1997) J Chr **B 698**, 161-170.

Triazolam.- LE; GC-ECD (deactivated metal col.; fused silica poor); sens. 0.5 ng/ml(Q). Nishioka, R. (1996) J Chr **B 681**, 401-404.

* **Valproate.**- (1): dialysis; micro-SPE on fibre, desorbed in GC-FID port; sens. 1 µg/ml. (2; **mtbs**. too; incl. rat): LE (extra clean-up for rat brain); deriv.; GC-MS(EI); sens. 0.5 ng/ml. (3; a **mtb**. too): LE; GC-FID; sens. 0.1 µg/ml(Q). (1) Krogh, M. et al. (1995) J Chr **B 673**, 299-305. (2) Darius, J. (1996) J Chr **B 682**, 67-72. (3) Carlin, A. & Simmons, J. (1997) J Chr **B 694**, 115-121; cf. **690**, 139-152 (Dills, R.L. & Shen, D.D.; brain) - minimization of interferences from making PFBB deriv.

Zolpidem.- Pptn.(MeOH); C-18: MeOH/phos./TEA/NaN_3 pH 6.8; fluor; sens. 1.5 ng/ml(Q). Method rapid & simple. Ptacek, P. et al. (1997) J Chr **B 694**, 409-413.

Zolpidem & zopiclone.- (1): LE; C-18: MeOH/THF/TEA pH 2,6; 305 nm; sens. 20 ng/ml. (2a, **zolpidem & mtbs.**; urine): CE: phos. pH 5.6, fluor (LIF - discussed); sens. 10 ng/ml. (2b, **zopiclone & mtbs**; urine & saliva): LE; **chiral** CE: β-cyclodextrin/triethanolamine/phos. pH 2.75, fluor; sens. 6 ng/ml. (2a,b) Hempel, G. & Blaschke, G. (1996) J Chr **B 675**, (2a) 131-137; (2b) 139-146.

Zonisamide (& Phenobarbital, Phenytoin etc.).- LE; CE with SDS, pH 8 phos.; 210 nm; sens. 3 µg/ml. TM use. Makino, K. et al. (1997) J Chr **B 695**, 417-425.

Zotepine.- LE; GC-NPD; sens. 1 ng/ml. PK use. Ulrich, S. et al. (1996) J PhBmAn **14**, 441-449.

A **dopamine D_4 receptor antagonist** [N-heterocyclic moieties] (incl. hydrol, urine).- LE; C-18: MeCN/acet. (urine: pH 4); MS-MS(CI); sens. 0.1(urine: 0.5) ng/ml. Chavez-Eng, C.M. et al. (1997) J Chr **B 691**, 77-85.

CAM 4515 & 4750 [tachykinin antagonists] (rat).- Pptn.(MeCN); [for CAM 4750: LE]; C-18: MeCN/phos. pH 7 [4750: citric ac. pH 3.5]; fluor; sens. 5 ng/ml(Q). Van Noord, T. et al. (1996) J PhBmAn **14**, 1709-1716.

CI 1007 mtbs., sulfates (monkey, incl. urine).- Pptn. (MeOH & DMSO); C-8: MeCN/acet. pH 6.5, grad.; fluor; sens. 50 ng/ml(Q). Feng, M.R. et al. (1997) J Chr **B 693**, 159-166.

(±)-**DU 124884** [5HT receptor agonist] & 'des' **mtbs**, (±)-KC 9048.- Pptn. (MeCN); **chiral** deriv.; LE; NP col.: solvent mixture; fluor; sens. 0.02 ng/ml. Naidong, W. et al. (1996) J PhBmAn **14**, 325-337.

Sch 39166 & mtb. 40853 [anti-psychotic; naphthazepine moiety].- LE; CN col.: MeCN/phos. pH 6.5; 205 nm; sens. 1 ng/ml. PK use. Accord with a GC method. Lin, C-C. et al. (1996) J Chr **B 682**, 309-313.

JOURNAL TITLES: Chromatog... - Chr; Liquid - Lq; Pharm. Biomed. Anal. - PhBmAn	Prefix * if notable, e.g. for a method study. Metabolites, **mtb(s)**.; invstg. if investigated. For (e.g.) conjugates: hydrolysis, hydrol. Pptn., protein precipitation. LE, liquid-liquid extraction. (Commonly - in SPE too - dry down and redissolve; unstated.) Packings: e.g. C-18, CN (cyano), AGP (α_1-acid glycoprot.) MeCN, acetonitrile; iPrOH, isopropanol; DEA, diethylamine. If gradient: grad. Enantiomer studies/agents: chir.... Automation implicit if stated on-line or switching (swtch). Derivatization/conversion: deriv (sometimes post-column). If nm stated, absorptiometric detection; EC, electrochemical. Approx. LOD or (Q) LLOQ: sens. Pharmacokinetic studies, PK.	*KEY (abridged) Generally human plasma/serum*

5. NEUROACTIVE AGENTS, continued

5b: stimulatory/various: AcCh-/dopamine-/-5HT-related; **anti**-depression/-hyperkinesis/-schioid/-MAO; nootropic; psychotic

* ANTIDEPRESSANTS (various; fibre-SPE, GC): Ulrich, S... (1997) J Chr **B 696**, 217-234.

* **Amisulpride** [sulpiride analog] (incl. urine).- LE; **Chiralpak AS** col.: hexane/EtOH/DEA; 280 nm; sens. 2.5 ng/ml(Q). PK use. Ascalone, V. et al. (1996) J Chr **B 676**, 95-105.

* **Amitryptline/Nortryptline**.- (1; & OH **mtbs**.): LE; GC-NPD; sens. 1.5 ng/ml. (2): SPE (polymer, 'HLB'; **novel** - discussed, & compared with LE); C-18 (35°): MeOH/phos. pH 7; 254 nm; sens. ~0.5 µg/ml. (3; incl. microsomal incubates): LE; C-18: MeCN/TEA/phos. pH 3; 240 nm; sens. 2 ng/ml(Q). (4) SPE (weak cation-exch.); CN col.: MeCN/MeOH/phos. pH 7; 215 nm; sens. 1 ng/ml. (1) Ulrich, S. et al. (1996) J Chr **B 685**, 81-89. (2) Cheng, Y-F. et al. (1997) J Lq Chr **20**, 2461-2473. (3) Ghahramani, P. et al.: as for (1), 307-313. (4) Miljkovic, B. et al. (1997) L Lq Chr **20**, 1067-1078.

Amphetamines & mtbs.: SEE **Metam...** **Carbidopa**: see **Levodopa**

Dihydroergotamine (incl. rabbit).- SPE (CN; MeCN eluted protein); C-18: MeCN/iPrOH/phos. pH 7; fluor; sens. 0.06(rabbit:0.4) ng/ml. Romeijn, S.G. et al. (1997) J Chr **B 692**, 227-232.

Doxepin & mtb., **N**-des, *cis & trans* (dog also; incl. urine).- LE; NP col.: solvent mixture; 254 nm; sens. 1 ng/ml(Q). Yan, J-H. et al. (1997) J Chr **B 691**, 227-232. SEE ALSO Amitryptline, ref. (2).

Fluvoxamine.- (1; enantiomers, **Fluoxetine**† & **Norfluoxetine**): LE; **chiral** deriv.; GC-MS(EI,SIM); sens. 2 ng/ml. (2; **Paroxetine** also): SPE (on-line; swtch, for guard col. also); CN col.: MeCN/MeOH/phos. pH 6.8; 215 nm; sens. 5 ng/ml. (3): LE; C-18: MeCN/phos./octylamine pH 6.4; 230 nm; sens. 5 µg/ml. (1) Eap, C.B. et al. (1996) J Chr **B 682**, 265-272. (2) Bagli, M. et al. (1997) J Lq Chr **20**, 283-295. (3) Gennaro, M.C. et al., AS FOR (2), 3017-3028.

Imipramine/Desimipramine & mtbs, -OH (incl. urine).- LE; C-18: MeCN/K_2HPO_4; EC; sens. 3 ng/ml. Chen, A.G. et al. (1997) J Chr **B 693**, 153-158.

* **Levodopa & mtbs**. incl. dopamine.- 'Microdialysis' by probe on sample (NOT *in vivo!)*; C-18: phos. 2.5/MeOH/octanesulfonate/EDTA; EC; sens. 0.2 ng/ml. TM use. Dethy, S. et al. (1997) Clin Chem **43**, 740-744. ALSO (**Carbidopa** too) Tolokan, A. et al. (1997) J Chr **B 698**, 201-207: pptn. (PCA); auto-FIA alumina step; C-18... pH 2.8; EC; sens. 0.3/1.5 ng/ml.

Lysergide (urine).- SPE; C-18: MeCN/TEA/acet. pH 8; MS(ESP; SIM); sens. 0.5 ng/ml(Q). White, S.A. et al. (1997) J Chr **B 689**, 335-340.

†Assay of F_3Me-phenol (**mtb.**; incl. rat liver & brain): Urichuck, L.J. et al. (1997) J Chr **B 698**, 103-109 (LE; GC-ECD). SEE also #7, Fen.../Fluox...

Methamphetamine/Amphetamine & mtbs. (-OH).- (1; incl. urine, hydrol.): LE (high salt); deriv.; GC-NPD; sens. 1 ng/ml. (2; urine): SPE(SCX); **chiral** col.(cyclodextrin): MeCN/MeOH/phos. pH 6; MS(TSP); sens. 50 ng/ml. (1) Cheung, S. et al. (1997) J Chr **B 690**, 77-87. (2) Katagi, M. et al. (1996) J Chr **B 676**, 35-43.

Naloxone.- SPE (CBA; pre-elution drying); CN col.: MeOH;/phos. pH 5.3; EC; sens, 1 ng/ml. Franklin, M. et al. (1996) J Chr **B 679**, 199-203.

Naltrexone & mtb..- LE; phenyl col.: MeOH/phos. ac.; EC; sens. 0.2 ng/ml(Q). Davidson, A.F. et al. (1996) J PhBmAn **14**, 1717-1725. Also Huang, W. et al. (1997) J Anal Tox **21**, 252-257: SPE; deriv.; GC-MS(nCI).

Pyrovalerone, & mtb., invstg. (rat, urine).- LE; GC-NPD or (**mtb.**) GC-MS; sens. 5 ng/ml (plasma). Lho, D-S. et al. (1996) J Chr **B 687**, 253-259.

Risperidone & mtb. (-OH).- LE; C-18: MeOH/acet. pH 7; 280 nm; sens. 1 ng/ml(Q). Olesen, O.V. & Linnet, K. (1997) J Chr **B 698**, 209-216.

Schizandrin.- SPE; GC-MS(EI); sens. 2 ng/ml(Q). Ono, H. et al. (1995) J Chr **B 674**, 293-297.

Sertaline & norsertaline.- LE; SCX col.: MeOH/ClO_4^- pH 7; 215 nm; sens. 10 μg/ml. Patel, J. et al. (1996) Biomed Chr **10**, 351-354.

Sultopride.- LE; NP col.(40°): MeCN/acet.; 240 nm, then MS(nCI); sens. 5 ng/ml(Q). Jitsufuchi, N. et al. (1997) J Chr **B 690**, 153-159.

Trimipramine, chiral eluant.- SPE; C-18: acet./EtOH/β-cyclodextrin; UV; sens. 10 ng/ml. Ameyibor, E. & Stewart, J.T. (1997) J Lq Chr **20**, 3107-3119.

* **A 82895** & (enantiomer) **A 79814** [cholinergic channel activator; tetrahydropyrrole moiety] (dog & monkey).- LE (add salt); deriv. (for fluor) [optimization discussed; UV or EC not apt]; LE; C-18: aq. MeCN; fluor; sens. 10 pg/ml. Hui, Y-H. et al. (1995) J PhBmAn **14**, 131-142.

ABT 089 [cholinergic; pyrrolidine & pyridine moieties] (animals; incl. brain).- LE; fluor deriv.; LE; YMC basic col.: MeCN/MeOH/$Me_4N^+ClO_4^-$ /TFA; fluor; sens. 0.1 ng/ml(Q). Hui, Y. et al. (1997) Chr **B 695**, 337-347.

BMS 181101 & mtbs., hydroxy [antidepressant] (rat & monkey).- SPE(CN); phenyl col.: MeCN/MeOH/phos./Me_4N^+ pH 3; 287 nm; sens. 50 ng/ml(Q). Shah, V.P. et al. (1996) Biomed Chr **10**, 135-138.

DU 124884 & mtb. (des; KC 9048) [a 5HT receptor agonist].- Pptn.(MeCN) & dry down; **chiral** deriv. on EtOH extr.; NP col.: solvent mixture; fluor; sens. 20 pg/ml. Naidong, W. et al. (1996) J PhBmAn **14**, 325-337.

E2011 & distereoisomeric **mtb.**- LE; SPE (C-18 then SiO_2; **2 Chiralpak AD** cols.: solvent mixture; 275 nm; sens. ~1 ng/ml(Q), Naitoh, T. et al. (1997) J Chr **B 694**, 153-161.

GTS 21 & mtb. (rat).- SPE; C-18: MeCN/phos. pH 2.8; 405 nm; sens. 1 ng/ml(Q). Azuma, R. (1996) J Chr **B 686**, 229-234.

HSR 609 [H_1-receptor antagonist] (incl. urine).- Pptn. (for plasma; MeCN); LE, & deriv. (for fluor) on aq. layer; C-18(TSK gel): MeCN/citrate pH 2/SDS; fluor; sens. 0.5(urine:5) ng/ml(Q). Hasegawa, H. et al. (1996) J Chr **B 687**, 419-425.

LY 292356 [muscarinic agent; thiadiazine moiety] (rat, rabbit & monkey).- LE; CN col.: MeCN/acet.; MS-MS(ESP,SRM); sens. 0.1 ng/ml(Q). Cornpropst, J.D. et al. (1995) J Chr **B 673**, 67-74.

U 89968E [$5HT_{1a}$ agonist; valine...] (rat).- SPE (weak cation exch.); deriv.; CN col.: MeCN/TFA; fluor; sens. 4 ng/ml(Q). PK use. Rees, S.A. & Austin, P.M. (1997) J PhBmAn **15**, 739-748.

6. ANTI-INFECTIVE/ANTI-CANCER (or cytostatic) AGENTS

6a: some structural groups [A-to-Z within each], firstly: #cef/ceph/cillin/penem/β-lactamase-inhibitory (then #oxacin; #nucleoside-type)

Amoxicillin (rabbit, incl. middle ear fluid & whole blood).- SPE; C-8: MeCN/phos./TBA^+ pH 6.5 [quicker with a pH 2.8 variant]; 210 nm; sens. 0.2 μg/ml(Q). Yuan, Z. et al. (1997) J Chr **B 692**, 361-366.

Ampicillin.- (1; rabbit): pptn.(PCA); pre-col. (swtch) then C-18: MeOH/acetic ac./heptansulfonate; 225 nm; sens. 0.2 μg/ml. (2): similar to (1), but human. (3; incl. middle ear fluid): SPE; C-18: MeCN/TBA^+ pH 6.5; 210 nm; sens. 0.1 μg/ml(Q). (2) Muth, P. et al. (1996) J Chr **A 729**, 259-266. (3) (1) Tolhurst, T.A. et al. (1996) Chr'ia **42**, 223-226. (2) Yuan, Z. et al. (1995) J Chr **B 674**, 93-99.

Cefepime.- Pptn.(MeCN); C-18: MeCN/acet. pH 4; 254 nm; sens. 0.5 μg/ml. Storage loss in serum. Elkhaili, H. et al. (1997) J Chr **B 690**, 181-188.

Cefuroxime (lung lavage fluid).- SPE; C-18: MeCN/phos. pH 3.2; 280 nm; sens. 1 ng/ml. Rosseel, M.T. et al. (1997) J Chr **B 689**, 438-441.

Imipenem.- Ultrafilter; C-18: borate pH 7.2; 300 nm; sens. 30 ng/ml. PK use. Garcia-Capdevila, L. et al. (1997) J Chr **B 692**, 127-132.

* **Meropenem**.- (1): pptn.(MeCN); wash by LE; C-18: MeCN/ammon. acet.; 298 nm; sens. 0.25 μg/ml(Q). PK & TM use. (2; rat, 50 μl sample): C-8 pre-col. (swtch; back-flush); C-18: MeCN/phos. pH 7; 300 nm; sens. 0.1 μg/ml(Q). (3): bioassay (microbial); PK use. (1) Elkhaili, H. et al. (1996) J Chr **B 686**, 19-26. (2) Lee, H.S. et al. (1996) Chr'ia **42**, 405-408. 3) Mattie, H. et al. (1997) AtMcAgCh **41**, 2083-2088.

Penicillin V (phenoxymethylpenicillin).- SPE; C-18: MeCN/phos. pH 6; 269 nm; sens. 35 ng/ml. Krauwinkel, W.J.J. et al. (1996) J Chr **B 679**, 129-135.

Piperacillin.- (1; **tazobactam** also; incl. urine): Pptn.(MeCN); C-18: MeCN/amm.acet. pH 5; 218 & 222 nm; sens. 3 & 0.5 μg/ml. (2; **Ceftazidime** also): Pptn.(MeOH); C-18: MeCN/phos. pH 3.8; 254 nm; sens. 5 μg/ml(Q). (1) Augey, V. et al. (1996) J Chr **B 682**, 125-136. (2) Campanero, M.A. et al. (1997) Chr'ia **46**, 374-380.

Ticarcillin epimers (incl. urine).- SPE(SAX); C-18: MeOH/phos. pH 7; 254 nm; sens. <5 μg/ml. PK & binding studies. Li, Y.H. et al. (1997) J Chr **B 694**, 145-152.

DA 1131 [a carbapenem] (rat; incl. urine & water homogenates of liver etc.).- Pptn.(Zn^{2+}); C-18: MeCN/phos. pH 5; 320 nm; sens. 0.1(urine:0.5) μg/ml. Kim, S-H. et al. (1997) J Chr **B 688**, 95-99.

* **FCE 22101** [a penem] (urine; near-similar for plasma).- LE (high salt); C-18(40°) [elaborate swtch]: MeCN/pH 3.2 buffer, grad.; 320 nm; sens. 500(plasma:5) ng/ml. Ultrafiltration suits instead of LE, but acid-lability precludes PCA pptn. Matsuoka, M. et al. (1996) J Chr **B 685**, 273-279. Column procedure: '2-sided bracketed injection' [Claessens, M.A. & Keyken, M.A. (1987) Chr'ia **23**, 331.]

[Out of A-Z order:-]

BMS 180680 [a monobactam] (dog & rat).- Pptn.(MeCN); C-18: MeCN/phos. Bu_4HSO_4/TFA pH 4.1; 252 nm; sens. 0.2 μg/ml. Simple method. PK use. Srinivas, N.R. et al. (1997) J Lq Chr **20**, 1091-1101.

L-749,345 [a penem; guanidine groups] (primates, incl. urine).- Load direct (swtch); C-18: MeCN/MeOH/TFA, grad.; 307 or (urine) 280 nm; sens. <1 μg/ml. PK use. Sundlof, J.G. et al. (1997) AtMcAgCh **41**, 1743-1748.

For abridged Key, turn back 2 pp.
For amplification see start of #ABC.

AtMcAgCh *denotes* Antimicrobial Agents & Chemotherapy

Entries marked * are of notable interest, e.g. a method study

'Oxacin'-type

Ciprofloxacin & mtbs.- Pptn.(MeCN); MeCN/MeOH/TCA pH 3; on-line photoreaction; fluor; sens. 0.05 µg/ml(Q). Krol, G.J. et al. (1995) J PhBmAn **14**, 181-190.

* **Levofloxacin** (incl. urine).- LE; C-18: aq.MeOH with Cu^{2+} & L-isoleucine (**chiral** eluant); 330 nm; sens. 0.08(urine:23) µg/ml(Q). Wong, F.A. et al. (1997) J PhBmAn **15**, 765-771 [cf. Chien, S.C. et al. (1997) AtMcAgCh **41**, 1765-1769].

Norfloxacin (0.1 ml sample).- LE; C-18 (3 µm, in cartridge): MeCN/TEA/phos. pH 2.5; 279 nm; sens. 0.1 µg/ml(Q). Wallis, S.C. et al. (1995) J Chr **B 674**, 306-309.

Ofloxacin.- (1; aqueous humour): C-18: MeOH/MeCN/citric ac.; fluor; sens. 20 ng/ml. (2): Pptn.(MeCN) ; C-18(SGX) THF/phos. pH 2.6; fluor; sens. sens. 8 ng/ml. (1) Basci, N.E. et al. (1997) J PhBmAn **15**, 663-666. (2) Macek, J. (1995) J Chr **B 673**, 316-319.

Rufloxacin: SEE #1, Felbinac & Flurprofen entries

Temfloxacin (incl. urine).- LE; deriv. & wash; NP col.: solvents/ammonia; 280 nm; sens. 5 ng/ml. Matsuoka, M. et al. (1996) J Chr **B 676**, 117-124.

Trovafloxacin (incl. urine).- SPE; C-18: MeCN/TBA^+ etc.; phos. pH 3; 275 nm; sens. 0.1 µg/ml. Teng, R. et al. (1996) J Chr **B 675**, 53-59 [also (1996) AtMcAgCh **40**, 561-566: PK use].

Nucleoside-related

Acyclovir.- (1; also carboxy-Me **mtb.**; incl. urine): add NaCl; SPE; C-18: MeCN/SDS/phos. pH 2.1; fluor; sens. 0.05(urine:0.3) µg/ml. TM use, (2): ptn. (PCA); C-18 phos. pH 3.5; 254 nm; sens. 0.1 µg/ml(Q). (3) similar but fluor; sens. 30 ng/ml. (1) Svensson, J-O. et al. (1997) J Chr **B 690**, 363-366. (2) Boulieu, R. et al. (1997) J Chr **B 693**, 233-236. (3) Feh, K.K. et al. (1997) J Chr **B 693**, 214-244. ALSO Smidovnik, A. et al. (1997) J HR Chr **20**, 503-506 (LE; C-18).

Ateviridine & N-des mtb. (saliva & CSF).- SPE; C-18: MeCN/iPrOH/acet. pH 4; fluor; sens. <1 ng/ml. Howard, G.M. et al. (1997) J Chr **B 693**, 431-436.

* **Bromo-/Chloro-uracil.**- Pptn. (ammon. sulfate); LE; C-18: phos. pH 3; 260 nm; sens. 12 ng/ml(Q). Simple; optimized conditions. PK use. Gamelin, E. et al. (1997) J Chr **B 695**, 409-416.

* **Cytosine derivs.**- (1; e.g. **Cidofovir**, a CMP analog; monkey): pptn. (MeCN/ acetic ac.); deriv. for fluor (phenacyl bromide); C-18(45°): MeCN/TFA; fluor; sens. 5 ng/ml. (2; **N-alkyl-1-β-D-arabinofuranosyl...**; red cells also, sonicated): add urea and stand (strong binding!); SPE (losses if PCA or LE); C-18 (2 cols.; 2nd at 45°): MeOH/formate pH 2.7; 275 nm; sens. 20(cells:40) ng/ml. (3; cultured cells): **[³H]-cytosine arabinoside mtbs.** invstg. - C-18; CaF_2 in radio flow-cell. (1) Eisenberg, E.J. & Kundy, K.C. (1996) J Chr **B 679**, 119-127. (3) Rentsch, K.M. et al. (1995) J Chr **B 673**, 259-266. (3) Braess, J. et al. (1996) J Chr **B 676**, 131-140.

* **2' Deoxy-3'-thiacytidine mtb., Lamivudine** (urine).- SPE by disc (SDB-RPS); C-18: MeOH/acet./acetic; 270 nm; sens. 0.5 µg/ml(Q). Plumb, R.S. et al. (1996) J Chr **B 687**, 457-461.

Didanosine.- Ultrafilter; phenyl col.: iPrOH/citrate pH 5; 250 nm; sens. 25 ng/ml. PK/TM use. Rosell-Rovira, M.L. et al. (1996) J Chr **B 675**, 89-92.

2'-Fluoro-2',3'-dideoxy-adenosine, & (mtb.) -inosine (dog; incl. urine).- SPE; C-8: a glycol ether/phos.(K_2H & KH_2); 257 nm; sens. 0.1(urine:2) µg/ml. Campbell, D.A. et al. (1996) J Pharm Sci **85**, 890-892.

* **5-Fluorouracil.**- (1): LE (high salt); C-18(40°): MeOH/phos. pH 5.5, grad.; 266 µm; sens. 0.02 µg/ml(Q). Method rugged. (2; incl. prodrugs - **4-deoxy & tegafur**; rat): 2-stage pptn.(MeCN; $MgSO_4$); PRP col.: MeCN/TBA^+ pH 11; 254 & 313 nm; sens. 0.1 µg/ml. (3; incl. whole blood & red cells): LE (high salt); C-18: phos. pH 3; 270 nm; sens. 10 ng/ml. (4; **uridine mtbs.** also): SPE & LE; C-18: aq.MeOH; 268 or 275 nm; sens. 10 ng/ml. (1) Coe, R.A. et al. (1996) J PhBmAn **14**, 1733-1741. (2) Jarugula, V.R. & Boudinot, F.D. (1996) J Chr **B 677**, 199-203.- Cf. rat-liver studies, incl. **tegafur**: Fuse, E. et al. (1996) Biochem Pharmacol **52**, 561-568. (3) Wattanatorn, W. et al. (1997) J Chr **B 692**, 233-237. (4) Joulia, J.M. et al. (1997) J Chr **B 692**, 427-435.

5-Fluorouracil & derivs./dihydro mtb.- (1; **fluorouridine** also; rat): LE (sonicate); filter; C-18: phos. pH 3; 210 nm; sens. 0.2-0.8 µg/ml. (2; **5'-deoxy** also): LE (high salt); deriv.; GC-MS(EI); sens. 20 ng/ml. (3), **5-chloro-2',3'-dideoxy-3'-fluorouridine** (935U83): SPE (anion-exch.); C-18: MeCN/amm. acet.; GC-MS(nCI,SIM); sens. 4 ng/ml(Q). For glucuronide, hydrol. also. (1) Fruitwala, M.A. & Shanghavi, N.M. (1997) J Lq Chr **20**, 471-480. (2) Zambonin, C.G. et al. (1996) Anal Chim Acta **329**, 143-152. (3) Dunn, J.A. & Savina, P.M. (1996) J Chr **B 686**, 241-248.

Ganciclovir (incl. urine).- Pptn. (serum; PCA, removed as $KClO_4$ ppt.); filter; C-18: MeCN/phos. pH 2,5; 254 nm; sens. 0.05(urine:3) µg/ml. Page, T. et al. (1996) J Chr **B 675**, 342-346.

Guanine derivs.- (1; benzyl at position 6): LE; C-18: MeCN/MeOH/phos. pH 3.2, grad.; 280 nm; sens. 25 ng/ml(Q). (2; **BCX 34**, variant at 9; urine also): Pptn. (PCA); SPE; ISRP col.: MeCN/ammon. sulfate, pH 6; 233 nm; sens. 0.03 µg/ml(Q). (1) Stefan, T.L. et al. (1996) J Chr **B 681**, 331-338. (2) Yan, J. et al. (1997) J Chr **B 690**, 295-303.

Lamivudine: SEE 2'-Deoxy-3'-thia... entry

6-Mercaptopurine mtb. (Me- & -OH, invstg.).- PCA then K^+ as for ganciclovir, ABOVE; deriv.; GC-MS(EI). Keuzenkamp-Jansen, C.W. et al. (1996) Clin Chem **42**, 380-386.

Uracil-related 'conservers' of uridine: SEE 'TAU'... entry in #7

* **Zidovudine & phos. mtbs.** (mononuclear blood cells, BBMC).- (1): Extract (60% MeOH, -20°); C-18: MeCN/TBA phos. pH 7.5; 254 nm; RIA on fractions after alk. phosphatase treatment and SPE; sens. 0.2 ng/ml(Q). (2): similar (RIA), but SPE(SAX) cartridge conveniently served as separatory column; sens. <1 pg/10^5 cells. (1) Peter, K. et al. (1996) J PhBmAn **14**, 491-499. (2) Robbins, B.L. et al. (1997) AtMcAgCh **41**, 2651-2654.

SDZ WAG 994 [adenosine deriv.] (dog, monkey & rat).- SPE steps (phenyl & cation-exch.); C-8 (YMC Basic): MeCN/phos. pH 6; 266 nm; sens. 1 ng/ml(Q). Cramer, J.A. et al. (1997) J PhBmAn **15**, 749-758.

6b: Anti-infectives (not of a #6a type)

Clindamycin is in #6c

* **Amikacin.**- Ultrafilter; deriv. (for fluor); CE (MEKC; 40°): SDS/pH 7 borate/phos.; fluor; sens. 0.5 µg/ml. Oguri, S. & Miki, Y. (1996) J Chr **B 686**, 205-210.

Amnosidine [an aminoglycoside].- Deriv. (for UV detection); LE (wash); C-18: MeOH/phos.pH 3; 350 nm; sens. 0.2 µg/ml. PK use. Kanyok, T.P. et al. (1997) AtMcAgCh **41**, 982-986.

Amphotericin B. (1). pptn.(MeOH); C-18: MeCN/EDTA; 405 nm; sens. 50 ng/ml. (2; dog): similar; PK use. (1) Lopez-Galera, R. et al. (1995) J Chr **B 674**, 298-300. (2) Beno, P. et al. (1997) J Lq Chr **20**, 1857-1866.

Artemether or (3) **Artesunate, & mtb., dihydroartemisin.**- (1) LE; CN col.(35°): MeCN/acet. pH 5; EC; sens.10 ng/ml(Q). (2) similar; **mtb.** has α & β isomers - discussed. (3) SPE; C-8(35°): MeCN/acetic ac.; EC; sens. 12 ng/ml(Q). PK use. (1) Sandrenan, N. et al. (1997) J Chr **B 691**, 145-153. (2) Karbwang, J. et al. (1997) J Chr **B 690**, 259-265. (3) Navaratnam, V. et al. (1997) J Chr **B 692**, 157-162.

Atovaquone.- Pptn.(MeCN/acetic ac.); C-6: MeOH/TFA/TEA pH 2; 254 nm; sens. 0.5 μg/ml. TM use. Hansson, A.G. et al. (1996) J Chr **B 675**, 180-182.

Benflumetol.- (1): LE; C-18: MeOH/DEA/acetic ac.; 335 nm; sens. 5 ng/ml(Q). (2): MeCN/DEA/acet. pH 4.5; otherwise similar. (1) Zeng, M-Y. et al. (1996) J Chr **B 681**, 299-306. (2) Mansor, S.M. et al. (1996) J Chr **B 682**, 321-325.

N-Benzoylstaurosporine (CGP 41251) & oxid. **mtbs.** (0.1 ml sample; incl. rat).- LE; C-18(end-capped): MeCN/TEA/phos. pH 3.6, grad.; fluor; sens. 0.1 ng/ml. van Gijn, R. et al. (1995) J PhBmAn **14**, 165-174.

Buthionine sulfoxime diastereomers [for GSH study].- CE(MEKC): SDS/phos. pH 7; 190 nm; sens. 4 μg/ml. Sandor, V. et al. (1995) J Chr **B 673**, 123-131.

Chlorguanide: SEE **Proguanil**; **Clindamycin**: SEE #6c entry

* **Chloroquine & des-Et-.**-'Perfusion HPLC', cross-linked polymer col. (fast, & allows high pH; 60°): borate/DEA/THF, grad.; fluor; sens. 30 ng/ml(Q). Augustijns, P. (1997) J Lq Chr **20**, 1103-1113 [for work-up SEE **13**, 1203].

Clofazimine & other **phenazines** (rat; incl. tissues).- LE; C-18: THF/ acetic ac./hexanesulfonate; 285 nm; sens. 10 ng/ml or /g(Q). O'Connor, R. et al. (1996) J Chr **B 681**, 307-315.

Clotrimazole & mtbs. (incl. whole blood).- SPE; for free/bound value, ultrafilter; C-8: MeOH/phos. pH 6.3; 210 nm; sens. 0.03 μg/ml. PK use. Rifai, N. et al. (1995) Clin Chem **41**, 387-391. [An anti-sickling agent.]

Dapsone.- Published LC method adapted to 50 μl samples (can be on filter paper). TM use. Mberu, E.K. et al. (1996) J Chr **B 677**, 385-387.

(+)Dihydrocalanolide A (anti-HIV; non-nucleoside).- SPE; C-18: aq. MeCN; 318 nm; sens. 20 ng/ml(Q). Newman, R.A. et al. (1997) J Lq Chr **20**, 651-659.

Distamycin A deriv., FCE 26644 [polysulfonated] (monkey).- LE; C-18: MeCN/TBA$^+$/phos. pH 7; 323 nm; sens. 0.3 μg/ml(Q). PK use. Frigerio, E. et al. (1996) J Chr **A 729**, 237-242.

Doxycycline (turkey).- LE; C-18: MeCN/EDTA/oxalic ac./$HClO_4$; 350 nm; sens. 0.1 ng/ml. Santos, M.D. et al. (1996) J Chr **B 682**, 301-308.

* **Droloxifene.**- SPE (cation-exch.; discussed); C-18(40°): MeCN/phos. pH 3.5; photo-cyclization; fluor; sens. 10 pg/ml. Tess, D.A. et al. (1995) J Chr **B 674**, 253-260.

* **Erythromycin** (incl. urine).- Filter; C-18: phos. pH 7/MeCN plus a ruthenium complex for post-col. electro-generated chemiluminescence; sens. 0.05 μg/ml. Ridlen, J.S. et al. (1997) J Chr **B 694**, 393-400.

Ethambutol (incl. urine).- Deriv. for fluor; LE; CN col.: MeCN/phos. pH 2.5; fluor; sens. 10 ng(urine 10 μg)/ml(Q). Breda, M. et al. (1996) J Chr **A 729**, 301-307.

Fenvalerate & mtbs., acidic - invstg. (rat tissue homogenates).- LE; C-18: aq. MeCN; 230 nm. IR after HPLC clean-up. Misra, S. & Sharma, C.D. (1997) Biomed Chr **11**, 50-53.

* **Fluconazole.**- (1): Variants (discussed): #1 no treatment, #2 pptn.(MeCN), #3 LE, #4 SPE; CE(MEKC): SDS/borate/phos. pH 9.2; 190 nm; sens., μg/ml: #1 & #2, 5; #3, 1; #4, 0.1. Results confirmed by HPLC. (2): pptn.(MeCN+salt); C-8: aq.MeCN; 260 nm; 0.15 μg/ml. (1) von Heeren, F. et al. (1996) J Chr **A 745**, 165-172. (2) Cociglio, M. et al. (1996) J Chr **B 686**, 11-17.

Fosfomycin (urine).- Deriv.; GC-FID; sens. 10 μg/ml. PK use. Dios-Vieitez, M.C. et al. (1996) Chr'ia **43**, 293-295.

* **Geneticin** [a 2-deoxystreptamine] (mouse, incl. tissues).- Pptn. (MeOH or TCA; invstg.); deriv. (for UV detection); C-18: aq.MeCN, grad.; 340 nm; sens. 0.08 μg/ml. PK use. Bethune, L. et al. (1997) AtMcAgCh **41**, 661-664.

Genistein & mtb., 4-ethylphenol (mouse; incl. urine).- LE; C-8: MeCN/ formate pH 4; 260 nm; sens. 10 ng/ml. Supko, J.G. et al. (1995) J Chr **B 666**, 157-167.

Gentamycin variants.- SPE(CN - CBA); deriv. for fluor; C-18: aq. MeCN; fluor; sens. 0.2 μg/ml(Q). Stead, D.A. et al. (1996) J Chr **B 675**, 295-302.

Isoniazid (& SEE **Pyrazinamidum**).- (1; active **mtbs.** also: acetyl & hydrazine (incl. urine): pptn.(TCA); ±hydrol.; deriv.; C-8(50°): MeCN/iPrOH/ KH_2PO_4, grad.; 340 nm; sens. 1 μg/ml(Q). (2; acetylniazid & hydrazine also): pptn.(TFA; 0°, slick!); deriv.; C-18: MeCN/ TEA/ acetic ac. pH 5; 340 nm; sens. 0.2 μg/ml. (1) Seifart, H.I. et al. (1995) J Chr **B 674**, 269-275. (2): Sadeg, N. et al. (1996) J Chr **B 675**, 113-117.

Isepamicin [an aminoglycoside].- RIA (with kit); sens. 1 ng/ml(Q). Radwanski, E. et al. (1997) AtMcAgCh **41**, 1794-1796.

Itraconazole & mth., -OH.- LE; C-8: MeCN/DEA; 258 nm; sens. 10 ng/ml. Congas, D. et al. (1996) J Chr **B 687**, 453-456.

Ketoconazole: SEE below, SC 52151 entry

* **Letrozole & mtb.**, CGP 44645 (incl. urine).- LE by 'ASPEC'; C-18: MeCN/ phos. pH 7; fluor; sens. 1 ng/ml(Q). Marfil, F. et al. (1996) J Chr **B 683**, 251-258.

* **Melarsoprol** stereoisomers - invstg.; unstable (incl. whole blood - strong binding - & urine).- LE; C-18: MeCN/phos. pH 3; 283 nm; sens. 3 ng/ml(Q). MS and ^{13}C-NMR to identify. Ericsson, O. et al. (1997) J Chr **B 690**, 243-251.

Miconazole.- SPE; C-18: MeOH/MeCN/phos. pH 7; 230 nm; sens. 5 ng/ml. Kobylinska, M. et al. (1996) J Chr **B 685**, 191-195.

Mitoguazone & other **bis-amidinohydrazones** of alkylglycols.- Ultrafilter; C-18: MeOH/acet. pH 4.3, grad.; 282 nm; sens. 40 ng/ml. Koskinen, M. et al. (1996) J Chr **B 685**, 141-149.

* **Mitomycin C** (small animals†).- Sample (small) loaded direct onto 'CAPCELL PAK MF Ph 1' col. [hydrophilic **and** hydrophobic groups, on silica beads]: water [better than MeCN/phos. for separating interferants]; 365 nm; sens. 5 ng/ml. PK use. Sens. claimed better than methods with LE at start; yet Fuse et al. [Biochem Pharmacol **52**, 561] reported 3 ng/ml (LE; NP col.): SEE #6a, Fluorouracil. Song, D. et al. (1996) J Chr **B 676**, 165-169.

Moxidectin (cow).- Pptn.(MeCN); SPE; deriv.; C-18: MeCN/MeOH/acetic ac.; fluor; sens. 0.1 ng/ml. Alvinerie, M. et al. (1995) J Chr **B 674**, 119-124.

Nelfinavir [HIV protease inhibitor].- LE; C-18: MeCN/phos. pH 3.4; 220 nm; sens. 0.05 μg/ml. Stability in plasma studied; PK use. Wu, E.Y. et al. (1997) **695**, 373-380.

Nitrofurantoin (incl. urine).- LE (for plasma); C-18: MeCN phos. pH 3; 370 nm; sens. 0.01(urine:0.4) μg/ml(Q). PK use. Muth, P. et al. (1996) J Chr **A 729**, 251-258.

* **Oracine & mtbs.**, invstg. (incl. urine, hepatocytes & microsomes).- LE; TLC; colour/fluor, reagent (indicative of identity - also from UV spectra and NMR on eluates after C-18 re-run). Wsol, V. et al. (1996) J Chr **B 681**, 160-175.

Paramycin (incl. urine).- Pptn. (PCA); deriv.; LE; C-18: aq.MeOH pH 3; 350 nm; sens.0.2 μg/ml. Lu, J. et al. (1997) J Chr **B 695**, 329-335.

* **Pentamidine.** (1; incl. urine): Ultrafilter; MEKC: pH 8.4 borate/SDS; 214 nm; sens. 0.3 μg/ml. (2, **mtbs.** invstg.; urine, incl. rat): Optionally SPE (cf. low concn. of mtbs. in human urine; discussed); C-18: MeCN/acet./ TEA pH 6; MS-MS(CI,SIM) with minimization of interface-capillary deterioration (sensitivity loss). (1) Garzon, M.J. et al. (1997) J Chr **B 688**, 135-142. (2) Nordin, J. et al. (1997) J Chr **A 777**, 73-79.

---

†Humans: Joseph, G. et al. (1997) J Chr **B 698**, 261-267.- Pptn. (MeCN; freeze-dry, vac. centrifuge); C-18: MeOH/phos. pH 6.5; 365 nm; sens. 1 ng/ml; PK use.

* **Praziquantel.**- (1): SPE(discs); **Chiracel** OJR col.: MeCN/ClO_4^-; 210 nm; sens. 5 ng/ml. (2): SPE; C-18: aq.MeCN; 217 nm; 0.06 µg/ml. PK use. (3): LE; **Chiracel** OD-H col.: hexane/EtOH; 220 nm; sens. 12 ng/ml(Q). (1) Liu, J. et al. (1997) J Chr **B 692,** 141-147. (2) Jung, H. et al. (1997) AtMcAgCh **41,** 1256-1259. (3) Jabor, V.A.P. et al. (1997) J Chr **B 696,** 301-311.

Proguanil (Chlorguanide) & mtbs (filter-paper samples).- SPE; C-18: MeCN/phos. pH 2/SDS; 254 nm; sens. 0.02 µg/ml(Q). Kolawole, J.A. et al. (1995) J Chr **B 674,** 149-154.

Pyrazinamide & mtbs. (rat urine).- Ultrafilter; C-18: phos.pH 5.2; UV; sens. 0.4 µg/ml(Q). Mehmedagic, A. et al. (1997) J Chr **B 695,** 365-372.

Pyrazinamidum (Pyrazinamide), Isoniazid (Rimifon) & Rifapentine [anti-TB drugs] (incl. urine & CSF).- Filter; C-18: aq. MeOH; 230 nm (pyrazinamidum; phenyl col. & heptanesulfonate); sens. 0.1 µg/ml. TM use. Zhen, Q.P. et al. (1997) J Lq Chr **20,** 459-469.

Pyrimethamine.- LE(MeCN, then CH_2Cl_2; C-18: MeOH/octanesulfonate pH 2.2; 240 nm; sens. 3 ng/ml. Na-Bangchang, K. et al. (1997) J Chr **B 689,** 433-437. ALSO (**Sulfadoxine** too) Astier, H. et al. (1997) J Chr **B 698,** 217-223.- Auto-SPE; C-18: MeOH/MeCN/TEA/phos. ac. pH 5.6; 240 nm; **sens.** 10 ng/ml.

Pyronaridine.- LE; C-18: MeCN/phos. pH 6; fluor; 10 ng/ml. PK use. Jayarman, S.D. et al. (1997) J Chr **B690,** 253-257.

Quinine (rat, incl. CSF & perilymph).- LE; C-18: MeCN/TEA/phos. pH 2.5; 254 nm; sens. 10 ng/ml. Chmurzynski, L. (1997) J Chr **B 693,** 423-429.

* **Rifabutin** [& see above, Pyrinaz...].- (1): SPE; C-8: MeCN/acet./phos. pH 4; 275 nm; sens. 5 ng/ml(Q). (2): pptn.(MeCN); CN pre-col. (swtch); C-8: MeCN/PBS; 278 nm; sens. 33 ng/ml. (1) Lau, Y.Y. et al. (1996) J Chr **B 676.** 125-130. (2) Bartels, H. & Bartels, R. (1996) J Chr **B 686,** 235-240.

Rifapentine [a rifamycin].- LE; C-18: MeOH/phos. pH 5.5; 336 nm; sens. 0.5 µg/ml(Q). He, X.B. et al. (1996) J Chr **B 681,** 412-415.

Saquinavir.- LE; C-8: MeCN/sulfuric ac./Bu_4N^+ pH 3.5; 240 nm; sens. 1 ng/ml. Ha, H.R. et al. (1997) J Chr **B 694,** 427-433.

Secnidazole.- LE; C-18: MeCN/acet. pH 4; 276 nm; sens. 0.5 µg/ml. Ravi, S.K. et al. (1997) J Chr **B 691,** 208-211.

Sinefungin (rat).- Pptn.(MeCN); C-18: MeCN/phos. pH 6; 210 or (more specific) 260 nm; sens. 1 µg/ml(Q). Tharasse-Bloch, C. et al. (1995) J Chr **B 674,** 247-252.

Sulfadiazine (incl. urine).- LE; C-18: MeCN/phos. pH 4.7; 270 nm; sens. 0.03(urine:0.6) µg/ml(Q). PK use. Metz, R. et al. (1996) J Chr **A 729,** 243-249.

[**Sulfadoxine**: SEE Pyrimethamine.

* **Suramin & analogs.**- Add TBA^+; pptn.(MeCN - better than MeOH; keep 2 h at 4°); C-8: MeOH/ TBA^+/phos. pH 6.5, grad.; 238 nm; sens. 7 ng/ml. Kassack, M. et al. (1996) J Chr **B 686,** 275-284.

* **Thiacetazone** (incl. rat).- Filter (centrifugally); phenyl ISRP col.: aq.MeCN; 322 nm; sens. 63 ng/ml. Col. survives >200 injections. Song, D. et al (1997) J Chr **B 690,** 289-294.

* **Tinidazole.**- Ultrafilter; MEKC with C-18 guard col. & CN col.: aq. PrOH SDS; 320 nm; sens. 0.1 µg/ml. No better with (OPTION) protein pptn. and conventional HPLC. PK use. Habel, D. et al. (1997) Biomed Chr **11,** 16-18.

BILA 2185 BS [HIV protease inhibitor; a piperidine carboxamide] (rat).- LE; C-8: MeCN/diMe-octylamine/phos. pH 3, grad.; 198/262 nm; sens. ~5 ng/ml. Liard, E. et al. (1997) J PhBmAn **14,** 151-154.

CGP 61755 [protease inhibitor] (incl. urine).- Pptn.(MeCN); LE; hydrol.; fluor deriv.; C-18: MeCN/borate pH 9; fluor; sens. 5 ng/ml. PK use. Flesch, G. et al. (1997) J Chr **B 696,** 123-130.

* **EM 523 & mtbs.**, active [erythromycin deriv.] (incl. urine).- LE; C-18: Na_2HPO_4/MeCN, grad.; on-line generation (oxidative; with ruthenium) of chemiluminescence; sens. 1(urine;10) ng/mL Monji, H. et al. (1997) J Chr **B 690**, 305-313.

GG 167/GR 121167 [sialic acid analog] (urine).- Diol col. (swtch; heart-cut) & SCX col.: MeCN/phos. pH 2.5; 238 nm; sens. 0.3 µg/ml(Q). Morris, D.M. et al. (1995) J PhBmAn **14**, 191-201.

L-696229 [non-nucleoside reverse transcriptase inhibitor] (animals incl. monkey).- LE; C-8 (base-deactivated): MeCN/TEA/phos. pH 3; 319 nm; sens. 50 ng/ml(Q).

* **L-702007** [as for L-696229] & -OH **mtb.**- LE; 2 C-18 cols., uncoupled after analyte had entered 2nd (then late components removed from 1st): MeCN/MeOH/NaH_2PO_4; fluor; sens. 10 ng/ml(Q). PK use. Woolf, E.J. et al. (1996) J Chr **A 729**, 211-217.

LY 303336 [lipopeptide, antifungal] (dog).- SPE(phenyl); C-8: MeCN/ phos. pH 4.5; 300 nm; sens. 20 ng/ml(Q). Zomes, L.L. & Stratford, R.E. (1997) J Chr **B 695**, 381-387.

* **PNU 103017** enantiomers [HIV protease inhibitor] (incl. rat & dog).- Pptn. (MeCN & EtOH); SPE(phenyl); Pirkle-type **chiral** col. ('Whelk-O1'; Brownlee CN guard col.): EtOH/hexane/acetic ac.; 295 nm; sens. 0.05 µg/ml. Williams, M.G. & Zhong, W-Z. (1997) J Chr **B 694**, 169-177.

SC 52151 [HIV protease inhibitor] & **Ketoconazole** (dog).- Pptn. (phos. ac.); SPE(CN); C-18: MeCN/TEA/phos. pH 3; 239 nm; sens. 0.02 µg/ml(Q). Yuan, T.H. et al. (1997) Xenobiotica **27**, 489-497.

TOP 53 & glucuronide [podophyllotoxin deriv.] (rat, incl. urine, ±hydrol.).- LE; C-18(35°): MeOH/phos. pH 7; EC; sens. 2(urine:20) ng/ml(Q). PK use. Kitamura, R. et al. (1997) J Chr **B 690**, 283-288.

6c: Anti-cancer agents & cytostatics ['hived off' from the former #6b; immunosuppressives now in #2]

Amidox & mtb. N-des (rat, incl. bile).- LE; C-18: MeOH/phos./heptane-sulfonate pH 4, grad.; 264 nm; sens. 0.1 µg/ml(Q). Vielnascher, E. et al. (1997) J Chr **B 696**, 267-274.

9-Aminocampothecin.- For lactone: LE; C-18: aq.MeOH pH 2; fluor; sens. 50 pg/ml(Q). For total drug: hydrol./pptn.; C-18 etc. Loos, W.J. et al. (1997) J Chr **B 694**, 435-441. Cf. a Topotecan entry, BELOW.

Aminopterin deriv. (10-deaza) & **mtb.**- Pptn.(MeOH); C-18: MeOH/phos./ Bu_4N^+ pH 7.5, grad.; fluor; sens. 0.2 ng/ml(Q). Su, H. et al. (1997) J Chr **B 695**, 401-408.

* **Amprolium** (chicken).- Pptn. (PCA); C-18: MeCN/KH_2PO_4/hexanesulfonate; on-line deriv.; fluor; sens. 2 ng/ml. Hamamoto, K. et al. (1997) J Chr **B 693**, 489-492.

Busulphan (Busulfan).- Deriv.; LE; C-18: aq. MeOH; 254 nm; sens. 20 ng/ml. Heggie, J.R. et al. (1997) J Chr **B 692**, 437-444

Campothecin-type agents: SEE 9-Amino..., Hydroxy..., Topotecan & GI...

* **N-2-Chloroethylaziridine [cyclophosphamide mtb.**, volatile; cytotoxic] (rat).- LE, & Kuderna-Danish concentrator with Snyder col.; GC-MS(CI); sens. 5 ng/ml. Lu, H. & Chan, K.K. (1996) J Chr **B 678**, 219-225.

Cisplatin.- Pptn.(TCA); deriv.; C-18: MeOH/MeCN; 254 nm; sens. 1 µg/ml. TM use. Khukawar, M.Y. et al. (1997) J Chr **B 693**, 175-179.

Clindamycin.- SPE; C-18: MeCN/phos./THF pH 5; 204 nm; sens. 0.05 µg/ml(Q). Liu, C-M. et al. (1997) J Chr **B 696**, 298-302.

Colchicine (incl. whole blood & urine).- LE; C-18: MeCN/ammon. formate pH 3; MS(ESP); sens. 50 ng/ml. Hoja, H. et al. (1996) Analusis **24**, 391-394.

Cyclophosphamide.- (1): SPE; C-8 then (heart-cuts) **chiral** col.(AGP); 195 nm; sens. 1 µg/ml(Q). (2; **mtbs.** showed between-person diffs.): LE; GC-NPD; sens. 0.2 µg/ml. (1) Corlett, S.A. & Chrystyn, H. (1996) J Chr **B 682**, 337-342. (2) Sessink, P.J.M. et al. (1995) J Chr **B 673**, 205-212. SEE ALSO **N-2-Chloroethylaziridine** entry.

Docetaxel & -OH **mtbs.**- SPE (CN, end-capped); C-8: MeCN/acet. pH 5; 227 nm; sens. 10 ng/ml. Rosing, H. et al. (1997) J Chr **B 696**,89-98.

Dolastatin 10 [linear pentapeptide].- SPE; CZE, pH 2.9; 200 nm; sens. 25 ng/ml. Huie, W.R. et al. (1997) J Chr **B 693**, 451-461.

Etoposide, unbound.- Ultrafilter ('Centrifree' device); phenyl col.: MeOH/acetic ac.; fluor; sens. 50 ng/ml(Q). TM use. Robieux, I. et al. (1996) J Chr **B 686**, 35-41.

* **Finasteride.**- SPE (graphitized carbon black; C-18: MeCN/phos. pH 4; 215 nm; sens. 5 ng/ml. Carlucci, G. & Mazzeo, P. (1997) J Chr **B 693**, 245-248. Cf. (1975; Guarna...) **674**, 197-204: GC-MS, isotope dilution.

10-Hydroxycampothecin, acid & lactone (rat; incl. urine, faeces; liver & kidney homogenates).- LE; C-18: MeCN/acet./TBA phos. pH 6.4; fluor; sens. 2 ng/ml(Q). Li, Y-F. & Zhang, R.H. (1996) J Chr **B 686**, 257-265.

4-Hydroxyifosfamide (incl. urine).- Add citric acid to stabilize; deriv.; C-18: MeOH/TMAB/heptanesulfonate pH 3; fluor; sens. 12 ng/ml(Q). Kaijser, G.P. et al. (1997) J PhBmAn **15**, 773-781 [cf. Ifosfamide entry].

Hydroxyurea (incl. peritoneal fluid).- Pptn. (MeCN); C-18: MeOH/ClO_4^-; EC; sens. 25 ng/ml(Q). Pujari, M.P. (1997) J Chr **B 694**, 185-191.

Idarubic|in & |ol (100 µl only).- LE; CE: phos./MeCN/spermidine pH 5; LIF; sens. 0.5 ng/ml(Q). Hempel, G. et al. (1997) J Chr **B 698**, 287-292.

Ifosfamide & mtbs.- (1; various; rat): LE; deriv.; GC-MS(CI); sens. 0.1 µg/ml(Q). (2; incl. urine): LE; **Chirasil**-L-val GC col.; NPD; sens. 20-100 ng/ml. (3; mustard active **mtb.**): deriv. (invstg.); SPE; C-8: MeCN/phos. pH 7; 280 nm; sens. 0.2 µg/ml. (1) Wang, J.J-H. & Chan, K.K. (1995) J Chr **B 674**, 205-217. (2) Kaijser, G.P. et al. (1997) J Chr **B 690**, 131-138. (3) Kaijser...as for (2): (1996) **686**, 249-244. ABOVE: 2 cognate entries.

* **Irinotecan** [CPT 11] & active **mtb.**- LE (lactone only) or pptn.(PCA + MeOH; total drug); C-18: MeOH/acet./TBA^+ pH 5.5; fluor; sens. 2 ng/ml(Q; 0.5 for lactone). de Bruijn, P. et al. (1997) J Chr **B 698**, 277-285.

Lobaplatin diastereomers (D-19466).- Ultrafilter; SPE; C-18: MeCN/TEA/phos. pH 6.4; 210 nm; sens. 20 ng/ml. PK/TM use. Welink, J. et al. (1996) J Chr **B 675**, 107-111.

Lometrexol.- SPE (cation-exch.); phenyl col.: MeCN/acet. pH 4; EC; sens. 5 ng/ml. Synold, T.W. et al. (1996) J Chr **B 683**, 245-249.

Melphalan.- (1): Pptn.(MeOH); C-18: aq. acetic ac.; 261 nm; sens. 8 ng/ml. (2; -OH **mtbs.** also; incl. tissues & limb perfusates): Pptn. (MeOH); phenyl col.: MeOH/octanesulfonate/acetic ac. pH 2.7; fluor; sens. 2 µg/ml(Q). (1) Pinguet, F. et al. (1996) J Chr **B 686**, 43-49. (2) Wu, Z-Y. et al. (1995) J Chr **B 673**, 267-279.

* **Methotrexate.**- (1): SPE; C-18: MeCN/TMA^+/phos. pH 2.5; 313 nm; sens. 5 ng/ml. (2): Pptn.(acetone); on-line (col.) deriv. for fluor; C-18: MeCN/phos. pH 6.6; fluor; sens. 3 ng/ml. (3; -OH **mtb.** also): C-8 diol pre-col. (swtch); C-18... [ref. cited]; on-line deriv.; fluor: better sens. (×10) than UV. (4, urine; -OH **mtb.** also): SPE; C-18: MeCN/phos. pH 5.3; 303 nm; sens. 2 ng/ml. TM use. (1) Aboleneen, H. et al. (1996) J Chr **B 681**, 317-322. (2) Emara, S. et al. (1997) Biomed Chr **11**, 42-46. (3) Yu, Z. et al. (1997) J Chr **B 689**, 379-386 [cf. Chr'ia **44**, 589-594: optimized use of restricted access media]. (4) Hirai, T. et al. (1997) J Chr **B 690**, 267-273.

Mitozantrone (mouse; whole blood & tissues).- Add hexanesulfonate and ascorbic ac.; LE; C-18: MeCN/formate pH 2.7; 658 nm; sens. 2 µg/ml. PK use. Rentsch, K. et al. (1996) J Chr **B 679**, 185-192.

Mofarotene [an arotinoid] (incl. rat & dog).-Pptn.(EtOH); SPE; C-18: MeCN/ammon. acet.; 200 nm; sens. 10 ng/ml(Q). PK use. Wyss, R. et al. (1996) J Chr **A 729**, 315-322.

Mycophenolic acid & glucuronide.- SPE; RP('BDS') col.: MeCN/phos./TBA^+ pH 6.6; 254 nm; sens. 0.1 µg/ml(Q). Huang, J.J.Z. et al. (1997) J Chr **B 698**, 293-300.

Oryzalin & mtb., N-des (mice; also rat-liver incubates).- SPE; C-18: aq.MeCN; MS-MS(CI); sens. 0.1 µg/ml. PK use. Dvorakova, K. et. al. (1997) J Chr **B 696**, 275-281.

* **Paclitaxel.**- (1; incl. -OH **mtbs.**): SPE(CN); C-8: MeCN/MeOH/acet. pH 5; photodiode array; sens. 10 ng/ml(Q). (2; incl. urine): LE; MEKC: MeCN/pH 8.5 buffer; 230 nm; sens. 20 ng/ml(Q). PK use. (1) Huizing, M.T. (1995) J Chr **B 674**, 271-278. (2) Hempel, G. et al. (1996) J Chr **A 748**, 173-179.

Prospidin (papilloma homogenates).- Pptn.(MeOH); deriv.; CE; MeOH/ borate pH 11 (sample at anode; separation buffer minimized wall adsorption); 254 nm; sens. 1 µg/ml. Okun, V.M. et al. (1996) J Chr **B 675**, 313-319.

Tamoxifen & mtbs., invstg. (cattle & rat).- LE; CE (MeOH/MeCN/acet.); 214 nm. Sanders, J.M. et al. (1997) J Chr **B 695**, 417-425. Cf. **698**, 261-267.

Thalidomide.- (1): Add sulfuric ac. (stabilizes); SPE; C-18: MeCN/phos. pH 3, grad.; 220 nm (phenacetin i.s. 248 nm; photodiode array); sens. 12 ng/ml. (2; **mtbs**/enantiomers in microsomal incubates): **chiral** CE with a cyclodextrin, pH 6; 225 nm. (1) Simmons, B.R. et al. (1997) Anal Chim Acta **339**, 91-97. (2) Weinz, C. et al. (1995) J Chr **B 674**, 287-292.

* **Topotecan** lactone & acid [campothecin analog].- (1; ±hydrol.; rat & dog; quick red-cell removal): Pptn.(MeOH, -30°); C-18 (19° best for separating an interferant, N/A to human plasma): MeOH/hexanesulfonate/a diamine; fluor; sens. 0.2 ng/ml(Q). (2): A simple general approach; lactone & acid actually separated.- (2a; also a **mtb.** and **10-Aminocampothecin & GI 147211**): C-18: MeCN/TEA/acet. pH 5.5; fluor. (2b, for **mtb.**): Similar, but whole blood; higher % MeCN. (1) Rosing, H. et al. (1996) J PhBmAn **15**, 279-285. (2) Warner, R.L. et al. (1997) J Chr **B 691**, 161-171. Cf. 9-Amino... & GI 147211 entries.

Turosteride [5α-reductase inhibitor].- LE; SPE(CN); C-18: aq. MeCN/MeOH; 210 nm; sens. 5 ng/ml(Q). Basileo, G. et al. (1997) J Chr **B 688**, 117-125.

Vinorelbine.- LE; C-18: MeCN/phos. pH 2.7; fluor; sens. 1 ng/ml. Robieux, I. et al. (1996) J Chr **B 675**, 183-187.

AG 331 [thymidylate synthetase inhibitor] (urine also).- LE <u>or</u> SPE; C-18: MeCN/phos. pH 3; 230 nm; sens. 20 ng/ml. Li, W.Y. et al. (1995) J Chr **B 673**, 281-288.

CGP 48664 [inhibits polyamine biosyn.] (incl. dog & rat).- SPE (C-2; by ASPEC); C-18: MeCN/phos. acid/SDS; 230 nm; sens. 10 ng/ml(Q). Degen, P.H. & Zhinden, P. (1996) J Chr **B 681**, 339-345.

CI 980 [a carbamic ester].- LE; Pirkle-type col. (non-chiral use): hexane/iPrOH/MeCN; fluor; sens. 0.2 ng/ml(Q). Markoglau, N. & Wainer, I.W. (1997) J Chr **B 693**, 493-497.

* **CI 1010** [prodrug for a radiosensitizer; a 2-nitroimidazole] (rat).- SPE, at 0-4° & with other stratagems to combat rogue peaks from the C-18 packing; **ChiralPak** (amylose) cols.: hexane/EtOH/DEA, grad.; 325 nm; sens. 0.2 µg/ml(Q). Kagel, J.R. et al. (1997) **A 763**, 129-137.

* **GI 147211** [campothecin analog]. Precipitative LE ; C-8: MeCN/acet. pH 5.5; fluor; sens. 0.15 ng/ml(Q). Selinger, K.A. et al. (1995) J PhBmAn **13**, 1521-1530. SEE ALSO Topotecan entry, ABOVE.

JM 216 & mtbs., invstg. in plasma incubates [a novel Pt cpd.].- Ultra-filter; polymer col. (PLRP-S): aq.MeOH, grad.; MS(API-ESP). Poon, G.K. et al. (1995) J PhBmAn **13**, 1493-1498.

LY 191704 [5α-reductase inhibitor; a benzoquinolinone] (rat & dog).- SPE(SCX); **Chiracel OD-R** col.: aq.MeCN/MeOH, grad.; 220 nm; sens. 0.1 µg/ml(Q). Farid, N.A. & Coleman, N.L. (1997) J Chr **B 691**, 212-216.

MK 0386 & mtb, carboxylic [a 4-azasteroid] (incl. semen).- LE; C-8: aq. MeCN, acidified; MS-MS(CI); sens. 0.2 ng/ml(Q). Constanzer, M.L. et al. (1997) J Chr **B 693**, 117-129.

NB-506 & mtb. (desformyl, active) [anti-cancer] (incl. urine).- Pptn. (pH 2.8); SPE; C-18(40°): aq. MeCN/MeOH [then, if urine, wash col. with 80% MeOH]; 305 nm; sens. 10(urine:100) ng/ml(Q). Tanenaga, N. et al. (1995) J Chr **B 674**, 111-117.

NU/ICRF 505 & mtb., conjugate [topoisomerase I inhibitor; tyrosine moiety] (incl. mouse tumour, saline homogenate).- Pptn. (DSMO & MeCN); C-18(40°): MeOH/acet./TFA pH 3, grad.; 545 nm; sens. 50 ng/ml. Cummings, J. et al. (1996) J Chr **B 685**, 159-164.

* **SR 4233 & mtbs.** [a benzotriazine] (rat muscle/tumour microdialysis mini-samples).- C-18 (microbore): MeCN/phos.; pH 2.5; EC (optimized); sens. 0.02 µg/ml(Q). Liang, X-Z. et al. (1996) J PhBmAn **15**, 113-119.

TAS 103 & mtb., N-des (monkey).- Pptn.(PCA); SPE; C-18: MeCN/TFA, grad.; 460 nm; sens. 5 ng/ml(Q). Azuma, R. & Urakawa, A. (1997) J Chr **B 691**, 179-185.

* An **imidazoacridinone, CI 1311** (mouse, incl. tumour homogenate).- Pptn. (MeCN); C-18: MeCN/phos./citrate pH **4**; fluor; sens. 0.1 µg/ml(Q). Drug, strongly protein-bound, sticks to membranes; so ultrafiltration/dialysis precluded. Calabrese, C.R. & Loadman, P.M. (1997) J Chr **B 690**, 275-281.

7. VARIOUS AGENTS, mostly in 'Unclassified' sub-section

[#2 now 'houses' immuno-affecting]

Anti-diabetics

Glibenclamide (Gliburide).- (1): C-8: MeCN/PCA; 230 nm; sens. 5 ng/ml. Spiking by adding sample to dried-down std. (2; incl. rat): LE; C-18: MeCN/phos. pH 4; sens. 20 ng/ml. PK use. (1) Valdes Santurio, J.R. & Gonzalez Porto, E. (1996) J Chr **B 682**, 364-370. (2) Al-Dhawailie, A. et al. (1995) J Lq Chr **18**, 3981-3990. Cf. art. #B-9n.

* **Lithium.**- Dilute, then capillary isotachophoresis, pH 5.4; sens. 15 µM. Valaskova, I. et al. (1995) J Chr **B 674**, 310-313.

Pioglitazone & mtbs.- SPE; C-8: aq.MeCN/acetic ac. pH 5.5; 269 nm; sens. 20 ng/ml(Q). Zhong, W.Z. & Williams, M.G. (1996) J PhBmAn **14**, 465-473.

* **Sulfonylureas: acetohexamide, tolbutamide, etc.** (envisaged matrix: urine).- SPE, on-line (sleeve insert; MeCN in elution buffer) with CZE: borate/phos. pH 8.4; 200 nm. Aim: sens. improvement by pre-CZE concentration step. Strausbuch, M.A. et al. (1995) J Chr **A 717**, 279-291.

AY 4166 & mtbs. [phenylalanine deriv.] (incl. urine).- SPE; C-18 cols. (50°): MeCN/phos. pH 6.6, + EtOH for some mtbs. (swtch); 210 nm; sens. 0.1(urine:0.5) ng/ml. Ono, L. et al. (1997) J Chr **B 692**, 397-404.

JOURNAL TITLES: Chromatog... - Chr; Liquid - Lq; Pharm. Biomed. Anal. - PhBmAn

KEY (abridged) Generally human plasma/serum

Prefix * if notable, e.g. for a method study. Metabolites, mtb(s).; invstg. if investigated. For (e.g.) conjugates: hydrolysis, hydrol. Pptn., protein precipitation. LE, liquid-liquid extraction. (Commonly - in SPE too - dry down and redissolve; unstated.) Packings: e.g. C-18, CN (cyano), AGP (α_1-acid glycoprot.) MeCN, acetonitrile; iPrOH, isopropanol; DEA, diethylamine. If gradient: grad. Enantiomer studies/agents: chir.... Automation implicit if stated on-line or switching (swtch). Derivatization/conversion: deriv, (sometimes post-column). If nm stated, absorptiometric detection; EC, electrochemical. Approx. LOD or (Q) LLOQ: sens. Pharmacokinetic studies, PK.

Retinoids

- mostly retinoic acid (R.a.) variants: *cis/trans* (c/t), **mtbs.** of **retinol** (Vit. A; most assays **embrace** it; exclude light! SEE p. 80 (Wyss) in Vol. 22, this series; also p. 271 in Vol. 18

Retinoyl-β-glucuronides (rat urine).- SPE; C-18: MeOH/Et acet., grad.; 363 nm; sens. 10 ng/ml. Li, S. et al. (1996) J Chr **B 683**, 155-162.

Tretinoin (all-t-R.a.) & Isotretinoin (13-c-R.a.).- Pptn.(iPrOH); SPE(C-1); deriv.; 2 C-18 cols.: MeCN/acet. pH 6, grad.; MS(SIM); sens. 0.2 ng/ml(Q). Lehman, P.A. & Franz, T.J. (1996) J Pharm Sci **85**, 287-290.
* **SAME, & 4-oxo mtbs.**- (1): Pptn.(MeCN); C-18(30°) with on-col. concentration: MeOH/acet. pH 5.75, grad.; 340 nm; sens. 1 ng/ml(Q). (2; **9-c-R.a.** also; rat, rabbit): LE; C-18: MeOH/MeCN/THF, grad.; 350 nm; sens. 2 ng/ml. (3; **9-c-R.a.** also): SPE (on-line); back-flushing onto 'pKb' col.: MeCN/BuOH/MeOH/acet.; 360 nm; sens. 2 ng/ml. (4): Pptn.(EtOH +$AmSO_4$); LE [preferable to SPE]; NP col.: hexane/iPrOH/acetic ac., grad. [obviates complexity of a swtch procedure]; 350 nm; sens. 0.5 ng/ml(Q).
(1) Teerlink, T. et al. (1997) J Chr **B 694**, 83-92 [cf. **681**, 153-160 (Dimitrova...): quite similar; mouse keratinocytes etc., developmental toxicology]. (2) Disdier, B. et al. (1996) J Chr **B 683**, 143-154.
(3) Gundersen, T.E. et al. (1997) J Chr **B 691**, 43-58. (4) Lanvers, C. et al. (1996) J Chr **B 685**, 233-240. SEE ALSO **677**, 225-231 (Hagen): use of EC.

* **Am 80** [a retinobenzoic acid].- SPE; C-18: MeCN/TFA, grad.; RIA on dried-down fractions; sens. 0.04 ng/ml(Q), better than by direct RIA. Itoh, M. et al (1996) Anal Lett **29**, 1377-1388 (context: topical application, PK).
[For **Mofarotene** SEE entry in #6c]

Unclassified

Acetylcysteine: SEE #2

Amifostine [WR 1065; a phosphorothioctic ac.] & (DTT step) disulfides.- Pptn.(PCA); C-18: MeOH/$ClCH_2COOH$/hexanesulfonate/cysteamine pH 2.5; EC; sens. 0.05 µg/ml(Q). Korst, A.E.C. et al. (1997) J Chr **B 693**, 167-174.

Chlorobutanol [preservative] (mouse, incl. urine & embryo homogenate).- LE; GC-ECD; sens. 1 µg/ml(Q). Amiss, T.J. & Smoak, I.W. (1995) J Chr **B 673**, 59-66.

* **Deferiprone** [chelator; hydrophilic!] (rat, rabbit).- Pptn.(TCA); graphitized-C col.: MeCN/TEA/NaH_2PO_4/EDTA; 270 nm; sens. 5 µg/ml. PK use. Whole blood intractable. Skinner, T.L. et al. (1996) J Chr Sci **34**, 52-57.

Dorzolamide & mtb..- LE; base-deactivated CN col.: MeCN/acet./TFA; MS-MS(CI); sens. 0.5 ng/ml. Accord with UV-detection. Constanzer, M.L. et al. (1997) J PhBmAn **15**, 1001-1008.

* **Ebrotidine & mtbs.** [H_2 receptor antagonist] (urine).- LE (optimized; SPE worse); C-18: MeCN/hexanesulfonate pH 3; 235 nm; sens. 26 ng/ml. Rozman, E. et al. (1997) J Chr **B 688**, 107-115.

* D-**Fenfluramine (& Nor-), Fluoxetine** (rat, incl. brain & off-line micro-dialysate).- LE; deriv. (SAX to clean up); C-18: MeCN/phos. pH 4.5, grad.; fluor; sens. 1 ng/mg brain. Clausing, P. et al. (1997) J Chr **B 692**, 419-426. SEE also entry in #5b.

Fomepizole (4-methylpyrazole) [poison antidote] (incl. plasma dialysate).- Pptn.(TCA); C-18: MeCN/phos. pH 6; 220 nm; sens. 0.3 µg/ml. Jobard, E. et al. (1997) J Chr **B 695**, 444-447.

Frusemide (Furosemide; & SEE Piretanide entry).- Ultrafilter; SPE (on-line; swtch, +TBA, to ion-pair); C-18 (40°): MeCN/phos. pH 7; 271 nm; sens. 5 ng/ml(Q). TM use. Okuda, T. et al. (1996) J Chr **B 682**, 343-348.

Hydroxyflutamide.- SPE; C-18: MeCN/MeOH; 300 nm; sens. 10 ng/ml(Q). Leibinger, J. et al. (1996) J PhBmAn **14**, 1377-1381.

Lansoprazole (1): LE; C-18: MeCN/octylamine; 283 nm; sens. 12 ng/ml. (2; PK use): LE; TLC; fluor; sens. 20 ng/ml. (1) Borner, K. et al. (1997) Chr'ia **45**, 450-452. (2) Pandya, K.K. et al. (1997) Chr **B 693**, 199-204.

Melanotan II [sun-tan agent, peptide; MSH analog] (rabbit urine).- SPE; C-8: MeCN/phos. pH 3.6; 217 nm; sens. 0.1 µg/ml(Q). Walters, A.E. et al. (1997) J Chr **B 690**, 99-103.

* **Melatonin.**- (1): 'SPE'/cols.: C-18 & 'Carbograph' (3-stage); aq.MeCN; fluor; sens. 3 pg/ml. Accord with RIA. (2; rat, pineal): add alumina/HCl & homogenize; LE; NP col.: Et acet. [apt for]: fluor; sens. 3 pg/g, similar to RIA. (1) Lagana, A. et al. (1995) Anal Chim Acta **316**, 375-385. (2) Vitale, A.A. et al. (1996) J Chr **B 681**, 381-384.

Norethindrone (Norethisterone) & mtbs.- Hydrol.; SPE; deriv.; GC-MS (SIM); sens. 250 pg/ml. Pommier, F. et al. (1995) J Chr **B 674**, 155-165.

* **Omeprazole.**- (1): LE; C-18: MeOH/phos. pH 7; 302 nm; sens. 10 ng/ml(Q). (2; -OH **mtb.** also): direct onto **chiral** (AGP) col.: MeCN/phos. pH 5.7-7.2; 302 nm. A chemometric optimization study, esp. solvent choice (opalescence problem with MeCN, not MeOH) & col. temp. (1) Macek, J. et al. (1997) J Chr **B 689**, 239-243. (2) Karlsson, A. & Hermansson, S. (1997) Chr'ia **44**, 10-18.

Ondansetron & mtbs., -OH.- (1): SPE(CN); **Chiracel** OD-R col.: MeCN/ClO_4^-; 210 nm; sens. 15 ng/ml(Q). (2): LE; C-18: MeCN/phos. pH 3; 305 nm; sens. 0.5 ng/ml. (1) Liu, J. & Stewart, J.T. (1997) J Chr **B 694**, 179-184. (2) Depot, M. et al. (1997) J Chr **B 693**, 399-406. SEE also #5a (the true 'home').

Oxabolone (K 520) mtbs., invest. (cf. sport testing; urine).- Deriv.; GC-MS (EI & CI). Barbarulo, M.V. et al. (1995) J HRChr **18**, 705-708.

* **Pamidronate** (urine).- Co-pptn. with Ca phos.; SPE (R_4N^+); deriv.; ion-pair LE to clean up, then H_2O_2; C-18: MeCN/phos./R_4N^+/etidronate, pH 7.7; fluor; sens. 3 ng/ml. Sparidans, R.W. et al. (1997) J Chr **B 696**, 137-144.

Pilocarpine (incl. urine & saliva).- LE; deriv. for fluor; CN col.(37°): solvents/phos. ac.; fluor; sens. 1 ng/ml. PK use. Aromdee, C. et al. (1996) J Chr **B 677**, 313-318.

Piretanide & Furosemide (Frusemide) (urine).- LE; C-18: MeCN/phos. pH 7; EC; sens. 0.01 & 0.5 µg/ml(Q). Barroso, M.B. et al. (1996) J Chr **B 675** 303-312.

Ractopamine (monkey & pig).- SPE(-COOH); C-18: MeCN/phos. pH 4.5; EC; sens. 2 ng/ml(Q). Turnberg, M.P. et al. (1996) J Chr **B 675**, 279-285.

Ranitidine.- (1): LE; SPE(CN); phenyl col.: MeCN/TEA/phos. pH 6.0; 228 nm; sens. 10 ng/ml. (2) LE; TLC; fluor quenching; sens. 5 ng/ml. PK use. (3; 100 µl sample): LE; C-18: MeCN/phos. pH 3.8; 313 nm; sens. 2 ng/ml. Rugged method; PK use. (4; oxides too; incl. urine): LE; C-18: MeCN/phos. pH 6; deriv.; fluor; sens. 0.03 µg/ml. (1) Farthing, D. et al. (1997) J Chr **B 688**, 350-353. (2) Mody, V.D. et al (1996) J Chr **B 676**, 175-179. (3) Lopez-Calull, C. et al. (1997) J Chr **B 693**, 228-232. (4) Vinas, P. et al. (1997) J Chr **B 693**, 443-449.

* **Sinistrin** [GFR marker, better than inulin] (incl. urine).- SPE; 'HPX-87C' col.: water ['ion-moderated partition chromatography']; on-line deriv. for EC; sens. 0.8 µg/ml. PK use. Sechaud, R. et al. (1996) J PhBmAn **14**, 483-490.

Spironolactone & mtbs., thio-Me & canrenone.- LE; 2 C-18 cols. (long; 5°): MeCN/phos. pH 3.4; 245 nm; sens. 10 ng/ml(Q). PK or TM use. Jankowski, A. et al. (1996) J PhBmAn **14**, 1359-1365.

Stanozolol, mtbs. invstg. (cattle, hydrol. urine).- SPE; deriv.; GC-MS. Ferchaud, V. et al. (1997) J Chr **B 695**, 269-277.

Tamsulosin (incl. plasma dialysate & urine).- LE; C-18: MePH/acet. pH 4; MS-MS(ESP); sens. 0.01(urine:0.1) ng/ml. Matsushima, H. et al. (1997) J Chr **B 695**, 317-327.

'TAU'/'BBBA': uracil derivs. to aid actions of uridine as e.g., a CV drug or toxicity-reducing adjunct to anti-cancer/anti-AIDS/immuno-suppressive agents [TAU: uridine prodrug; BBBA: uridine phosphorylase inhibitor with barbiturate moiety] (mice).- Pptn. (TCA; washing step gave neutral supernatant); C-18: MeCN/acet. pH 4.8, grad.; 254 & 268 nm - disclosing uridine & related mtbs. but negligible TAU or BBBA. Ashour, O.M. et al. (1996) Biochem Pharmacol **51**, 1601-1611.

Thalidomide: SEE #6c

* **Tiopronin & mtb.** (urine).- C-18: octanesulfonate/acet./TBA^+/cetrimonium$^+$ pH 5-5.7; FIA, chemiluminescence; sens. 0.1 µg/ml. Method (optimized) simple/selective/sensitive. Zhao, Y. et al. (1997) Chr'ia **44**, 31-36.

* **CI 988** [peptoid; CCK-B antagonist].- (1; incl. monkey): SPE steps (robotics); C-18: MeCN/citrate pH 3.5; fluor; sens. 0.2 ng/ml(Q). (2; rat): a follow-up, with trial & validation of kindred methods incl. LC-MS: sens. 2 ng/ml(Q). Hinton, J.P. et al.: (1, 1995): Biomed Chr **9**, 94-97; (2, 1996): J PhBmAn **14**, 815-824.

* **E 3810 & mtbs.** [a benzimidazole; anti-ulcer].- (1; dog): direct loading: LC with trapping in 1 cm affinity (avidin) col. (phos. pH 7.5); **chiral** flavoprotein col.: MeCN/phos. pH 5.5; 290 nm; sens. 0.03 µg/ml(Q). (2; dog & rat): LE; **Chiracel** col.: MeCN/ClO_4^- pH 7.8, grad.; 290 nm; sens. 0.03(rat:0.1) µg/ml. (1) Mano, N. et al. (1996) J Pharm Sci **85**, 903-907; cf. J Chr **572**, 133 & **623**, 221. (2) Takakuwa, S. et al. (1995) J Chr **B 673**, 113-122.

MK 677 [GH secretagogue].- LE; C-18: aq.MeCN/acet., acid pH; MS-MS(CI); sens. 0.1 ng/ml(Q). PK use. (2): Constanzer, M.L. et al. (1997) J Chr **B 693**, 131-137 [cf. **A 773**, 311-320 (Zhou, L. et al.): **chiral** CZE on pure cpd.].

Ro 24-2090 [Vit. D analog].- LE; deriv.; NP col.: solvent mixture; MS (nCI); sens. 25 pg/ml(Q). Wang, K. et al. (1996) Anal Biochem **243**, 28-40.

S 1033 [prostaglandin deriv.; antiglaucoma] (incl. rabbit).- SPE [and, if human plasma, C-18 clean-up: MeCN/acetic ac.]. RIA or, ALTERNATIVELY (good accord): 2-stage deriv. with intervening TLC; GC-MS(nCI); sens. 0.1 ng/ml(Q) but worse (rabbit) or better (human) if RIA. Yamauchi, A. et al. (1996) J PhBmAn **15**, 175-182.

SDZ PSC 833 [a non-immunosuppressive cyclosporin] (whole blood).- Lysis; pptn. (MeCN/MeOH); C-18(75°): aq.MeCN, grad.; 203 nm; sens. 0.8 µg/ml. TM use. Scott, M.G. et al. (1997) Clin Chem **43**, 505-510.

JOURNAL TITLES: Chromatog... - Chr; Liquid - Lq; Pharm. Biomed. Anal. - PhBmAn

See start of Compendium for amplification

KEY (abridged) Generally human plasma/serum

Prefix * if notable, e.g. for a method study.
Metabolites, mtb(s).; invstg. if investigated.
For (e.g.) conjugates: hydrolysis, hydrol.
Pptn., protein precipitation. LE, liquid-liquid extraction.
(Commonly - in SPE too - dry down and redissolve; unstated.)
Packings: e.g. C-18, CN (cyano), AGP (α_1-acid glycoprot.)
MeCN, acetonitrile; iPrOH, isopropanol; DEA, diethylamine.
If gradient: grad. Enantiomer studies/agents: chir....
Automation implicit if stated on-line or switching (swtch).
Derivatization/conversion: deriv, (sometimes post-column).
If nm stated, absorptiometric detection; EC, electrochemical.
Approx. LOD or (Q) LLOQ: sens. Pharmacokinetic studies, PK.

Section #C

STRATEGIES FOCUSED ON ANALYTE DISTINCTION OR DETECTION

The suffix **'n'** as in #C-3n distinguishes a 'NOTE' from main articles. COMMENTS (starting on p. 190) comprise discussion remarks and Editor's 'annotations'.

Art. E-5n, on a novel PCR approach for ultra-sensitive immunoassay, is pertinent to Section #C.

Also pertinent to Section #C: art. D-7, some Section #B articles, and some of the #ABC entries marked *.

PERTINENT 'LORE' in past vols. [some allusions in present text]

A dip into each General Index (near-constant in style) and list of Contents may be rewarding to readers who can consult a shelf run.
- See list facing title page; Plenum published 4 'A' vols. (#12, #14, #16 & #18). [Earlier vols.: #5 and (Horwood) #7 & #10.]

A recommended 'guide-book': *The Analysis of Drugs in Biological Fluids* - J. Chamberlain (2nd edn., 1995); CRC Press; 351 pp. [1525 refs.] It explains basics such as the biomatrix, features and metabolism of drugs, and choosing and validating analytical methods. Even experienced bioanalysts will find useful tips, many culled from the present book series and especially useful in the 'Pitfalls and Practical Solutions' chapter which draws strongly on the out-of-print Vol. 10 ('Trace-Organic Sample Handling', ed. E. Reid).

#C-1

APPLICATIONS OF HPLC COMBINED WITH NMR AND MS *

I.D. Wilson†, S. Taylor, B. Wright & E. Clayton

Dept. of Safety of Medicines, Zeneca Pharmaceuticals,
Mereside, Alderley Park, Macclesfield, Cheshire SK10 4TG, U.K.

The combination of HPLC with NMR and MS in a single integrated system for the purposes of efficient metabolite identification is described. The major practical problem associated with HPLC-NMR-MS is the need for careful solvent selection so as not to compromise the performance of either spectrometer. The utility of the system is illustrated using the example of ibuprofen/metabolites extracted from human urine.

The routine use of HPLC with MS for bioanalysis, including quantification, confirmation of identity and structure determination, is now well established. The HPLC-NMR combination is more recent but is also gaining adherents. Indeed although HPLC-NMR is not as widely used as HPLC-MS it shows much promise and will inevitably become routinely used in structure determination. The applicability of HPLC-NMR, with emphasis on drug metabolism, has been reviewed in this series [1] and elsewhere [2-4].

In metabolic studies it is rare that the identity of metabolites can be determined solely by either HPLC-MS or HPLC-NMR as both techniques have limitations. Thus, in the case of positional isomers, e.g. hydroxylation of an aromatic ring, MS alone may be able to show that the biotransformation has occurred, but not the exact location. With NMR this is a trivial problem. However, NMR is similarly limited when 'NMR-invisible' groups have been introduced. Thus, having located the site of ring hydroxylation, NMR may not be able to say whether it is present free or conjugated to sulfate, as MS would readily show. Hence the complementary nature of the data derived from NMR and MS is a strong incentive to construct a system in which MS and NMR are integrated. Such LC-NMR-MS systems have been put together, as firstly reported in 1995 [5]. Other examples soon appeared [TFMA = 4-tri-fluoromethyl-aniline], e.g. HPLC-NMR-ion-trap MS for paracetamol [6] and 2-bromo-TFMA [7], HPLC-NMR-MS for paracetamol metabolites [8], 2-bromo-TFMA [9], 2-chloro-TFMA [10] and ibuprofen metabolites [11].

The time saving achievable with this combination is clear: the characterization of drug metabolites in complex matrices, e.g. biofluids, in a single run can eliminate totally the time-consuming and tedious isolation procedures required by traditional methods.

---

†addressee for any correspondence

**Abbreviations*.- TFA, trifluoroacetic acid; S/N, signal-to-noise; NOESY, nuclear Overhauser enhancement spectroscopy.

Here we briefly describe the practicalities of the HPLC-NMR-MS approach, with illustrative examples from our recent work.

EXPERIMENTAL

Sample preparation.- So as to obtain a concentrated sample, 20 ml human urine representing the 3-h period following the oral administration of 400 mg of ibuprofen were taken to pH ~2 with 1 M HCl and applied to a 3 ml SPE cartridge containing 500 mg of C-18 sorbent (BondElut; Jones Chromatography, Hengoed, UK). The cartridge had previously been conditioned using methanol (5 ml) and 0.1 M HCl (5 ml). After sample application 0.1 M HCl (2 ml) and methanol (5 ml) were applied sequentially. The methanolic eluate was taken to 1 ml (N_2, ambient temp.). For analysis, 40 µl of the concentrate was mixed with 160 µl of D_2O and the whole of the sample injected onto the column.

Chromatography.- Separations of ibuprofen metabolites were obtained using a 250 × 4.6 mm Hypersil BDS column with a linear gradient from 20% acetonitrile/80% D_2O (containing 0.2% formic acid) to 60% acetonitrile over 45 min, delivered *via* a Bruker LC22 HPLC pump. Immediately after monitoring (254 nm), the eluate was split 95:5, with the bulk of the flow directed to a Bruker DRX500 500 MHz NMR spectrometer (Bruker Spectrospin, Coventry) and the remainder to a 'Platform LC' single quadrupole mass spectrometer (Micromass, Wythenshawe). The splitter was a stainless steel 'T' piece (Valco). All transfer lines were of polyethylethylketone (PEEK). Transfer times, at 1 ml/min flow-rate, were: UV to NMR, 52 sec; UV to MS, 36 sec.

NMR spectroscopy.- The spectrometer had a dedicated $^{1}H/^{19}F$-flow probe, with 120 µl cell volume. Stopped-flow NMR experiments were done using 1D-NOESY pulse sequence for double solvent suppression (acetonitrile and water). Data were generally acquired with between 32 and 160 transients into 16K data points with a pulse repetition time of ~3 sec. Chemical shifts were referenced to acetonitrile at 1.93 ppm.

Mass spectroscopy.- MS data were acquired over the range 200-600 amu with 1 sec scan time and 0.1 sec interscan delay. The needle voltage was set to 3.5 kV. The nebulizing gas was N_2 at 100 psi. Positive and negative ion modes were used to obtain MS data with cone voltages between 25V and 90V, switching between scans. Data acquisition was by Micromass MassLynx NT version 2.3.

RESULTS AND DISCUSSION

Instrument layout (see [1] for some points of detail)

Fig. 1 illustrates the configuration of the experimental HPLC-NMR-MS system. Chromatography was performed on a conventional HPLC system, controlled from the NMR console. The system also incorporated a 'sampling unit' where UV-observed peaks can be collected and stored for subsequent NMR analysis etc. if direct in-line spectroscopy is not required immediately. The NMR and MS detectors were connected in parallel, with the solvent flow

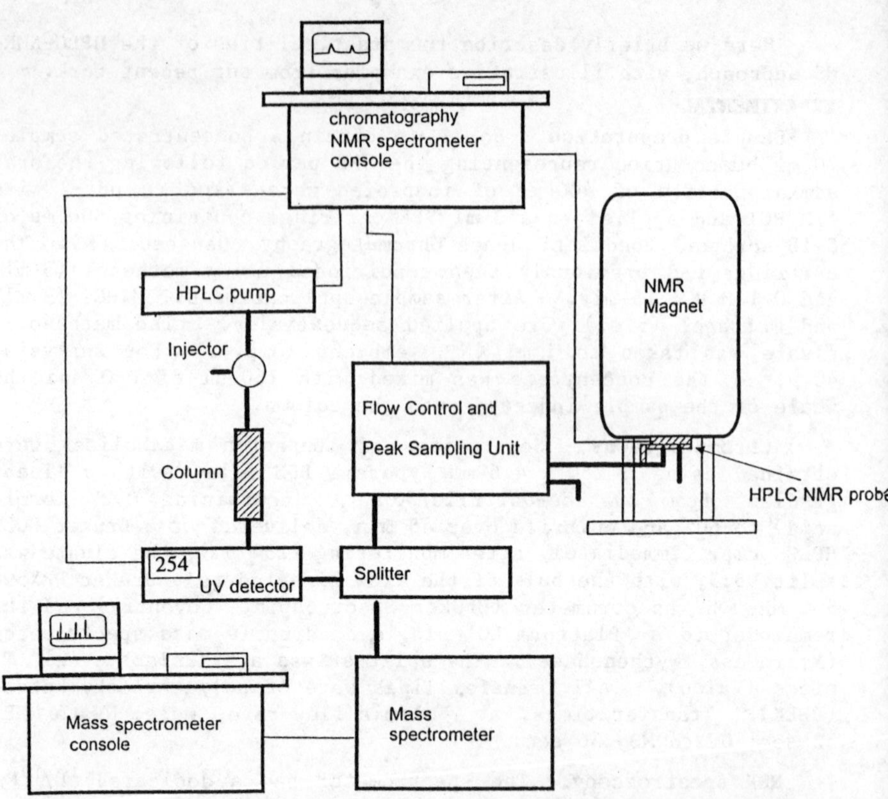

Fig. 1. The HPLC-NMR-MS system described here (schematic).

split post-detector (see above). For drug metabolism studies a radioactivity monitor (using a solid scintillant) can be inserted in-line with the UV detector.

It might seem unusual that the NMR and MS are not connected in-line. There are, however, several practical reasons for this. In particular, placing the MS after the NMR causes the NMR flow probe to become pressurized, resulting in solvent leakage and loss of sample. This is a technical problem which could easily be eliminated by modification of the probe design. However, a benefit of the in-parallel configuration is compatibility with the different time scales over which NMR and mass spectra are obtained. These do not differ if concentrated samples are being analyzed, allowing on-flow NMR; but if the concentration is too low for this, stopped-flow operation will be requisite, enabling spectra to be acquired over periods ranging from a few minutes up to several hours so as to get good S/N (or obtain 2-D NMR data).

In such circumstances the ability to easily disconnect the MS and use it for other tasks rather than leaving it idle is a distinct advantage. Thus, the MS could be set to work on alternative problems, re-examining the sample with non-deuterated solvents (see below) or perhaps setting up the next chromatographic run.

An additional difficulty that may arise when trying to juxtapose NMR and MS instruments in the in-line configuration results from the adverse effects of the strong magnetic field from NMR on MS performance. The in-parallel layout allows NMR to be kept well separated from the MS, which in this study was placed outside the 1 Gauss line, markedly reducing the undesirable effects of the stray magnetic field. In future such problems will be greatly minimized by reductions in the stray field through better design and the 'active' shielding of NMR magnets (e.g. the 'Ultrashield' NMR magnets from Bruker).

Whilst connecting the HPLC, NMR and MS instruments together is a relatively trivial exercise, the biggest problem with the operation of HPLC-NMR-MS systems is currently the absence of integrated data acquisition and control systems enabling all three instruments to be controlled from a single console. The best that can be done with the current state-of-the-art is to control either HPLC and NMR, or HPLC and MS, from one console - which is clearly inefficient. However, this problem is being addressed, and 'off-the-shelf' systems should shortly become commercially available.

Practical considerations

The most obvious problem that arises in HPLC-NMR-MS is the need for solvent systems that are compatible with both types of spectrometer. Thus acetonitrile/water*/D_2O mixtures with the pH modified using 'NMR-invisible' inorganic buffers, e.g. phosphate, are ideally suited to HPLC-NMR. Few HPLC-MS systems will work for long with 0.1 M phosphate buffers introduced directly into their ion sources, and an alternative is essential. An obvious first choice for NMR would be TFA, with no interferences in the ^1H NMR spectrum; but in practice TFA causes severe suppression of the ionization of many analytes which are then not detected by MS (e.g. see [7, 11]). Formic acid offers an acceptable compromise where an acidic modifier is needed for the HPLC. Whilst it does give rise to a signal in the ^1H NMR spectrum, this is a singlet which usually does not interfere with the resonances of compounds of interest. Although somewhat restrictive, the resulting range of mobile phases that can be prepared has proved to suit quite well for HPLC-NMR-MS applications in drug metabolism.

Whilst use of deuterated solvents in HPLC-NMR is not essential, as solvent suppression systems are now very efficient, in practice the use of D_2O does simplify the solvent suppression problems. D_2O can, however, result in difficulties in interpreting the mass spectra due to the exchange of labile protons by

*presence incidental, not intentional

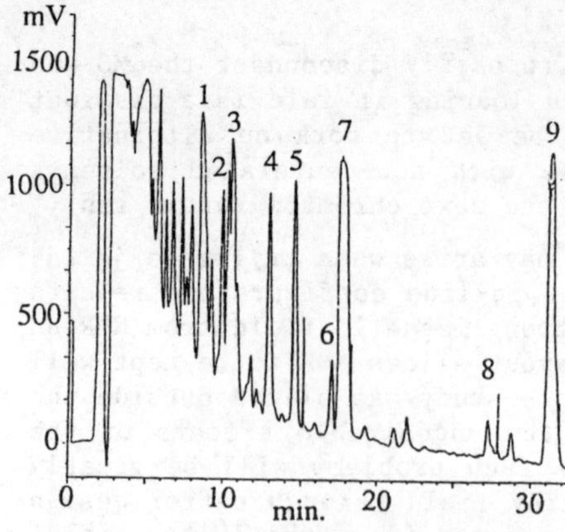

Fig. 2, *above*. **HPLC-UV** (254 nm) of the SPE extract of ibuprofen metabolite-containing urine. Numbers indicate metabolite peaks.

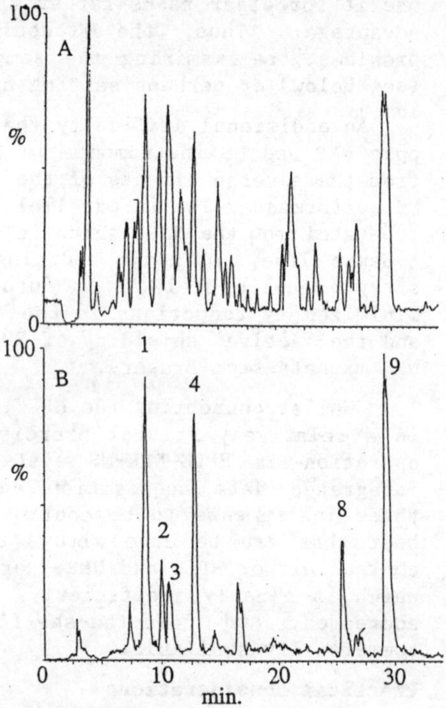

Fig. 3, *right*. HPLC-MS at a (**A**) low cone voltage, (**B**) high cone voltage selecting for the glucuronide-derived ion at m/z 196. Numbers of peaks correspond to those in Fig. 2.

deuterium. If necessary this problem can be eliminated by re-exchange (by mixing with H_2O or methanol *via* a T-piece) prior to the MS [12]. Alternatively, simply re-running the MS with a non-deuterated solvent system can provide the same result. Indeed, having both sets of MS data can be informative concerning the number of exchangeable protons on a metabolite.

Applications of HPLC-NMR-MS

Successive published studies [6, 7, 9, 10] were cited above. The present demonstration of identification of urinary ibuprofen metabolites illustrates well the strengths and weaknesses of the individual MS and NMR techniques and the capabilities of the combination [12]. Fig. 2 shows HPLC-UV for the SPE-derived sample. HPLC-NMR analysis of the major UV-absorbing peaks enabled the major drug-related components in the sample to be identified [cf. 1]; in most cases MS provided confirmatory data.

However, by setting the MS to look for losses characteristic of the presence of glucuronides, some additional, minor, peaks could be selected for further NMR study. The mass chromatogram obtained at low cone voltage, i.e. little fragmentation, is shown in Fig. 3A and, with a higher cone voltage, that for the diagnostic glucuronide fragment at m/z 196 is shown in Fig. 3B. MS, by

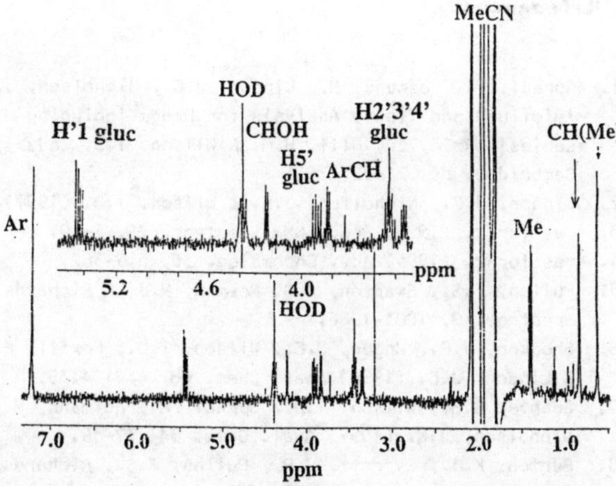

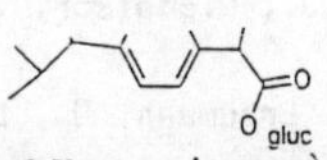

(1:1 mixture of diastereoisomers)

Fig. 4. HPLC-NMR of the mixture of diastereoisomers glucuronides of a hydroxy metabolite of the ibuprofen metabolite peak 4 in Fig. 2. HOD denotes $^1H^2HO$.

highlighting the presence of these glucuronides, allowed several minor components that might have been missed by HPLC-NMR alone to be fully characterized with, in this example, the NMR complementing the MS.

However, it was also the case that several of the non-glucuronidated metabolites, readily detectable by both UV and NMR, were not observed at all by MS; moreover, MS was obviously unable to distinguish between the isobaric mono-hydroxylated glucuronides and the diastereomers of the various *R* and *S* glucuronides. An example of two diastereomeric glucuronides co-eluting under these chromatographic conditions is shown in Fig. 4. Thus, the combination of NMR and MS analysis of the HPLC eluate gave a markedly more complete picture of the metabolic fate of ibuprofen than either would have furnished alone.

CONCLUSIONS

HPLC-NMR-MS is readily implemented experimentally and needs no specialized equipment or knowledge beyond that required to operate the individual systems. It can be used to provide rapid metabolite profiling and identification, and in favourable circumstances may completely eliminate the need for the isolation of the sample components.

A major requirement to enable HPLC-NMR-MS to be easily used and applied in the future is an integrated system for controlling all of the various components of the combination from the same console. Indeed, as a result of the specialization and 'compartmentalization' of HPLC, NMR and MS as separate disciplines, it will probably be more difficult to obtain appropriately trained operators for the instruments than to build them. In respect of this, 'intelligent' methods of data processing that could undertake some of the method development and spectral interpretation would be of great benefit.

References

1. Spraul, M., Hofmann, M., Lindon, J.C., Nicholson, J.K. & Wilson, I.D. (1994) in 'Biofluid and Tissue Analysis for Drugs including Hypolipidaemics' [Vol. 23, this series] (Reid, E., Hill, H.M. & Wilson, I.D., eds.), Royal Society of Chemistry, Cambridge, pp. 21-32.
2. Lindon, J.C., Nicholson, J.K. & Wilson, I.D. (1997) Drug Metab. Rev. **29**, 705-746.
3, as for 2. (1996) Prog. NMR Spectros. **29**, 1-49.
4. as for 2. (1995) Adv. Chromatog. **36**, 315-382.
5. Pullen, F.S., Swanson, A.G. Newman, M.J. & Richards, D.S. (1995) Rapid Comm. Mass Spectrom. **9**, 1003-1006.
6. Shockor, J.P., Unger, S.E., Wilson, I.D., Foxall, P.J., Nicholson, J.K. & & Lindon, J.C. (1996) Anal. Chem. **68**, 4431-4435.
7. Scarfe, G.B., Wilson, I.D., Spraul, M., Hofmann, M., Braumann, U., Lindon, J.C. & Nicholson, J.K. (1997) Anal. Comm. **34**, 37-39.
8. Burton, K.I., Everett, J.R., Pullen, F.S., Richards, S. & Swanson, A.G. (1997) J. Pharm. Biomed. Anal. **15**, 1903-1912.
9. Scarfe, G.B., Wright, B., Clayton, E., Taylor, S., Lindon, J.C. & Nicholson, J.K. (1998) Xenobiotica **28**. 373-388.
10. as for 9. (1998) [Drug Metab. Disp., intended] - unpublished work.
11. Clayton, E., Taylor, S., Wright, B. & Wilson, I.D. (1998) Chromatographia **45**, in press.
12. Holt, R.M., Newman, M.J., Pullen, F.S., Richards, D.S. & Swanson, A.G. (1997) J. Mass Spectrom. **32**, 64-70.

#C-2

APPROACHES TO INCREASING SENSITIVITY AND SPECIFICITY IN THE DETERMINATION OF DRUG ENANTIOMERS IN BIOFLUIDS

[1]M.J. Hemsley,† [1]H.M. Hill, [1]T.A.G. Noctor, [2]A.S. Prangle & [3]D. Thomas

[1]Covance Laboratories Ltd., Dept. of Biopharmaceutical Analysis, Otley Road, Harrogate, N. Yorks. HG3 1PY, U.K.

[2]School of Health Sciences, University of Sunderland, Wharncliffe Street, Sunderland SR2 3SD, U.K.

[3]Perkin Elmer, Post Office Lane, Beaconsfield HP9 1QA, U.K.

Enantiomeric separations of compounds of pharmaceutical interest often fail when they are applied to samples of biological origin. Poor sensitivity and/or inadequate specificity are the main reasons for this. A modification to the chromatographic system, such as the introduction of column switching or peak compression, and the use of a more selective detection method such as MS, are possible solutions to these problems. The chiral bioanalysis of warfarin, leucovorin, a primary aromatic amine and a partial 5-HT$_{1D}$ agonist exemplify the various solutions available to the bioanalyst.

The chiral separation of pharmaceutical compounds by HPLC can be effected by direct or indirect approaches. A direct enantiomeric separation relies on the formation of transient diastereomeric complexes, whereby the chiral selector forms part of the chromatographic system, either as a chiral mobile phase additive or a CSP*. The use of CSP's is particularly attractive, since, as we outlined in Vol. 24 [1], a wide range of types is available which can resolve most classes of compound. An indirect approach may also be used, whereby the enantiomers are first derivatized with a chiral, optically pure reagent, and the resulting disastereomers resolved on a conventional HPLC column.

When these techniques are applied to the chiral separation of drugs in biological fluids, two main problems arise. The first is poor specificity, due to interferences from endogenous sample components in the biological matrix, and often coupled with poor sensitivity. CSP's can be of low efficiency, and their performances (as plate numbers) and loadabilities are often low. The situation can be made worse by inadequate molar absorptivity of many compounds.

---

†addressee for any correspondence

**Abbreviations (see also Preface).*- AP(c)I, atmospheric pressure (chemical) ionization (context: MS-MS, tandem MS); CSP, chiral stationary phase; i.s., internal standard; LLOQ, lower limit of quantification; LV, leucovorin; MRM/SRM, multiple/single reaction monitoring; MeCN, acetonitrile; PCA, perchloric acid; TEA & DEA, tri- & di-ethylamine. QC, quality control (sample).

These problems can often be resolved through a modification to the chromatographic system. The use of peak compression, column switching and coupled column systems can all improve the sensitivity and specificity of an assay. Alternatively, a highly specific detection technique can be used. Detection methods based on chiroptical properties such as circular dichroism or optical rotation have been found to be too insensitive for bioanalytical needs. However, the use of MS-MS detection in chiral bioanalysis can bestow striking sensitivity and specificity upon an assay.

Chiral bioanalysis of warfarin utilizing peak compression

The continuing need to monitor the pharmacokinetics and metabolism of the enantiomers of warfarin has driven a search for improved methodology for determining its enantiomers in biofluids [2-6]. Here a peak compression effect has been used to allow 4 ng/ml to be detected by UV (surpassing fluorescence methods).

The enantiomeric separation is achieved using a chiral mobile phase additive and a non-chiral HPLC stationary phase. Plasma samples are first deproteinized using 10% v/v PCA. After centrifugation the supernatant is applied to a phenyl BondElut cartridge pre-conditioned with MeCN and 0.1% (v/v) acetic acid. MeCN/0.1% acetic acid is used to wash (20:80) and elute (50:50).*

The packing in the narrow analytical column (150 × 1.0 mm) is a porous graphitic carbon (Hypercarb, 5 μm), highly retentive and hence advantageous for peak compression systems. The mobile phase comprises MeCN/water/acetic acid/TEA (850:150:3:2.5 by vol.) containing 20 mM dimethylated-β-cyclodextrin as the chiral selector. A 20 μl injection of warfarin in mobile phase gives band broadening due to volume overload on the 1.0 mm i.d. column used; enantiomer resolution and peak shapes are poor (Fig. 1a). However, analyte is compressed into a very tight band at the head of the analytical column if 500 μl is injected in a weaker medium (lower % MeCN; Fig. 1b); this peak compression effect improves efficiency, resolution, peak shape and mass sensitivity.

Obviously the composition of the SPE elution solvent is very important. It must be strong enough to elute the analyte from the cartridge with a high recovery, but also weaker than the HPLC mobile phase to effect peak compression. For similar reasons, the choice of SPE sorbent should also be optimized. Ideally, the warfarin should be as weakly bound as possible, so that it can be eluted from the cartridge with a weak solvent. For warfarin the phenyl sorbent was found to be the optimal choice.

Chiral bioanalysis of LV using column switching/peak compression

A commonly used technique for resolving compounds of interest from interferants is to couple columns with phases of different selectivities by means of a valve switching system [cf. 1, 7].

*Details: 1 ml plasma; 200 mg Varian cartridges; ml: 5, 5, (wash) 10, (elute) 2 (20 μl injected). [Mobile phase variant in Fig. 1b.]

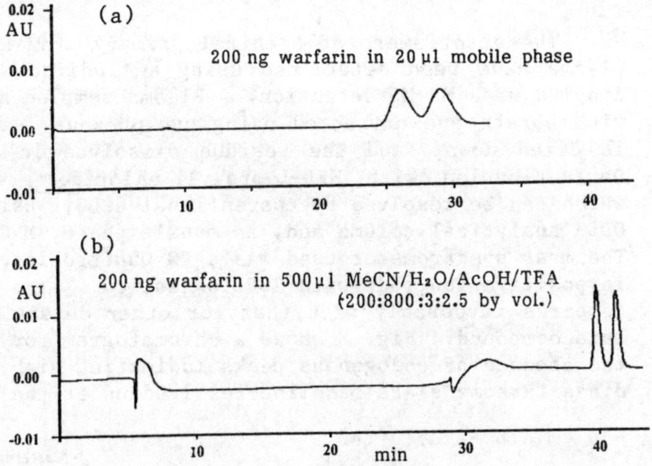

Fig. 1. HPLC peak compression with warfarin (pure standards; see text for some details). Loading was done in (**a**) normal mobile phase, (**b**) a weak version using a large vol. [see Fig.; AcOH, acetic acid]. Flow rate 24 µl/min; 305 nm for detection.

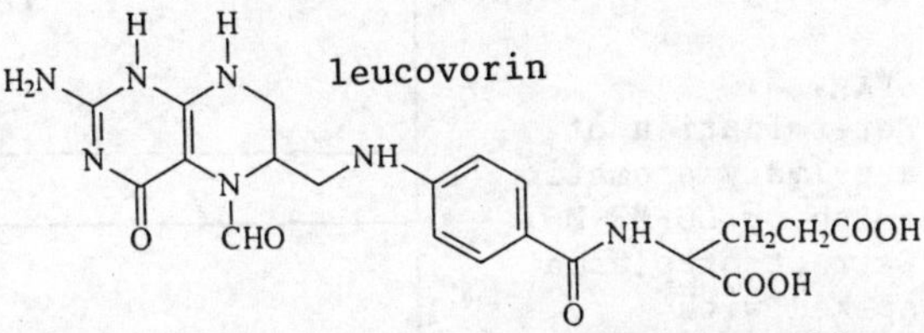

The enantiomers of leucovorin have been analyzed by coupling chiral and achiral stationary phases, in an HPLC system that also utilizes peak compression [8].* Firstly a BSA CSP is employed to separate (6*R*)-LV and (6*S*)-LV, with a purely aqueous mobile phase, 100 mM phosphate buffer pH 5.1. Unfortunately, the (6*R*)-LV and a major metabolite co-elute on the CSP; so further chromatographic selectivity is required. Hence the protein CSP is separately linked to 2 achiral columns *via* a valve switching system. Each of the enantiomers can thus be trapped on a separate C-18 column, and since the chiral system used is entirely aqueous, the enantiomers are concentrated into sharply defined bands at the head of each achiral column. Each enantiomer can then be eluted in turn merely by increasing the mobile phase strength, one C-18 column serving to free (6*R*)-LV from the metabolite. This coupled system with band compression allows enantiomer detection down to 5 ng/ml.

Chiral bioanalysis of a primary amine with MS-MS detection

Although the use of column switching and column coupling may offer a solution to a particular analytical problem, the correct HPLC system is often only arrived at after much time and effort spent in method development. Once a suitable system has been established, it is often not robust enough for routine sample analysis. The other alternative for increasing sensitivity and specificity in chiral bioanalysis is to use MS detection.

*For an analogous assay, see Amiodipine in #ABC, class 3.- *Ed.*

The enantiomers of a chiral primary aromatic amine in human plasma have been determined using an indirect chiral separation coupled with MS-MS detection. Plasma samples are first buffered with borate and extracted using cyclohexane. The organic phase is dried down, and the residue dissolved in carbonate buffer. Derivatization with HFB-L-prolyl chloride gives diastereomers which can be resolved by conventional HPLC, using a Prodigy 5 µm ODS3 analytical column and, as mobile phase, MeCN/1% acetic acid. The mass spectrometer used was a VG Quattro instrument, operated in positive ion mode with ESP ionization. The LLOQ, 1.0 ng/ml, compares favourably with that for other detection techniques for this compound. Fig. 2 shows a chromatogram for a plasma extract, the absence of endogenous peaks indicating high specificity; the diasastereomers are baseline-resolved under the conditions used.

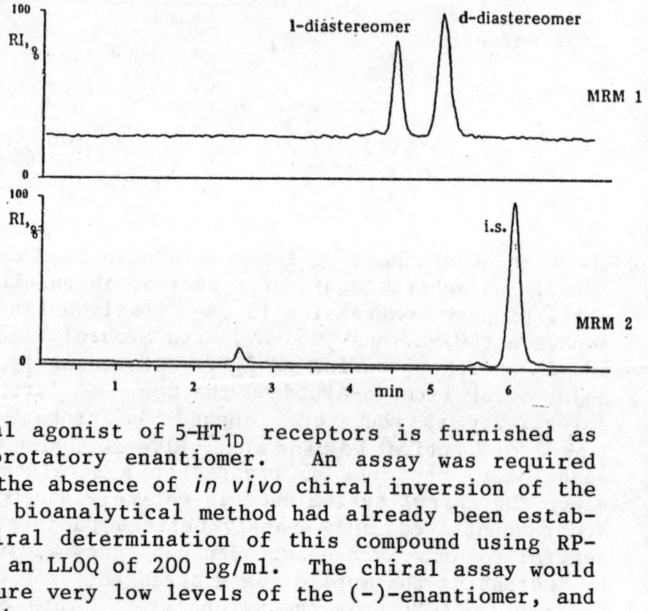

Fig. 2. Determination of a primary aromatic amine by LC-MS-MS. Extract of plasma spiked with 20 ng/ml.

Enantiomeric ratio of a partial 5-HT$_{1D}$ agonist in human plasma

A novel partial agonist of 5-HT$_{1D}$ receptors is furnished as the isolated dextrorotatory enantiomer. An assay was required that would confirm the absence of *in vivo* chiral inversion of the (+)-enantiomer. A bioanalytical method had already been established for the achiral determination of this compound using RP-LC-MS-MS, achieving an LLOQ of 200 pg/ml. The chiral assay would be required to measure very low levels of the (-)-enantiomer, and so an enantioselective system encompassing even greater sensitivity needed to be developed.

A satisfactory enantioseparation could be achieved with a direct approach using an amylose-based CSP. These columns operate only under NP conditions. The analytical column used is a Chiral-Pak AD [amylose tris(3,5-dimethylphenyl carbamate)], 250 × 4.6 mm (Daicel Chemical Industries, Tokyo); the mobile phase was hexane/ethanol/DEA (70:30:0.5 by vol.; 1 ml/min), run isocratically. The CSP column was coupled directly to a Sciex API III+ triple quadrupole mass spectrometer.

So far there has been little use of NP-HPLC systems with MS detection. This is due primarily to the enormous ascendancy of RP-HPLC in general applications. There is also perceived to be a small explosion hazard when highly volatile NP solvents such as hexane are introduced into an environment that may simultaneously employ high temperatures and high voltages. However, this drawback must be weighed against a major advantage of using NP systems with MS detection. In order that the analyte of interest can be detected by MS, analyte ions in the gas phase must be generated by an API interface. Theoretically, the production of such ions is more easily effected when highly volatile HPLC solvents are used. The resulting high sensitivity of NP-mode LC-MS-MS has enabled the enantiomers of the agonist to be determined with an on-column detection limit of 1 pg.

Procedure and assay performance

The agonist is extracted by SPE from 500 µl of human plasma buffered with 800 µl 10 mM ammonium acetate pH 11. The solution is allowed to pass through the bed of a pre-conditioned BondElut C-18 cartridge under minimal vacuum. After ammonium acetate washing the agonist is eluted with 400 µl 1% DEA in methanol. An Aspec XL (Gilson) sample processor is used to automate the SPE procedure. The eluate is blown to dryness (N_2) and reconstituted in 150 µl of analytical mobile phase; 75 µl is loaded.

The enantiomers are introduced into the mass spectrometer *via* an APcI interface (Heated Nebulizer probe, Sciex). APcI has proved a useful technique for introducing NP solvents into the MS ionization source, and the exposure risk is significantly reduced by various design features of the interface and by following a specific instrumental set-up [9]. This involves use of N_2 as both the nebulizer and the auxiliary gas, and the incorporation into the interface design of an over-pressure relief valve and a current-limiting resistor on the corona discharge needle. The probe is kept at 350° with 70 psi nebulizer gas pressure; the gas flows are 1.0 (curtain) and 4.5 (auxiliary) L/min, and the source pressure is 0.8 inches WC. Operation is in the positive ion SRM mode with 15 eV collision energy for the SRM transition of the agonist, m/z 244→213. Data are acquired using Sciex Standard API Software.

The validation procedure was designed to reflect the aim of determining the ratio of the agonist enantiomers rather than to quantitate the actual drug levels. A calibration curve is not required, but assay performance can be assessed by analyzing QC samples at different concentrations of (+)-agonist and at different enantiomeric ratios. QC's are prepared with (+)/(-) ratios 100:0, 95:5, 90:10, 75:25 and 50:50, this being done at nominal concentrations of (+)-agonist 16,000, 7500, 750 and 200 pg/ml. Fig. 3 shows LC-MS-MS of two such QC's. Essentially the LLOQ in respect of enantiomer ratio is 95:5 at 200 pg/ml, corresponding to ~10 pg/ml of (-)-agonist and having good accuracy. Precision is acceptable for the concentrations and ratios examined.

Concluding remarks.- Growing interest in the behaviour of enantiomers of potential medicinal agents has entailed a quest for improved sensitivity and specificity in chiral bioanalyses.

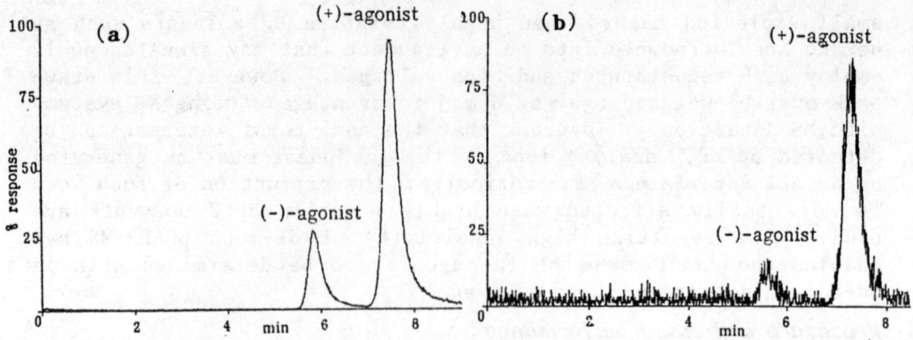

Fig. 3. LC-MS-MS of extracts from spiked plasma: (a) 750 pg/ml of (+)-agonist and at enantiomer ratio 75:25; (b) 200 pg/ml (the LLOQ) and at enantiomer ratio 95:5.

Chromatographic modifications such as column switching and peak compression have been employed with some success, but the use of a specific detection technique is the more favoured option. The elimination of possible interfering peaks and the associated high sensitivity of the technique have made MS-MS the ideal tool for measuring small quantities of enantiomers in biological fluids, and obviate the need for complicated chromatographic systems. In particular, NP-HPLC offers many benefits in enantioselective bioanalysis.

References

1. Noctor, T.A.G. (1996) in 'Biofluid Assay for Peptide-Related and Other Drugs' [Vol. 24, this series] (Reid, E., Hill, H.M. & Wilson, I.D., eds.), Royal Society of Chemistry, Cambridge, pp. 257-260.
2. Shibukawa, A., Nagao, M., Kuroda, Y. & Nakagawa, T. (1990) Anal. Chem. **62,** 712-716.
3. Erlandsson, P. & Nilsson, S. (1989) J. Chromatog. **482,** 35-51.
4. Thuaud, N., Sebille, B., Deratani, A. & Lelievre. G. (1991) J. Chromatog. **555,** 53-64.
5. Naidong, W. & Lee, J.W. (1993) J. Pharm. Biomed. Anal. **11,** 785-792.
6. Gariel, P., Grammond, J.P. & Guyon, F. (1993) J. Chromatog. **B 615,** 317-325.
7. Lindberg, C., Paulson, J., Blomqvist, A., Edholm, L-E. & Walhagen, A. (1992) in 'Bioanalytical Approaches for Drugs, including Anti-asthmatics and Metabolites' [Vol. 22, this series] (Reid, E. & Wilson, I.D., eds.), Royal Society of Chemistry, Cambridge, pp. 191-197.
8. Noctor, T.A.G. (1994) in 'A Practical Approach to Chiral Separations by Liquid Chromatography' (Subramanian, G., ed.), VCH, Weinheim, pp. 357-396.
9. Thomas, D., Crain, S.M., Sim, P.G. & Benoit, F.M. (1995) J. Chromatog. **A 695,** 1-9.

#C-3n

A Note on

COLUMN-SWITCHING WITH HEART-CUTTING: A USEFUL TOOL TO ACHIEVE SELECTIVITY IN BIOFLUIDS FOR DIFFICULT-TO-ANALYZE COMPOUNDS

A. Guenzi† & **D. Doppler**

Pharma Division, Preclinical Research, F. Hoffmann-La Roche, CH-4070 Basel, Switzerland

Very often the analysis of peptidomimetics in biofluids poses great challenges to the analytical chemist, due to the poor spectroscopic properties of the analytes and the high complexity of the media which need to be analyzed (plasma, urine etc.) [1]. In the course of a research project aimed at the discovery of vitronectin $\alpha_v\beta_3$ receptor antagonists, we had to characterize the oral bioavailability in the rat and the metabolic stability in rat hepatocytes of a number of such compounds, exemplified by two which presented UV maxima around 220 nm and no fluorescence:

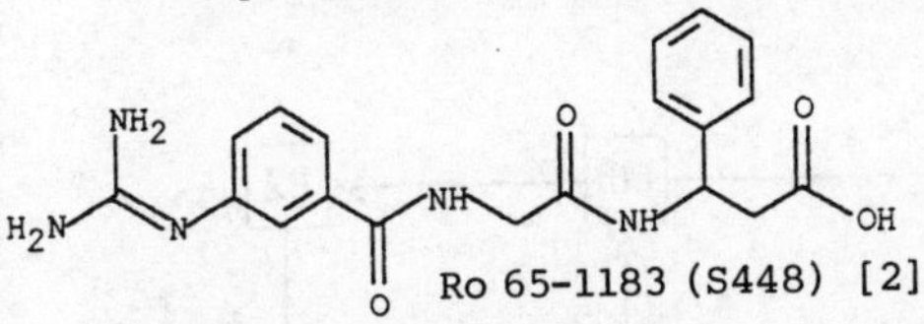

Ro 65-1183 (S448) [2]

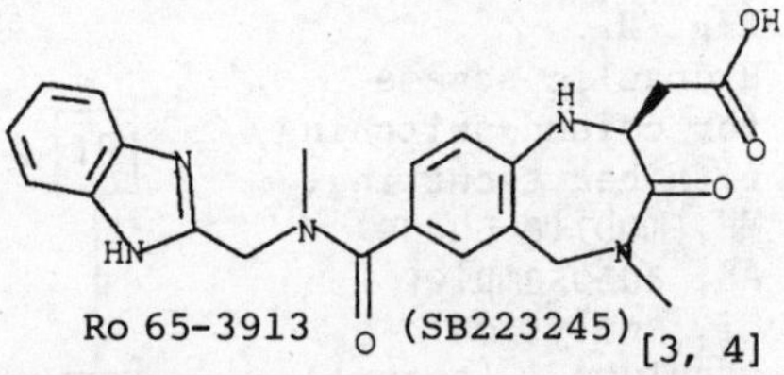

Ro 65-3913 (SB223245) [3, 4]

LC-MS or LC-UV?

An obvious approach to cope with these unfavourable properties would have been the use of LC-MS-MS, which is intrinsically selective. However, due to the unavailability of such an instrument for drug-discovery support, to the relatively small number of samples per compound (~50 overall, in plasma/urine/hepatocyte medium), and to priority problems, it was decided to attempt to solve the problem using the traditional UV detection, with the benefit of experience gained in several years of bioanalytical method development.

Suitable elution conditions were easily found for these compounds, on several columns (C-8, phenyl) from different manufacturers, using mixtures of MeCN* and 50 mM phosphate buffer, pH 3. Samples of plasma and hepatocyte incubation medium (containing 10% BSA) were worked up by protein precipitation (MeCN) because of the high water solubility and negative $\log P_{\text{octanol-buffer pH 7.4}}$ value of the analytes; this offered swift method development and high recovery. However, injection of extracts from blank media or diluted blank rat urine gave chromatograms such as Fig. 1 depicts - indicating inapplicability of these conditions to the analytes.

**Abbreviations*.- MeCN, acetonitrile; LLOQ, lower limit of quantification.

†addressee for any correspondence

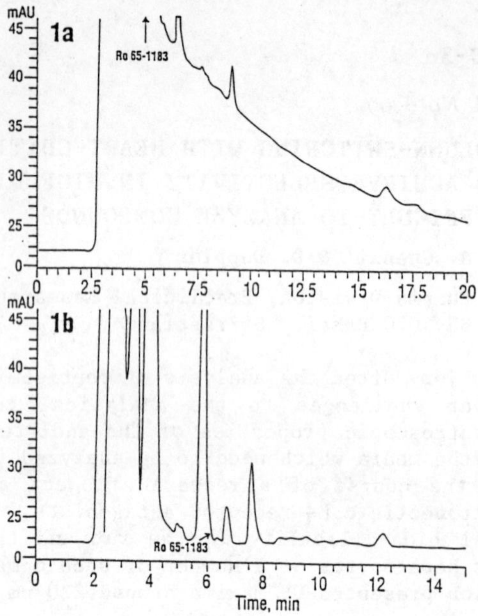

Fig. 1.
HPLC chromatogram of blank rat plasma after protein precipitation with MeCN. Column (Spherisorb S5): **a**, Phenyl; **b**, Octyl; 250 × 4.6 mm. Eluant: MeCN/50 mM phosphate buffer pH 3.0, 20:80; 1 ml/min. Detection: 220 nm.

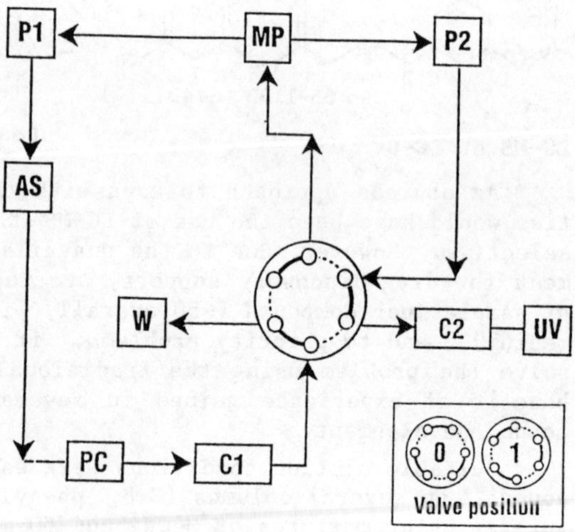

Fig. 2.
Hydraulic scheme for column-switching with heart-cutting.
MP, mobile phase
AS, autosampler
C1, analytical column 1 (phenyl)
UV, UV detector
P1 & P2, pumps
PC, guard column (phenyl)
C2, analytical column 2 (octyl)
W, waste

P1 #L6200A, P2 #L-6000; both Merck-Hitachi.
Rheodyne #7000E valve, modified for high-speed switching (ProLab, Reinach).

Solution: LC-UV with heart-cutting (Fig. 2)

We had some past success in analyzing hydrophilic compounds by column-switching with heart-cutting, whenever a single column couldn't separate the analyte from endogenous interferants. The combination of two different silica chemistries makes this '2-D' technique very easy and accessible. The only extra requisites are a second pump and an electrically operated switching valve.

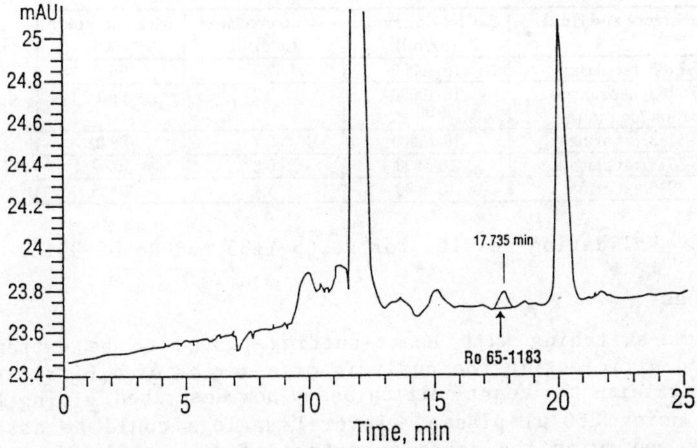

Fig. 3. Chromatogram for rat plasma spiked at 10 ng/ml (LLOQ) with Ro 65-1183, using column-switching and heart-cutting. Work-up: 200 µl plasma, 1 ml MeCN; vortex, centrifuge; dry down 1 ml supernatant, dissolve in 125 µl mobile phase, and inject 50 µl.

With the switching and heart-cutting (transfer from column 1 to 2 took 1.2 min), conditions evolved during column screening could now be exploited, and a good LLOQ achieved with 200 µl rat plasma for Ro 65-1183 (Fig. 3). Thereby we could characterize the kinetics after i.v. dosing and determine the oral bioavailability of this metabolically stable compound (Fig. 4). Although run time was 25 min, within 24 h one could extract and inject a pre-study run (8 calibrators and 4 QC's) and an unknown-sample run (same, plus 28 plasmas from i.v.-/p.o.-dosed rats) and get the results.

Similarly, the procedure served to analyze Ro 65-3913 (with the % MeCN altered: octyl column, 25%; phenyl, 20%) and several other analogs. Table 1 gives validation results.

Fig. 4. Plasma concentration-time profile of Ro 65-1183 following a 5 mg/kg i.v. (bolus) or 10 mg/kg p.o. (gavage) administration to rats.

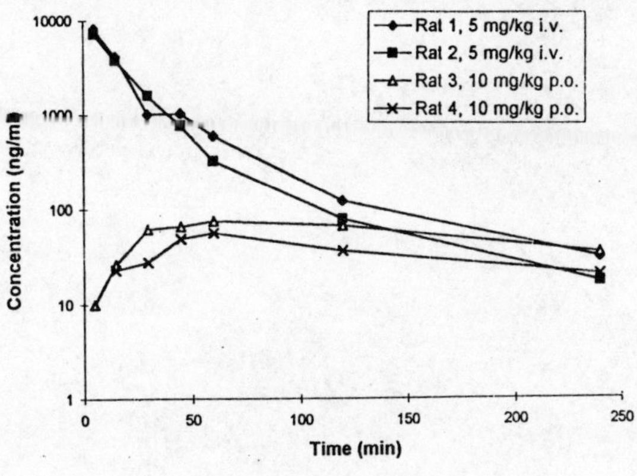

Substance and fluid	Calibration range (ng/ml)	Mean precision (n=4), %	Mean accuracy %
Ro 65-1183, rat plasma	10-10000	2.1	96.1
Ro 65-1183, hepatocyte incubation medium	10-10000	2.2	96.1
Ro 65-1183, rat urine	10-10000	0.7	98.0
Ro 65-3913, rat plasma	10-5000	0.8	102.7
Ro 65-3913, rat urine	10-5000	1.4	100.5

Table 1. Validation results for Ro 65-1183 and Ro 65-3913.

Conclusions

Column-switching with heart-cutting proved to be extremely useful and efficient in the analysis of a series of $\alpha_v\beta_3$ receptor antagonists. With the heart-cutting assay now described, giving LLOQ 10 ng/ml using 200 µl plasma, several analogs could be assayed merely by modifying the organic content of the mobile phase and the transfer time. This allowed the determination of their oral bioavailability and intrinsic clearance in a very short time.

References

1. Reid, E., Hill, H.M. & Wilson, I.D., eds. (1996) 'Biofluid Assay for Peptide-related and Other Drugs' [Vol. 24, this series], Royal Society of Chemistry, Cambridge, 350 pp.
2. Ruminski, G.P., Clare, M., Collins, P.W., Desai, B.N., Lindmark, R.J., Rico, J.G., Rogers, T.E. & Russell, M.A. (1997, March 6). World Patent WO 97/08145.
3. Ali, F., Bondinell, W., Huffman, W.F., Lago, M.A., Keenan, R.M., Kwon, C., Miller, W.H., Nguyen, T. & Takata, D.T. (1996, January 11). World Patent WO 96/00730.
4. Keenan, R.M., Miller, W.H., Kwon, C., Ali, F.E., Callahan, J.F., Calvo, R.R., Hwang, S., Kopple, K.D., Peishoff, C.E., Samanen, J.M., Wong, A.S., Yuan, C. & Huffman, W.F. (1997) **J. Med. Chem. 40,** 2289-2292.

#C-4

ELECTROCHEMICAL SENSORS: OPPORTUNITIES FOR HIGH-THROUGHPUT ANALYSIS DURING DRUG SELECTION

[1]S.A. Wring†, [2]J.P. Hart, [2]R. Pemberton¤, [3]R. Pittson, [4]L. Takahashi, [4]J. Tolan & [1]C. Serabjit-Singh

[1]Bioanalysis & Drug Metabolism Divn., Glaxo Wellcome Research Inc., 5 Moore Drive, Research Triangle Park, NC 27709, U.S.A.

[2]Sensors Group, Faculty of Applied Science, Univ. of the West of England, Coldharbour Lane, Frenchay, Bristol BS16 1QY, U.K.

[3]Gwent Electronic Materials Ltd., Monmouth House, Mamhilad Park, Pontypool, Gwent NP5 OHZ, U.K.

[4]Affymax Research Inc., 3410 Central Expressway, Santa Clara, CA 95051, U.S.A.

The throughput of many in vitro *systems employed for candidate drug selection could benefit from faster and simpler sensitive analytical techniques. EC* sensors and biosensors offer especial promise. This survey concerns trends in sensor research and how they may bear on drug analysis and selection, with emphasis on SPS's as they have met with the greatest commercial success, being easy to use, portable and easily made at low cost such that they can become disposable. Consideration is given to EC waveforms that may be able to achieve nM sensitivity and provide qualitative as well as quantitative data.*

Recent advances in combinatorial chemistry have afforded a marked increase in the number of compounds available for screening as possible drugs, calling for new analytical strategies to meet the needs of higher throughput and capacity. Improvements in analytical instrumentation and the use of API-MS have fortunately allowed us to meet current challenges. However, other technologies may be required to meet future needs.

Sensors and biosensors provide an exciting opportunity to realize marked improvements in analytical throughput, with the required specificity and sensitivity. These devices have proved valuable for biomedical and environmental applications but have yet to find widespread application in drug discovery or development. Sensors have featured in ~107000 *Chem. Abs.* citations, with ~8000 on biosensors - a notable growth area: the number of

---

†correspondent; e-mail: saw22503@glaxowellcome.com ¤U.K. contact

**Abbreviations*.- EC, electrochemical; SCE, saturated calomel electrode; SPS, screen-printed sensor (EC); SPCE, screen-printed carbon electrode. Ab, antibody; CyV, cyclic voltammogram [*Ed.*'s variant of authors' version, CV]; DMPC, dimyristoyl-L-α-phosphatidylcholine; LOD, limit of detection. [The term 'biosensor', here connoting enzyme use (not a biomatrix), is arguably 'unhappy'.- *Ed.*]

publications per year, only 1 or 2 during 1966-1979 and 9 in 1981, became 22 in 1982, 141 in 1985, 341 in 1988, 545 in 1991, 781 in 1994 and 1104 in 1997. Moreover, active research has borne fruit for several applications; e.g. inexpensive commercial devices have become acceptable for personal blood-glucose monitoring.

Sensor development is amenable to a broad range of detection technologies, although optical and EC detection have proven the most successful thus far. Here we focus on EC sensor development - design aspects, and approaches employed and the promise they hold in drug-selection applications. EC sensors are particularly exciting to consider because much research expertise has been gained that could now be applied to drug-discovery challenges. In an earlier volume [1] and elsewhere [2, 3] we have discussed EC sensor applications, and also problems, more thoroughly.

Commercial biosensors

The ExacTech (Medisense) biosensor for determining glucose in whole blood was probably the first inexpensive and successful commercial device offered as a mass-market tool. It comprised two electrodes screen-printed onto an inert plastic base, along with a reagent layer comprising glucose oxidase and a ferrocene derivative, which acts as an electron mediator to facilitate the transfer of an electron from the reduced enzyme to the working electrode [5]. The biosensor's working area is then coated with a filtration membrane to exclude proteins and cells (Fig. 1). The biosensor strip is inserted into a pen-sized device. A drop of blood is applied, a +400 mV potential (*vs.* Ag/AgCl) is applied by depressing a single button, and after 30 sec a digital read-out of glucose concentration is obtained. The device has also been used as the basis of sensors for 3-hydroxybutyrate [6], paracetamol [5, 7], salicylate [7, 8] and theophylline [9].

Biosensors based on oxidase enzymes

The ExacTech glucose device is generally similar in mode of action to most biosensors that employ oxidases. The electron mediator may be electrochemically oxidized as in the ExacTech sensor or already present in an oxidized state. The oxidase then passes the electron to the mediator, which in turn passes it to the sensor electrode, giving rise to the measured EC signal.

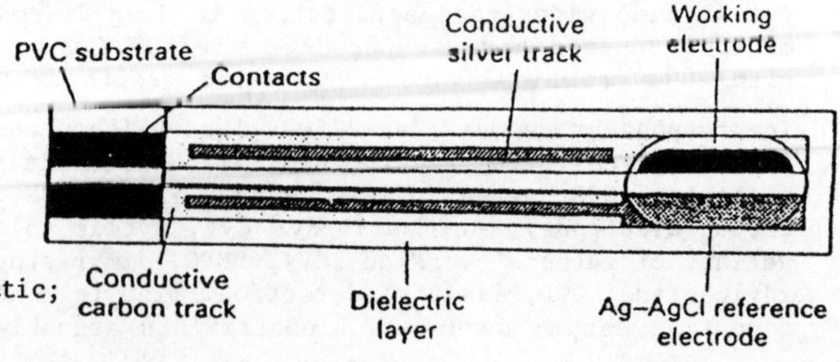

Fig. 1. The ExacTech glucose biosensor (schematic; from [5]). [Manufacturer: ExacTech.]

Fig. 2.
Electron transfer events for a glucose biosensor employing electrocatalytic detection of H_2O_2 with the electron mediator cobalt phthalocyanine. (SCE, SPCE: see * on first p. Source: ref. [10].)

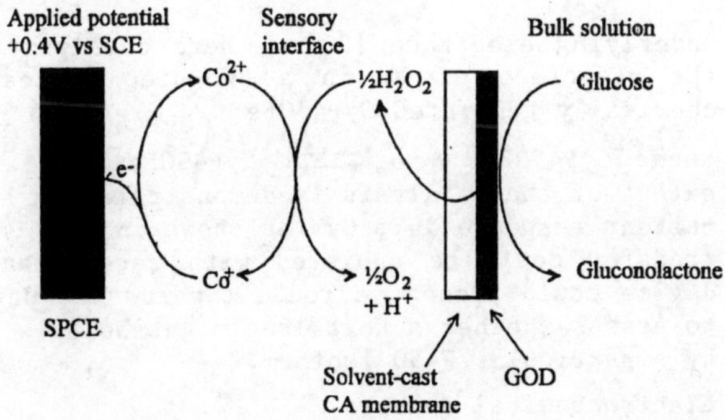

Other approaches for oxidase-based sensors have generally employed either the direct or electrocatalytic [10] measurement of H_2O_2 generated during the enzyme reaction (Fig. 2). It seems feasible that sensors could be developed using these approaches to screen for metabolism of drugs by monoamine oxidase, e.g. for sumatriptan [11].

Biosensors to follow NAD(P)H consumption

These devices may be useful for coupling with cytochrome P450 enzymes to make 'drug metabolism' sensors. The most promising of the approaches to follow NAD(P)H concentrations involve the use of phenoxazine, phenathiazine or phenazine electron mediators. In one approach Meldola's Blue, a phenoxazine, was incorporated into biosensors for ethanol [13] or lactate (Fig. 3) [14] to follow NADH generation by alcohol dehydrogenase or LDH respectively. Other analytes measured with NADH-linked sensors include glucose [15], hydroxybutyrate [16, 17] and creatinine [18].

Biosensors based on enzyme inhibition

Sensors based on attenuation of enzyme activity by specific inhibitors have also been successful. A similar approach may be feasible to screen for inhibitors of drug-metabolizing enzymes. This approach has been most widely employed for determination of organo-phosphorus pesticides [3, 19]. Thus, we [Hart & Hartley, in Bristol] have correlated the inhibition of acetylcholine esterase activity, measured by a fall in thiocholine generation, with levels of the pesticide paraoxon.

$$\text{acetylthiocholine iodide} \xrightarrow[\text{esterase}]{\text{acetylcholine}} \text{thiocholine} + \text{acetate} + I^-$$

The above-mentioned theophylline sensor works by drug-inhibition of alkaline phosphatase activity as measured by a fall in the production of *p*-aminophenol from *p*-aminophenylphosphate.

Biosensors based on direct electron transfer from enzymes

Direct electron transfer from an enzyme to the electrode may be achieved by conjugating a mediator to an enzyme or by transfer without a mediator. The latter was achieved for $P450_{cam}$ immobilized in a DMPC film that facilitated electron exchange with the underlying electrode [20]. Here catalysis was demonstrated for the reduction of O_2 in a two-step process involving electrochemically generated CyP450Fe: $CyP450Fe^{III} + e^- \rightleftharpoons CyP450Fe^{II}$ then $CyP450Fe^{II} + O_2 \rightleftharpoons CyP450Fe^{II} - O_2$. The quantitative extent of the catalysis is demonstrated by the marked increase in current response in a CyV as shown in Fig. 4. If direct electron transfer could be achieved with recombinant human CyP450's the device could present a route towards the development of a sensor to assess whether a test drug might be able to undergo metabolism by a particular P450 isoform.

Electrochemical sensors

Many drugs are electroactive [1] and these may be measured directly by EC sensors. This technique would find particular application for individual compounds, or libraries of compounds,

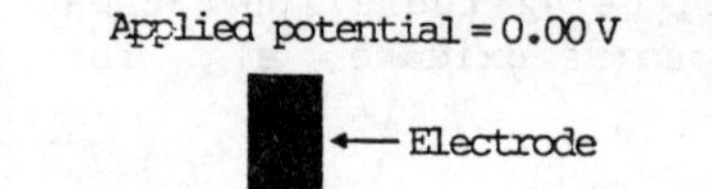

Fig. 3. *Above:* Sequence of events during operation of the NADH-linked ethanol biosensor (Meldola's Blue as electron mediator). From ref. [12].

Right: Amperograms, obtained at 23° in 0.5 M phosphate buffer pH 8, illustrating (a) reproducibility; (b) (c) & (d) calibration. Lactate concentrations (mM) as shown. From ref. [14].

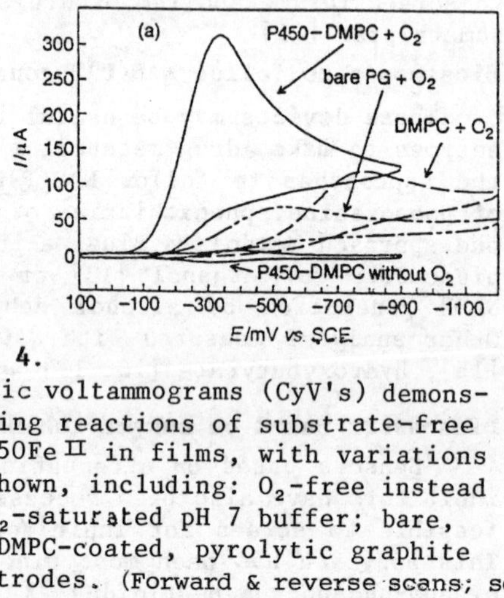

Fig. 4. Cyclic voltammograms (CyV's) demonstrating reactions of substrate-free $CyP450Fe^{II}$ in films, with variations as shown, including: O_2-free instead of O_2-saturated pH 7 buffer; bare, not DMPC-coated, pyrolytic graphite electrodes. (Forward & reverse scans; so dual lines.) Source: ref. [20].

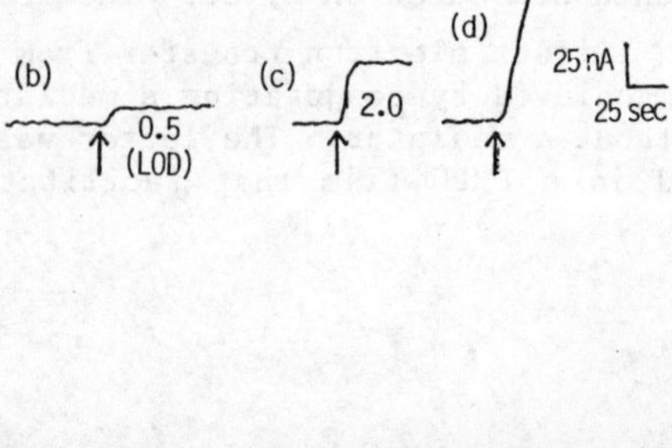

with known electroactivity. The strength of this approach is that arrays of sensors can be produced to make multiple parallel measurements. A simulation of this approach appears in Fig. 5, showing square-wave voltammograms for 5 related Glaxo Wellcome compounds. The CyV's can yield quantitative information from peak height or area and, by peak potential, qualitative information that may prove valuable during *in vitro* metabolism studies, as metabolites may well oxidize at different potentials to the parent drug. The technique is particularly useful for *in vitro* studies because samples may require little or no clean-up before analysis. We have found that square-wave voltammetry can typically afford µM levels of sensitivity.

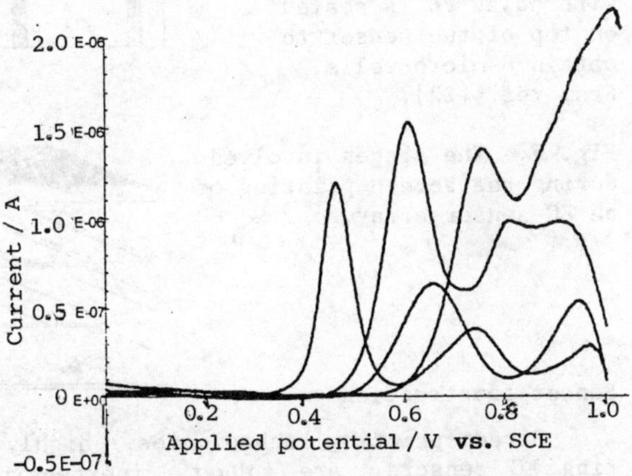

Fig. 5. Square-wave voltammograms of the oxidation of five structurally related candidate drugs (Glaxo Wellcome; heterocyclic, basic) at a glassy carbon electrode.

Immunosensors

Immunosensors, while unlikely to be used routinely during small-molecule drug discovery, may be useful for peptide drugs or therapeutic Ab's; they may also be adapted for sense/anti-sense detection of oligonucleotides.

An EC immunosensor array and a notable application have been described [21, 22]. The device comprised a 'strip' of 8 immunosensors formatted to the dimensions of a single column of a 96-well plate (Fig. 6). The herbicide 2,4-dichlorophenoxyacetic acid (2,4-D) was determined. Discs of nitrocellulose membrane with an immobilized monoclonal Ab to 2,4-D were put into the microwells. A competitive immunoassay was performed using a 2,4-D-horseradish peroxidase tracer. Tracer and test solution were incubated for 20 min, whereupon the disc was washed and an HRP substrate system (hydroquinone and H_2O_2) added; 30 min later, accumulated benzoquinone was detected amperometrically. The sensor could, in principle, be used as a generic format for any competitive immunoassay employing enzyme-conjugated tracers (e.g. peroxidase, alkaline phosphatase, cholinesterase or glucose oxidase). The base sensor devices are available commercially.*

*Source: Krejci Engineering, Tisnov, Czech Republic.

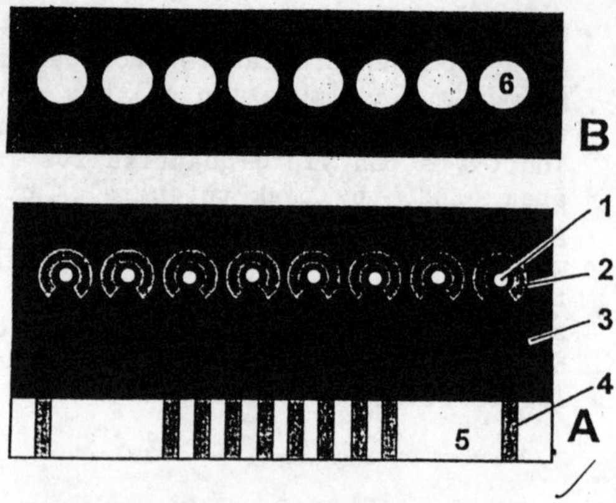

Fig. 6. (**A**) The SPS. The working electrodes #1 are prepared from gold paste, the reference #2 from silver paste. The isolating layer #3 covers conducting paths from electrodes to contacts #4. The sensor is prepared on a ceramic support #5. (**B**) The plastic sheet with holes #6 is sealed on top of the sensor to obtain 8 micro-wells. From ref. [22].

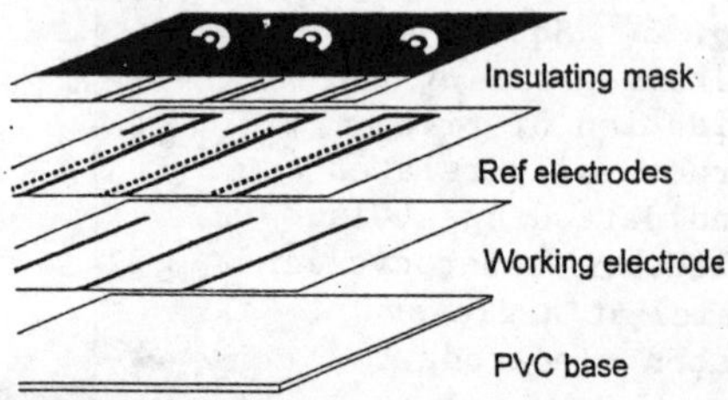

Fig. 7. The stages involved during the screen-printing of an EC sensor array.

Sensor construction

Screen-printing technologies, highly successful for constructing EC sensors, are robust, inexpensive and widely available. Fig. 7 illustrates the stages involved in screen-printing electrode arrays. Typically, a working electrode (e.g. gold or carbon) is printed onto an inert plastic or silica base material, followed by the reference one (e.g. Ag/AgCl). Enzymes and/or other reagents may also be printed. An insulating dielectric material may then be printed over both electrodes leaving a working area and an area for electrical contact exposed. An ion-selective or permselective membrane may also be layered over the working electrode prior to the application of the dielectric. Electrical contact is made with edge connectors.

CONCLUDING COMMENTS

We have overviewed, with examples, several strategies in sensor and biosensor research and suggested opportunities for applying the various approaches to drug research. Thus far, published applications have mostly targeted biomedical or environmental analyses. However, the increasing analytical requirements of early pharmaceutical research support could exceed the capacity of established technologies and bottlenecks may develop. Sensor technology has the potential of addressing these bottlenecks and should provide an opportunity to markedly improve our capacity to screen drugs. At present, expertise resides mostly in academia and it is likely that collaborations will be required to make best progress.

References

1. Wring, S.A., Ayres, D., Dayal, S. & Harrison, G. (1996) in 'Biofluid Assay for Peptide-Related and Other Drugs' [Vol. 24, this series] (Reid, E., Hill, H.M. & Wilson, I.D., eds.), Royal Society of Chemistry, Cambridge, pp. 262-274.
2. Hart, J.P. & Wring, S.A. (1994) Electroanalysis **6,** 617-624.
3. Hart, J.P. & Wring, S.A. (1997) Trends Anal. Chem. **16,** 89-103.
4. Matthews, D.E., Brown, E., Watson, A., Holman, R.R., Steemson, J., Hughes, S. & Scott, D. (1987) Lancet **1,** 778.
5. Green, M.J. & Hilditch, P.I. (1991) Anal. Proc. **28,** 374-376.
6. Batchelor, M.J., Green, M.J. & Sketch, C.L. (1989) Anal. Chim. Acta **221,** 289-294.
7. Hilditch, PI. & Green, M.J. (1991) Analyst **116,** 1217-1220.
8. Frew, J.E., Bayliff, S.W., Gibbs, P.N.B. & Green, M.J. (1990) Anal. Chim. Acta **224,** 39-46.
9. Foulds, N.C., Wiltshere, J.M. & Green, M.J. (1990) Anal. Chim. Acta **229,** 57-62.
10. Gilmartin, M.T., Hart, J.P. & Patton, D.T. (1995) Analyst **120,** 1973-1981.
11. Dixon, C.M., Park, G.R. &Tarbit, M.H. (1994) Biochem. Pharmacol. **47,** 1253-1257.
12. Wring, S.A. & Hart, J.P. (1992) Analyst **117,** 1215-1229.
13. Sprules, S.D., Hartley, I.C., Wedge, R., Hart, J.P. & Pittson, R. (1996) Anal. Chim. Acta **329,** 215-221.
14. Sprules, S.D., Hart, J.P., Wring, S.A. & Pittson, R. (1994) Analyst **119,** 253-257.
15. Appelqvist, R., Marko-Varge, G., Gorton, L., Torstensson, A. & Johansson, G. (1985) Anal. Chim. Acta **169,** 237-247.
16. Palleschi, G., Rathbone, H.S. & Mascini, M. (1989) Anal. Chim. Acta **209,** 223-230.
17. Palleschi, G., Rathbone, H.S. & Mascini, M. (1989) Electroanalysis **1,** 199-
18. Buch-Rasmussen, T. (1990) Anal. Chem. **62,** 932-936.
19. Trojanowicz, M. & Hitchman, M. (1996) Trends Anal. Chem. **15,** 38-45.
20. Zhang, Z., Nassar, A-E.F., Lu, Z., Schenkman, J.B. & Rusling, J.F. (1997) J. Chem. Soc. Faraday Trans. **93,** 1769-1774.
21. Skaldal, P. & Kalah, T. (1995) Anal. Chim. Acta **304,** 361-368.
22. Skaldal, P. & Kalah, T. (1995) Anal. Chim. Acta **316,** 73-78.

Acknowledgements

The authors thank colleagues at Glaxo Wellcome and Affymax for their interest and support of our research into potential applications of sensors for high-throughput drug screening. They also thank the research teams whose work is described in this report.

For Figs. that have appeared elsewhere, sources (now thanked) are evident from refs. cited in legends.

#C-5n

A Note on

DEVELOPMENT AND USE OF A DEEP-UV LASER-INDUCED FLUORESCENCE DETECTOR FOR HPLC: APPLICABILITY TO PHARMACEUTICAL BIOANALYSIS

[1]Richard F. Venn†, [2]Alan Howard & [2]Duncan Russell

[1]Pfizer Central Research, Sandwich CT13 9NJ, U.K. | [2]Chemistry Dept., Univ. of Southampton, Highfield, Southampton SO9 5NH

Laser induced fluorescence (LIF) has been used for detection in CE for a number of years [1] and also in HPLC [2, 3]. However, due to the difficulty and expense of obtaining deep-UV lasers, most work has been carried out with a few analytes having native excitation maxima at or above 325 nm (He-Cd laser line) or with compounds derivatized to generate fluorescence in the visible part of the spectrum [3, 4]. Derivatization, however, is not reproducible at low concentrations and is not a favoured approach for ultra-trace bioanalysis. The recent advent of a number of deep-UV lasers has opened the possibility of using the native fluorescence of analytes of interest (low-mol. wt. pharmaceuticals in our case).

Many compounds show fluorescence in the 220-300 nm excitation region, and LIF can offer a 10- to 100-fold increase in sensitivity over conventional fluorescence [5]. While HPLC-MS-MS offers excellent selectivity and sensitivity for many analytes, some are unsuitable for MS, sensitivity being poor due to poor ionization. Thus HPLC-LIF can offer an alternative approach. Fluorescence is an attractive solution for obtaining increased sensitivity, since (ideally) there is no background signal in the absence of analyte whereas with UV absorption there is a large background signal.- With UV detectors a small change in analyte concentration leads to a small change in a large signal, which limits sensitivity. For fluorescence, however, a small change on a very small background is more easily observed. Fluorescence intensity (F) is given by the equation $F = k.I_0.C\ l.\varepsilon.\emptyset$ where k is a constant; I_0, incident light intensity; C, analyte concentration; l, light path; ε, analyte's extinction coefficient at λ_{ex} ; ∅, fluorescence quantum efficiency. Thus fluorescence is directly proportional to incident light intensity (up to a limit, imposed by other factors such as excited-state half-life); hence the sensitivity of a fluorescence detector can be increased by increasing the incident light flux by, for example, the use of a laser.

We chose to investigate use of a Coherent Innova FreD laser*, which provides 6 fixed wavelengths in the UV region, from 229 up to 264 nm. This is a very stable system. The alternative, a

---

†correspondent

*#300; supplier: Coherent (UK) Ltd., Cambridge ('BeamBoost' from Technicol (UK) Ltd.)

Spectra-Physics MOPO 700 which is fully tuneable in the range 200 to 400 nm was considered to be a 'research instrument', was more expensive and, we judged, likely to be less stable in a routine environment than the Coherent Innova FreD 300 (layout: Fig. 1; all parts contained within the closed laser box). 'BBO Crystal' is an intra-cavity frequency-doubling crystal. We have found the configuration to be very stable, producing laser radiation of the chosen wavelength immediately on switch-on. (The possible scope and limitations of frequency doubling, e.g. 514→257 nm for the argon-ion laser, is discussed in [2].- *Ed.*)

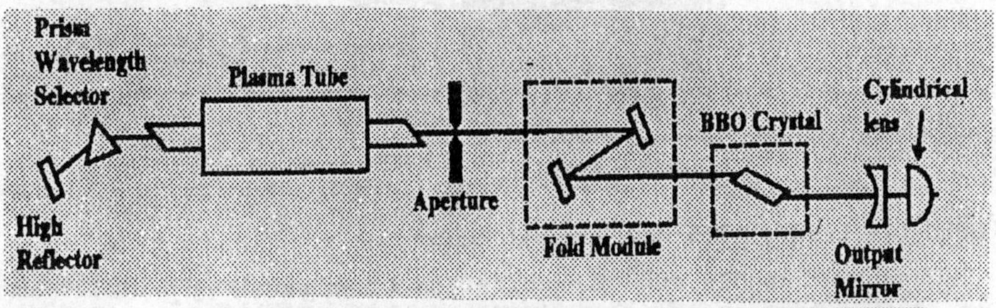

Fig. 1. Layout of the laser.

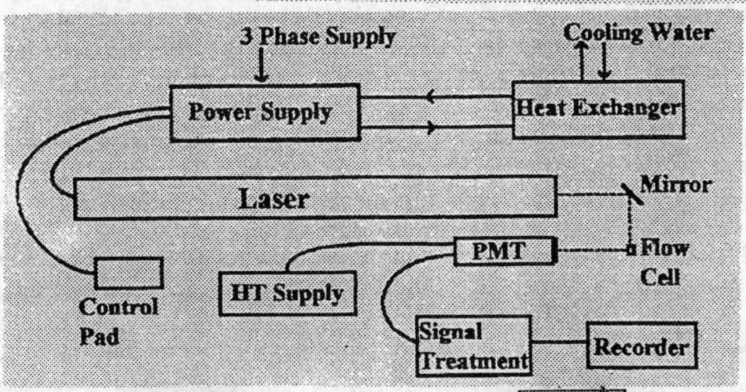

Fig. 2. Basic arrangement of parts for the LIF detector.

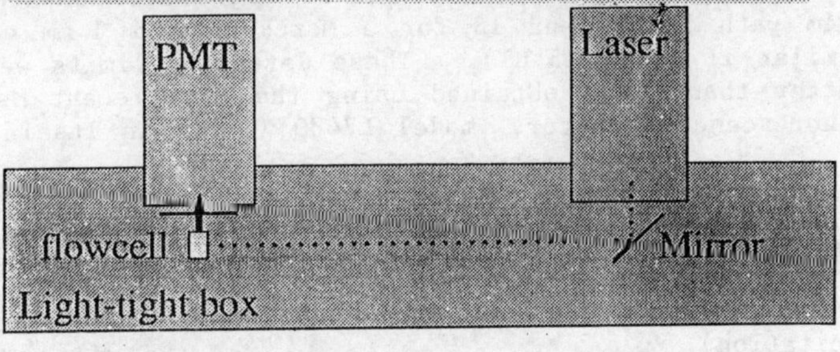

Fig. 3. Detector configuration adopted.

For the LIF detector (Fig. 2) we investigated a number of different flow cells and configurations [cf. 2 & 6] including fibre-optic collection of the emission as well as a commercial flow cell (Merck-Hitachi) with 12 or 40 µl volume. Fig. 3 shows the final configuration for the system. For the flow cell, two different configurations were found to be suitable (Fig. 4), one with fibre optics.

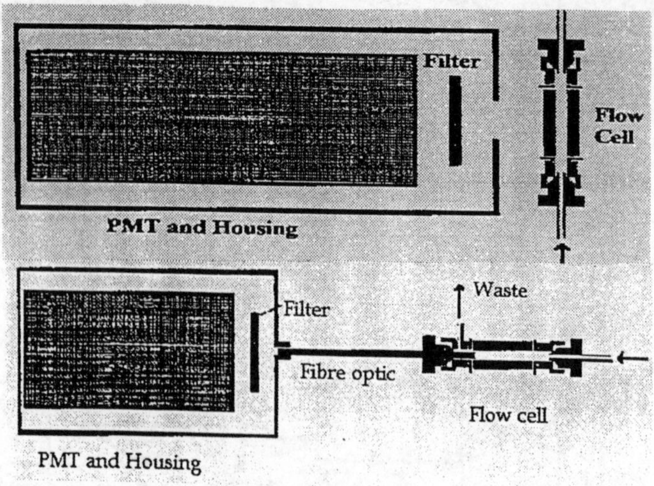

Fig. 4. Two effective flow-cell configurations.

HPLC-LIF trials with development compounds

Using an indole with an excitation laser wavelength of 229 nm at 12 mW output power, we could determine at little as 13 pg on-column. A high-pass cut-off filter (>300 nm) was used in front of the PMT to remove stray laser excitation light. In this configuration, we used a Marconi photomultiplier tube and power supply and associated amplification electronics. We did not find that a lock-in amplifier with a chopper improved performance.

HPLC traces obtained in this configuration, without (a) and with (b) a 'Beam-Boost' photochemical reaction unit in the eluent stream post-column, are shown in Fig. 5 for one indole and Fig. 6 for another indole. Flow-cell options include a 0.25 mm capillary placed in front of the PMT (optical fibre feasible), and cells for which indole detection limits (pg) were compared: 25 with 1 cm path length and 13 for a Merck-Hitachi 1 mm cell (40 µl; similar if 12 or 15 µl). These detection limits were slightly better than those obtained using the most recent Merck-Hitachi fluorescence detector, model L7480 (which in itself is 10-fold

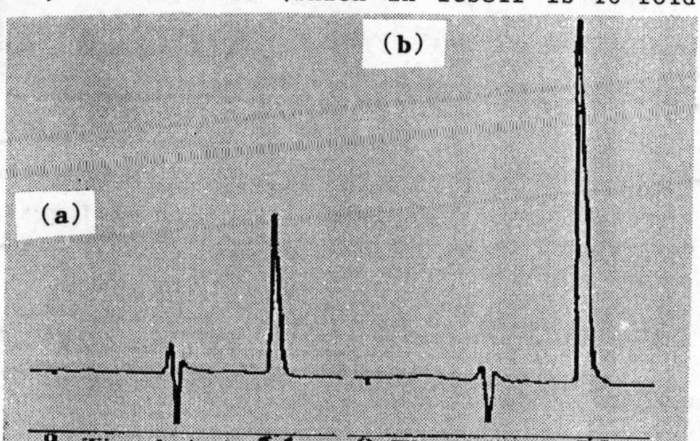

Fig. 5. HPLC-LIF on 60 ng of an indole (229 nm excitation). Configuration generally as in Fig. 6 legend. 'Beam-Boost' (a) OFF, (b) ON.

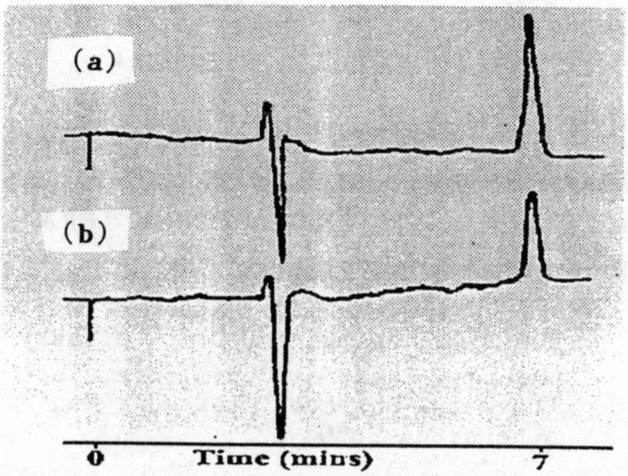

Fig. 6. As for Fig. 5 but 70 pg of a different indole.
Flow cell: 12 µl Merck-Hitachi; fibre-optic light collection.
Column: Hichrom Hi RPB, 150 × 3.2 mm; mobile phase: 35% acetonitrile with 0.1% TFA.

more sensitive for these indoles than the previous model). The performance of the LIF detector was limited by low and medium frequency noise from the continuous-wave laser. In particular, drift was a problem. Future work will centre on the use of a photodiode detector to provide automatic drift and noise compensation, as well as the use of other noise-reduction procedures,

References

1. Kok, S.J., Kristenson, E.M., Gooijer, C., Velthorst, N.H. & Brinkman, U.A.Th. (1997) J. Chromatog. **A 771**, 331-341.
2. Lingeman, H., van de Nesse, R.J., Brinkman, U.A.Th., Gooijer, C. & Velthorst, N.H. (1990) in 'Analysis for Drugs and Metabolites, including Anti-infective Agents' [Vol. 20, this series] (Reid, E. & Wilson, I.D., eds.), Royal Society of Chemistry, Cambridge, pp. 355-363.
3. Folestad, S., Johnson, L., Josefsson, B. & Galle, B. (1982) Anal. Chem. **54**, 925-929.
4. Rahavendran, S.V. & Karnes. H.T. (1997) Anal. Chem. **69**, 3022-3027.
5. van de Nesse, R.J., Velthorst, N.H., Brinkman, U.A.Th. & Gooijer, C. (1995) J. Chromatog. **A 704**, 1-25.
6. Brinkman, U.A.Th., de Jong, G.J. & Gooijer, C. (1988) in 'Bioanalysis of Drugs and Metabolites, especially Anti-inflammatory and Cardiovascular' [Vol. 18, this series] (Reid, E., Robinson, J.D. & Wilson, I.D., eds.), Plenum Press, New York, pp. 321-338.

#C-6

ENHANCED HPLC DETERMINATION OF SELECTED DRUGS IN SERUM USING POST-COLUMN IRRADIATION AND FLUORESCENCE/ELECTROCHEMICAL DETECTION

J.T. Stewart†, M. Siluveru, W.J. Bachman, T.G. Venkateshwaran, D.C. Delinsky & B.K. Matuszewski[¤]

Dept. of Medicinal Chemistry, Coll. of Pharmacy, University of Georgia, Athens, GA 30602-2351, U.S.A.

[¤]Merck Research Labs., West Point, PA 19486, U.S.A.

Subjecting analytes to a post-column reaction by irradiation on-line offers advantages of negligible effect on peak resolution and few practical restrictions on mobile phase composition. The primary advantage is the specificity of detection. The process entails no post-column addition of reagent, nor extra band broadening, and has often served to convert hard-to-detect analytes to species that can be detected with improved selectivity and specificity. The photochemical reactor is easy to build or can be bought.*

We have used such a system for analytes in human plasma/serum. ***Clofibric acid,*** *separated by SPE, was run (<3 min) on phenyl and C-8 columns and irradiated to allow EC§ detection.* ***Sulindac*** *and its sulfide and sulfone, separated by C-2 SPE and run on a C-8 column, were measured fluorimetrically (LLOQ 50 ng/ml) after irradiation under conditions that were optimized.* ***Fenbufen*** *and metabolites, separated by C-18 SPE and run on a CN column, were similarly measured (LOD 10 ng/ml), with optimized flow-rate through the reactor.*

Ondansetron, diazepam, oestrone-3-sulfate and clonazepam *are examples of other excellent candidates for so improving detection. In some cases we have found this approach (fluor) to rival MS-MS in terms of speed, specificity and sensitivity, allowing assay of some compounds not easily ionized and detectable by LC-MS-MS. A common feature of this way to improve drug assays is a* ***cis****-unsaturated moiety that undergoes a well-known photochemical, electrocyclic, concerted cyclization reaction, producing a dihydro analog which autoxidizes in the presence of traces of O_2 in the mobile phase to give a highly fluorescent cyclic product. Methods thus based have gained favour for selected new drug candidates in clinical trials.*

Post-column modifications using HPLC are reputed techniques for enhancing the detectability of analytes [1-10], as reviewed [11-13]. Both on-line and off-line techniques have been applied

---

†addressee for any correspondence

*For other literature on post-column photoreactions consult Index in this and earlier vols. and Borner's Vol. 24 art. (p. 147).-*Ed.*

§*Abbreviations (see also Preface)*.- EC, electrochemical; fluor, fluorimetry; i.s., internal standard; LLOQ, lower limit of quantitation; LOD, limit of detection.

to analytical problem-solving. We have been especially interested in the application of on-line photochemical reactions to extend the sensitivity and specificity of a given assay on biofluids using fluorescence or amperometric detection. With this reactor approach there is negligible effect on peak resolution and virtually no restriction on mobile phase composition, mixing or flow. The primary advantage of the technique is enhanced detection. The technique allows an analyst to convert non- or poorly fluorescent or amperometrically insensitive analytes to highly fluorescent or EC-responsive species. The post-column device ('reactor') is easy to build and can be purchased commercially.

Here we discuss enhanced detection of clofibric acid, sulindac and fenbufen in plasma/serum using EC/fluor detection. Recent work in our laboratory has indicated that the detection of many other analytes can be enhanced using the on-line post-column irradiation technique.

sulindac

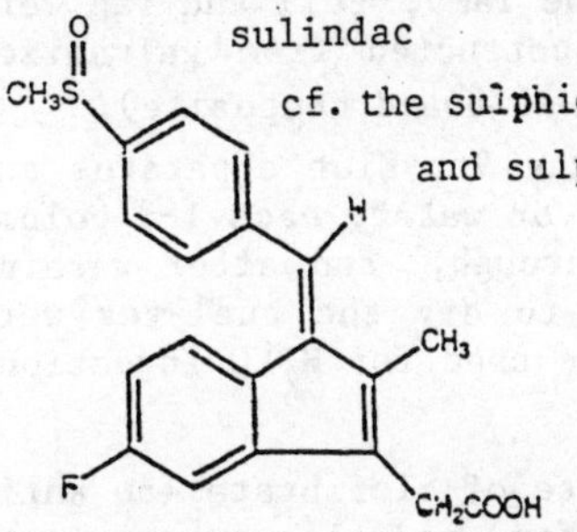

cf. the sulphide, CH_3S-

and sulphone, CH_3SO_2-

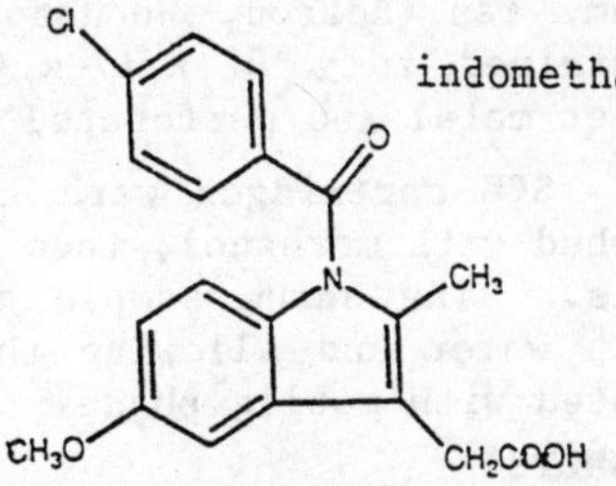

indomethacin

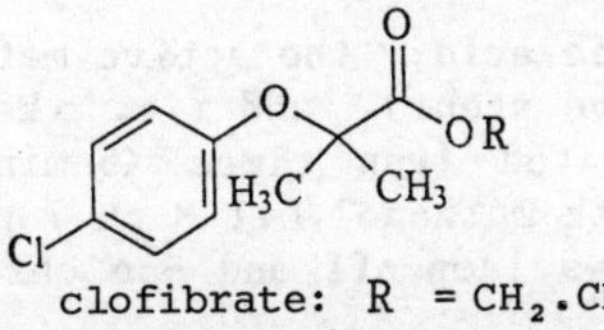

clofibrate: R = $CH_2 \cdot CH_3$

clofibric acid: R = H

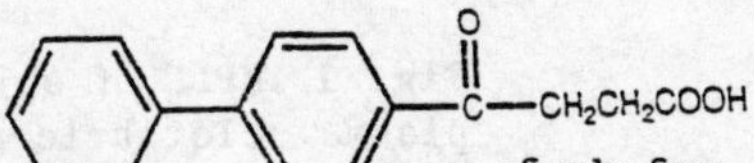

fenbufen (I)

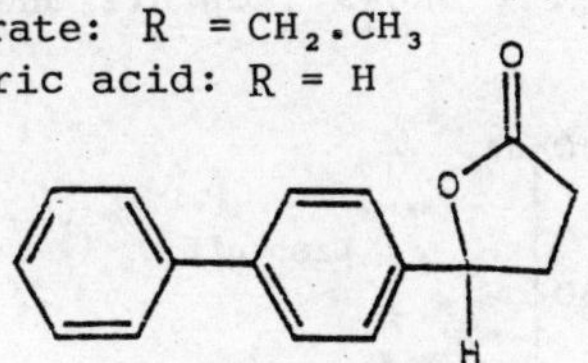

lactone of II (III)

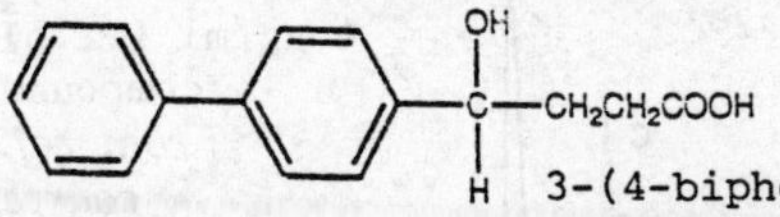

3-(4-biphenylhydroxymethyl-propionic acid (II)

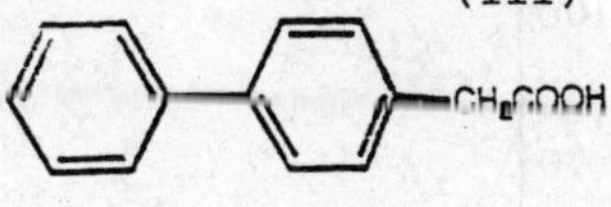

4-biphenylacetic acid (IV)

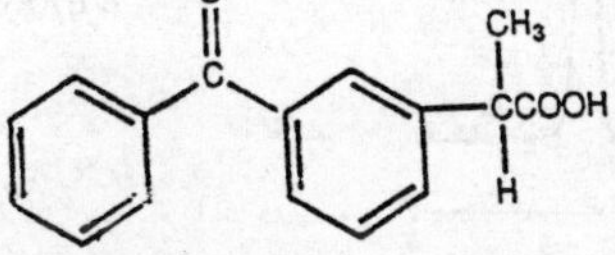

ketoprofen (V)

EXPERIMENTAL

Materials.- Some drugs were purchased from Sigma (St.Louis). Some were gifted by Lederle Labs (Pearl River, NY) or Merck Res. Labs. (Rahway, NJ). Drug-free human serum came from Biological Specialty Corp. (Colman, PA). Chemicals were top-grade (solvents HPLC grade).

Instrumentation.- A #110B HPLC pump (Beckman, Fullerton, CA) and #728 autosampler (Micromeritics Corp., Norcross, GA) equipped with a 100 µl loop were used. The columns were purchased from reputable suppliers and run at ambient temperature (22 ±1°). A #1045A programmable fluorescence detector (Hewlett Packard) was used with suitable excitation wavelengths and a 335 nm emission filter. A computing integrator was used to record each chromatogram and peak height responses. The EC detector comprised a BAS #4C-4B amperometric controller and #LC-17 thin layer transducer with a glassy carbon working electrode and Ag/AgCl reference electrode (Bioanalytical Systems, West Lafayette, IN).

The photochemical reactor employed a #SC3-9 UV lamp and #SCT-4 power supply (UVP, San Gabriel, CA). The irradiation coil was constructed from teflon tubing (Anspec, Ann Arbor, MI) 0.3 mm i.d. × 790 cm long. The coil fits directly onto the surface of the lamp in a sleeve-like manner. Air cooling was provided by a 3-in. diam. fan (Rotron, Woodstock, NY). The lamp, coil and fan were contained in a 32 × 17 × 16 cm box constructed from galvanized sheet metal and perforated masonite (a wallboard composite).

SPE cartridges were fitted onto the Vac-Elut apparatus and washed with methanol, then mobile phase or water, each 1-2 column vols. The serum sample was passed through, and after washing with water and allowing the cartridge to dry the analytes were eluted with mobile phase. Aliquots were used for HPLC injection.

RESULTS

Clofibric acid, the active metabolite of clofibrate (an anti-hypertensive agent), and i.s. (ketoprofen) and plasma components were separated (run times <5 min) on phenyl- or octyl-silane columns with methanol/0.1 M pH 7 phosphate buffer, 42:58 by vol. Fig. 1 shows lamp-off and -on chromatograms for a spiked sample.

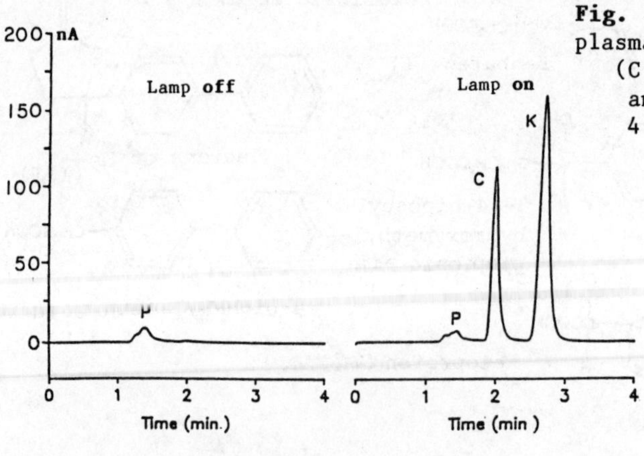

Fig. 1. HPLC of spiked plasma: clofibric acid (C; 90 ng injected) and i.s. (K), 90 & 4 µg/ml (P, plasma components). *From ref. [14], courtesy of Marcel Dekker.*

Analyte and i.s. were well resolved (R_s >3). Sample clean-up utilized a C-2 SPE procedure that was more efficient than previously reported liquid-liquid extractions, recoveries being >98% for both drug and i.s. The procedure featured a 10× dilution step and smaller injection volumes than in previous HPLC methods. Since the therapeutic range for clofibric acid is 65-100 µg/ml, recovery checks were run at 30, 90 and 180 µg/ml. The working electrode potential was based on a photo-induced hydrodynamic voltammogram obtained for clofibric acid in HPLC mobile phase. The analyte resided in the reactor for 33 sec. It was detected within the thin-layer cell at +1.1 V *vs.* Ag/AgCl. The LOD based on an S/N of 2 was 0.2 µg/ml (220 pg on-column), ~2.25 times better than for the same procedure with UV detection.

Sulindac (a widely used anti-inflammatory agent) and its two major metabolites, sulfide and sulfone, had fluorescence only if UV-irradiated, such that the post-column reactor could be of advantage [15]. The fluorescence response of all three analytes was studied at various flow rates (0.2-1.0 ml/min; so irradiation time 2.8 min to 33.4 sec in a 790 cm coil). Response was maximal at 0.3 ml/min; but excessive band broadening mandated that the method be run at 0.5 ml/min, this hardly affecting sensitivity.

Lamp-on chromatograms of blank and analyte samples are shown in Fig. 2. With use of a C-8 column (chosen for SPE also) and of 45:55 methanol/68 mM pH 2.5 phosphate, the run time was <21 min. For all 3 analytes recovery was >90% and LOD 10 ng/ml (S/N = 5). Indomethacin was selected as i.s. because it likewise produced a highly fluorescent species upon irradiation. Intra- and inter-day accuracy and precision data are listed in Table 1 (top half).

Table 1. Accuracy (% error) and precision (RSD, CV; %) for assay of sulindac, fenbufen and their metabolites in human serum. *Data from refs. [15, 16], courtesy of Elsevier Science Publishing.*

Analyte	Added, ng/ml	Found, INTRA-DAY (n = 3) ng/ml	error	CV	Found, INTER-DAY (n = 9) ng/ml	error	CV
Sulindac	100	95.5 ±2.2	-4.5	2.4	98.4 ±2.7	-1.6	2.7
	500	488.3 ±27.2	-2.4	5.6	489.6 ±1.9	-2.0	0.4
sulfone	100	94.7 ±5.1	-5.3	5.4	96.5 ±2.2	-3.5	2.3
	500	506.8 ±28.3	+1.4	5.6	510.2 ±15.2	+2.0	3.0
sulfide	100	97.2 ±4.2	-2.8	4.3	99.3 +2.5	-0.5	2.5
	500	511.9 ±16.8	+2.4	3.3	506.1 +16.4	+1.3	3.2
Fenbufen	60	64.5 ±2.6	+7.4	4.0	63.7 ±3.9	+6.1	6.5
	350	327.4 ±20.4	-6.5	6.2	332.0 ±16.1	-5.1	4.6
lactone	60	56.2 ±3.2	-6.3	5.8	57.0 ±3.8	-5.0	6.4
	350	367.1 ±20.6	+4.9	3.5	371.2 ±17.9	+6.1	5.1
4-biphenyl	60	63.6 ±3.8	+5.9	6.0	56.8 ±3.0	-5.4	5.4
acetic ac.	350	340.8 ±20.6	-2.6	6.1	329.0 ±25.9	-6.0	7.4

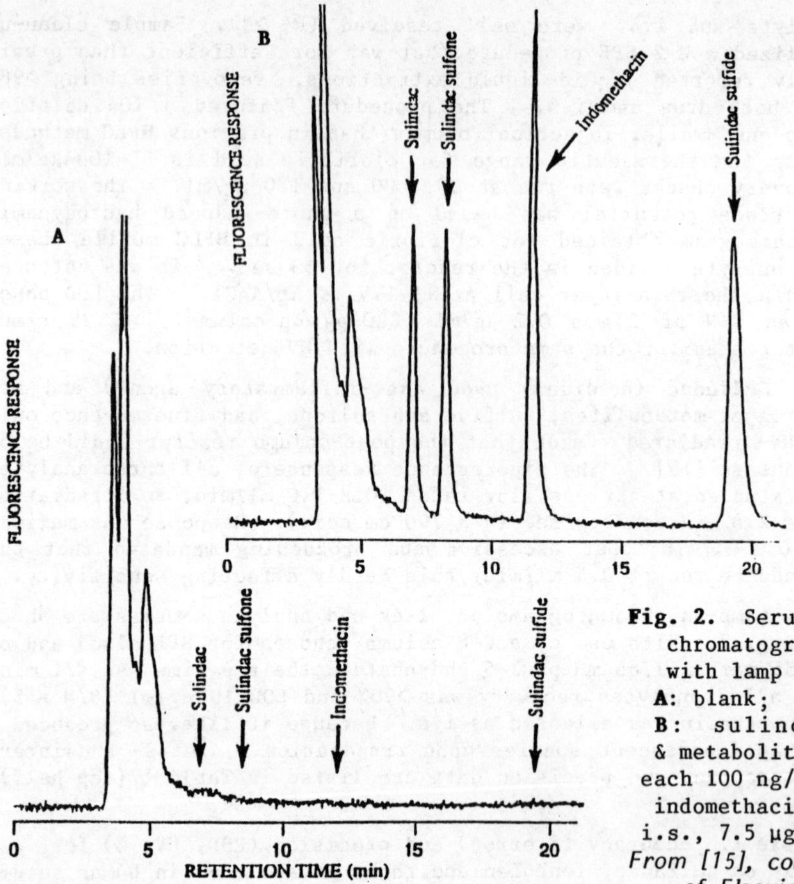

Fig. 2. Serum chromatograms with lamp ON. **A**: blank; **B**: sulindac & metabolites, each 100 ng/ml; indomethacin i.s., 7.5 µg/ml *From [15], courtesy of Elsevier.*

Fenbufen (another NSAID) has two metabolites, one of which quantitatively lactonizes *in vitro* [see above for formulae]. It has weak native fluorescence, whilst fenbufen has none; fluorescence increased after irradiation, whereas there was attenuation of the appreciable native fluorescence of the other metabolite, 4-biphenylacetic acid. For irradiation in the assay a flow rate of 0.5 ml/min (790 cm coil) was selected as a compromise for the 3 analytes, taking account of fluorescence/flow rate relationships [16]. Fig. 3 shows chromatograms for a blank (lamp on) and a spiked sample (lamp off or on). The use of a cyanopropyl column enabled the analytes to be separated in <11 min.

Sample clean-up was performed using a C-8 cartridge. The recoveries were 87 ±3% for fenbufen, 89 ± 3% for the lactone and 91.5 ±4% for 4-biphenylacetic acid. Ketoprofen was chosen as i.s. because its fluorescence was enhanced under the same analytical conditions as fenbufen and its SPE recovery was good (84 ± 3.6%).

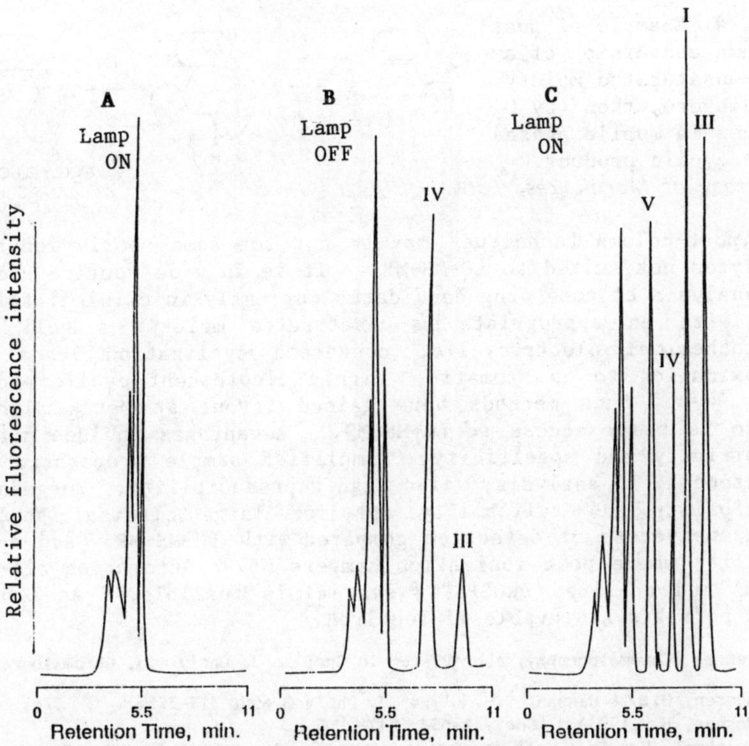

Fig. 3. HPLC of serum samples: **A,** blank; **B, C**: spiked with fenbufen (I; 220 ng/ml), lactone (III; 150 ng/ml), 4-biphenyl-acetic acid (IV; 100 ng/ml) and i.s. (V; 6 µg/ml). *From [16], courtesy of Elsevier.*

The LOD for fenbufen and metabolites was 10 ng/ml (1 ng on column). Accuracy and precision data are listed in Table 1. (For LOD, S/N = 3.)

FOLLOW-UP STUDIES

Further work in our laboratories has indicated that upon irradiation many medicinal agents of various therapeutic types will produce a highly fluorescent species that could be applied to the assay of a particular drug and maybe even its metabolites in biofluids. For instance, we have found that lamp-on compared with lamp-off increased fluorescence 2-fold for oestrone-3-sulfate, 45-fold for clonazepam and 14-fold for diazepam. We infer that assay methods much more sensitive than classical UV methods can probably be developed for clonazepam and diazepam. Other promising candidates for such enhancement include cyclophosphamide, trimethopterin, pargyline, ceftazidime, sulfamethizole, ondansetron, stanozolol and 19-nortestosterone.

Studies at Merck Research Labs. by one author (B.K.M.) have revealed the enhanced fluorescence technique to be very competitive with MS-MS in terms of speed, specificity and sensitivity.

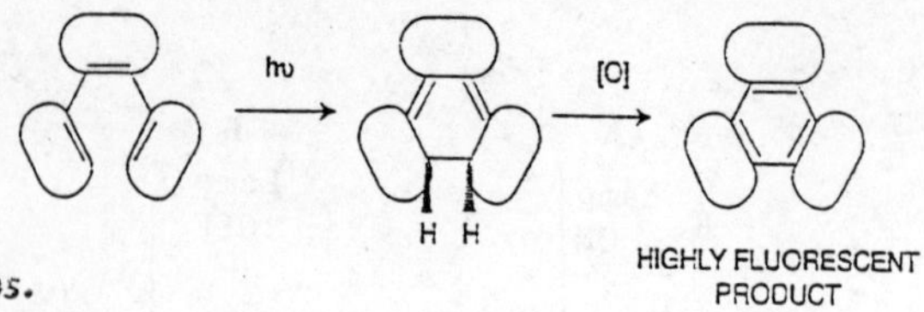

Fig. 4. Example of post-column conversion of a *cis*-unsaturated moiety to dihydro, then (by O_2 traces in mobile phase) to a cyclic product. *Courtesy of Merck Res. Labs.*

The post-column technique may be apt for some poorly ionizable analytes not suited to LC-MS-MS. It is in wide routine use for bioanalysis of some drug candidates currently in clinical trials, all with an appropriate *cis*-unsaturated moiety: a well-known photochemical electrocyclic, concerted cyclization leads with autoxidation to an aromatic, highly fluorescent cyclic product (Fig. 4). Such methods have gained favour at Merck although there is ready access to LC-MS-MS. Advantages include notable sensitivity and specificity, simplified sample preparation and shortened HPLC analysis; also high reproducibility, ruggedness, flexibility and reliability apt for large clinical studies, inexpensiveness of detection compared with LC-MS-MS, and applicability where poor ionization hampers MS. Automation off-line (SPE) and on-line (PROSPEKT® system) is feasible. An LLOQ of <100 pg/ml is achievable if required.

References [Chromatograph.. abbreviated to **Chr**; so **J. Chr'y = J. Chromatography**)]

1. Hansen, B.B. & Hansen, S.H. (1994) J. Chr'y **B 658,** 319-325.
2. Joshua, H. (1993) J. Chr'y **A 654,** 247-254.
3. Brinkman, U.A.Th., Welling, P.L.M., Devries, G., Scholten, A.H.M.T. & Frei, R.W. (1981) J. Chr'y **217,** 483-471.
4. Baustain, C.L. & Mikkelson, T.J. (1986) J. Pharm. Biomed. Anal. **4,** 237-246.
5. Lurie, I.S., Cooper, D.A. & Krull, I.S. (1993) J. Chr'y **A 629,** 143-151.
6. Miles, C.J. & Moye, H.R. (1987) Chr'ia **24,** 628-632.
7. Wolf, C.H. & Schmid, R.W. (1990) J. Liq. Chr'y **13,** 2207-2216.
8. Krull, I.S., Ding, S.D., Selavka, C. & Bratin, K. (1984) J. Forensic Sci. **29,** 449-463.
9. Selavka, C.M., Krull, I.S. & Bratin, J. (1986) J. Pharm. Biomed. Anal. **4,** 83-93.
10. Selavka, C.M., Krull, I.S. & Lurie, I.S.(1985) J. Chr'ic Sci. **23,** 499-508.
11. Krull, I.S. & LaCourse, W.R. (1985) Trends Anal. Chem. **4,** 118-124.
12. Birks, J.W. & Frei, R.W. (1982) Trends Anal. Chem. **1,** 361-367.
13. Krull, I.S. & LaCourse, W. (1986) in 'Reaction Detection in Liquid Chromatography' (Krull, I.S., ed.; Vol. 34), Marcel Dekker, New York, p. 303.
14. Bachman, W.J. & Stewart, J.T. (1989) J. Liq. Chr'y **12,** 2947-2959.
15. Siluveru, M. & Stewart, J.T. (1995) J. Chr'y **B 673,** 91-96.
16. Siluveru, M. & Stewart, J.T. (1996) J. Chr'y **B 682,** 89-94.

#C-7n

A Note on

PERSPECTIVES IN THE DEVELOPMENT OF FLUORESCENT AND ELECTROCHEMICAL DERIVATIZATION REAGENTS FOR ANALYSIS OF AMINES AND PEPTIDES IN PLASMA AND CELL CULTURE SYSTEMS

J.F. Stobaugh†, C. Ampasavate, M.J. Rose, R.G. Carlson & F. Kristjansson

Center for Bioanalytical Research, University of Kansas, Lawrence, KS 66047, U.S.A.

(1) REAGENT DEVELOPMENT ASPECTS

The trace determination of substances devoid of structural features that convey high detectability often represents a significant challenge to the bioanalyst. However, for analytes bearing a primary or secondary amine moiety, such as amino acids, certain drugs and their metabolites and numerous peptides, chemical derivatization techniques can frequently serve to transform the analytical target into a product exhibiting high detectability by fluorescence and/or EC* approaches. For non-peptide analytes bearing a primary amino group, reaction with oPA/RSH results in rapid formation of a substituted isoindole product that exhibits efficient fluorescence and is EC-active [1].

Unfortunately, it appears that any carboxy-terminus modification of an amino acid (ester or amide) furnishes an efficient non-fluorescent excited-state relaxation pathway for isoindoles derived from oPA, thus severely compromising the use of oPA in peptide analysis. A second, but less restricting, aspect of the oPA reagent is the instability of the isoindole product [1-4]. This is especially true since the introduction of auto-injectors that can perform derivatization reactions; however, this is an issue that one must be aware of when developing a method that incorporates the oPA derivatization chemistry.

In previous studies, we have examined and suggested a mechanism of isoindole formation [1], which allows one to realise the key steps leading to the transformation of a primary amine to the isoindole product. In short, this has allowed the development of the reagent system NDA/CN that has proven useful in the analysis of peptides [2-4]. The NDA/CN reagent reacts with primary amino groups to form efficiently fluorescent, stable CBI-peptide derivatives. Interestingly, this linear ring-extended (as compared to the oPA product) isoindole exhibits excitation/emission maxima (442/495 nm) that are ~100 nm longer than for the corresponding

---

†addressee for any correspondence *[Further abbreviations given later.*

**Abbreviations*.- CBI, 1-cyanobenz[f]isoindole; CN, cyanide (Na); EC, electrochemical (detection); (Im)NDA, (imidazole)naphthalene-2,3-dicarboxaldehyde; LIF, laser-induced fluorescence; MeCN, acetonitrile; oPA, *o*-phthaldialdehyde; RSH, mercaptan.

oPA-derived isoindoles. This appears to furnish efficient fluorescence with most mono-derivatized peptides (φ ~0.5); however, as with oPA, bis-derivatized analytes suffer a substantially reduced fluorescence.

After numerous studies, general guidelines can now be proposed for optimization of both rate and yield for derivatizing peptides with NDA/CN, the most important parameter being pH control. It is suggested that the reaction media be buffered to a pH equivalent to the pKa of the primary amine group targeted for derivatization. Specifically, for three types of analyte - α-amino group of a peptide N-terminus or of an amino acid, and ε-amino group of a peptide lys-residue - the reaction pH should be 7.5, 9.5 and 10.5 respectively. We have also established that peptide lys residues undergo extremely rapid derivatization (typical reaction time <2 min) as compared to N-terminus sites (~10 min), thus allowing selective labelling of a peptide lys residue and avoiding the quenching issue associated with bis-derivative formation [4].

CBI derivatives, as mentioned, have λ_{ex} 442 nm; this allows LIF detection using the He-Cd laser. Capillary separation systems (chromatography, CE, electrochromatography) that are coming into vogue call for robust LIF detection; for practical reasons (ion-tube lifetime, stability) an Ar^+-based (488 nm line) LIF system would be preferable to the He-Cd laser. With this goal in mind we modified the NDA molecule by linear fusion of an imidazole ring, producing ImNDA. When ImNDA is used to derivatize a primary amine, indeed the $\lambda_{ex,max}$ (490 nm) suits the Ar^+ 488 nm line. The peak for the ImNDA reaction product thus obtained is shown in Fig. 1.

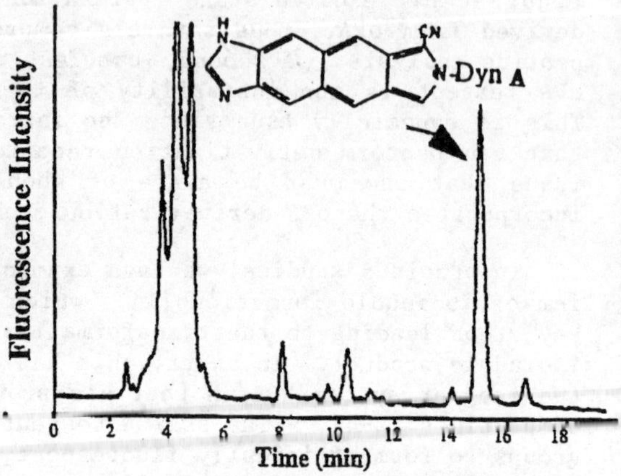

Fig. 1. Gradient-HPLC of dynorphin A_{1-11} (DynA; 30 μM initially) acquired from a BeWo cell culture transport experiment and derivatized with ImNDA/CN (product structure shown). Column: ODS-1 Spherisorb 5 μm, 250 × 4.6 mm. Mobile phase: **A**, MeCN/0.05% TFA (10:90 v/v) with 20 mM Na 1-heptanesulfonate; **B**, same but 90:10, increasing by 2%/min from 35% **B** start. Shimadzu RF 535 detector; ex, 467 nm; em, 510 nm. *For the experiment, see later in the text.*

Generation of EC responsiveness is an effective alternative to the prevailing fluorescence-derivatization approach. Recently we have begun a program to systematically develop reagents for oxidative EC detection, warranted because of the high sensitivity realizable and the ease, compared with optical methods, of scaling down in accord with the separation-system miniaturization trend. Moreover, the need to control reaction progress is minimized since with EC detection there is no signal quenching associated with bis- or higher derivatization. We first chose to evaluate THP-protected HLTE* as an acylating agent that could convey EC activity to a suitable primary or secondary amine analyte. HLTE was found to be a rather sluggish acylating agent, requiring rather stressful conditions (50°, 12-24 h). These evidently entailed notable oxidative degradation (polymerization) of the reagent, causing electrode fouling and the observation of numerous unidentified reaction products that interfered chromatographically. However, the BEA reaction product (Fig. 2) isolated in pure form was found to be optimally detectable in low fmol amounts with only +200 to +300 mV applied potential (Ag/AgCl ref. electrode).

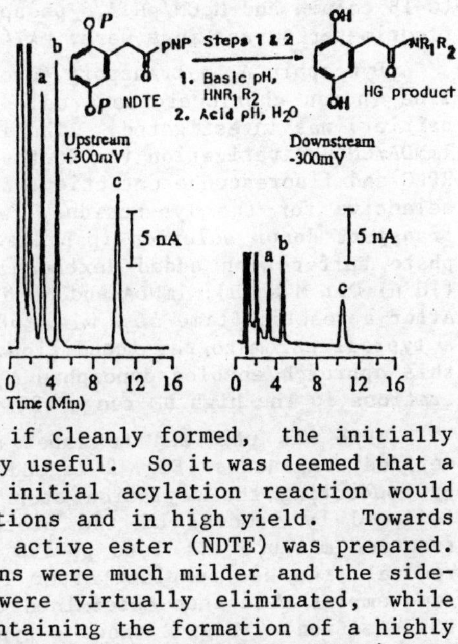

Fig. 2. RP-HPLC-EC determination of a 3-component mixture consisting of: **a**, glutamine-t-butyl ester; **b**, methionine methyl ester; **c**, isoleucine methyl ester (each ~50 µM). Hypersil ODS 5 µm column, 250 × 4.6 mm. Mobile phase: MeCN/pH 7.0 50 mM Na phosphate buffer (25:75). *For the experiment, see next page.*

Hence it was clear that, if cleanly formed, the initially selected product could be highly useful. So it was deemed that a reagent was needed where the initial acylation reaction would occur rapidly under mild conditions and in high yield. Towards this goal a bis-protected (THP) active ester (NDTE) was prepared. The required reaction conditions were much milder and the side-products previously observed were virtually eliminated, while achieving a high yield and maintaining the formation of a highly EC-detectable product (NDTE forms the same product as HLTE) that is quite stable in acidic media.

(2) APPLICATIONS

The NDA/CN reagent was successfully used to determine the undecapeptide Substance P (arg at position 1, lys at 3) in spiked plasma samples. After deproteinization (MeCN) and centrifugation, the supernatant was subjected to C-18 SPE. Due to the presence

**Further abbreviations*.- BEA, benzylethylamine; HLTE/NDTE, lactone/*p*-nitrophenol ester of 2',5'-dihydroxyphenylacetic acid (homogentisic acid), protectable by THP, tetrahydropyranyl group(s); SP, Substance P.

of both hydrophobic and polar (particularly arg) residues, mixed-mode retention occurred that permitted extensive washing of the loaded SPE unit. Elution was accomplished by a combination of high-organic/acidic media (pH 2.3 $HClO_4$/MeCN/THF, 58:22:20). The SPE eluate containing SP was then selectively derivatized (in pH 10.0 borate buffer/methanol, 20:80) with NDA/CN (equimolar amounts of the reagent components; 2 min reaction time), followed by RP-HPLC with fluorescence detection. With the spiked human plasma samples the procedure achieved quantitation of SP down to 5 nM (~200 fmol injected) with linear calibration over 3 decades.

NDA/CN has also been used to assay a peptide-nucleic acid trimer in cell sonicates after MeCN deproteinization and drying down [5]. After derivatization, essentially as above, and adding 100 nM n-butylamine, excess reagent was removed by ethyl acetate extraction, and gradient HPLC performed. Cell-derived samples were run with a PRP-X column using MeCN/pH 5.5 ammonium formate (C-18 column and MeCN/pH 2.6 phosphate for pure solutions). The fluorimeter nm settings were: ex 440, em 490 (5 ng detectable).

Dynorphin A_{1-11} transport across a monolayer of the BeWo cell line (human choriocarcinoma cell line - model for the placental barrier) was investigated. The intact peptide was determined by ImNDA/CN derivatization (product shown in Fig. 1) followed by RP-HPLC and fluorescence detection. The derivatization, which was selective for the lys-residue, was performed on 150 µl of the transport donor solution (peptide 30 µM at start; neutral phosphate buffer with added dextrose) after adjusting the pH to 10 (10 µl 0.1 M NaOH); ImNDA and NaCN (each 5 mM, 20 µl) were added. After a reaction time of 2 min, 50 µl was analyzed. Fig 1 shows a typical chromatogram (conditions stated in legend). Overall, this approach enabled dynorphin A_{1-11} to be determined at concentrations in the high nM range (30 pmol injected).

NDTE was utilized to label a mixture of three amino acid-derived components (Fig. 2 legend). The reaction was performed by conducting the acylation step for 30 min (room temp., pH 9) followed by removal of the protecting groups by diluting the reaction mixture 1:1 into pH 1.2 phosphoric acid. This group-removal step was complete after 1 h reaction time (room temp.), and samples were then determined by RP-HPLC using dual-electrode EC detection. Fig. 2 shows a typical chromatogram of the three components and the reaction scheme.

References

1. Sternson, L.A., Stobaugh, J.F. & Repta, A.J. (1985) Anal. Biochem. **144**, 233-246.
2. de Montigny, P., Stobaugh, J.F., Givens, R.S., Carlson, R.G., Srinvasachar, K., Sternson, L.A. & Higuchi, T. (1987) Anal. Chem. **69**, 1096-1101.
3. de Montigny, P., Riley, C.M., Sternson, L.A. & Stobaugh, J.F. (1990) J. Pharm. Biomed. Anal. **8**, 419-429.
4. Kristjansson, F., Thakur, A. & Stobaugh, J.F. (1992) Anal. Chim. Acta **262**, 209-215.
5. Nerurkar, M.M., Rose, M.J., Stobaugh, J.F. & Borchardt, R.T. (1997) J. Pharm. Biomed. Anal. **15**, 945-950.

#C-8

RAPID DEVELOPMENT OF REAGENT MONOCLONAL ANTIBODIES TO SUPPORT DRUG DISCOVERY AND DEVELOPMENT

[1]S.A. Wring†, [2]K.E. Kilpatrick, [3]I. Waterhouse, [3]R.M. Carr, [4]R.M. Hochel, [5]W.N. Jenner & [1]C. Serabjit-Singh

Glaxo Wellcome Research Inc., 5 Moore Drive, Research Triangle Park, NC 27709, U.S.A.: [1]Bioanalysis & Drug Metabolism Divn. and [2]Research Divn. Glaxo Wellcome Research & Development, U.K.: [3]Chemical Development Divn., Gunnels Wood Road, Stevenage, Herts. SG1 2NY; [5]Bioanalysis & Drug Metabolism Divn., Ware, Herts. SG12 0DP [4]Guilford College, 5800 W. Friendly Avenue, Greensboro, NC 27410

A novel and rapid technique is described for the production of reagent mAb's in ~30 days; this contrasts with conventional production techniques that typically require 3-9 months. Methods and data are presented from programs to produce Ab's to two drug haptens. We consider that the rapidity of this production technique will have a marked impact on increasing the value of reagent mAb's during drug research and development programs.*

Reagent Ab's* are employed widely in drug discovery and development. During early research they frequently have a pivotal role in target, receptor or gene-product localization. In later drug development they are needed for high-throughput IA's employed in support of clinical studies. Surrogate marker analysis is becoming increasingly important, and IA's are frequently the only analytical technique that is practical for measuring complex biological macromolecules. [Cf. P. End's art., #E-5n.] A further new and exciting area of application for reagent Ab's arises from DNA immunization techniques, wherein animals are immunized with genes that when expressed may give rise, to, e.g., human proteins in the immunized species. The recombinant protein can then elicit production of Ab's to human-specific epitopes.

Unfortunately, the applicability of reagent polyclonal Ab's and mAb's in drug development is often limited because of lengthy production times. To address this, we have developed a novel technique involving a 7-13 day, dual adjuvant regimen of repetitive immunizations at multiple sites proximal to draining lymph nodes (RIMMS) that markedly reduces the immunization phase of mAb production from ~3-6 months to only 7-13 days. This regimen is

†correspondent; e-mail: saw22503@glaxowellcome.com

**Abbreviations (see also Preface).*- Ab, antibody; mAb, monoclonal Ab. ALP, alkaline phosphatase; DMSO, dimethyl sulfoxide; (E)IA, (enzyme) immunoassay (e.g. ELISA); NSB, non-specific binding; RIMMS, Repetitive Immunizations at Multiple Sites; TEA, triethylamine. Other abbreviations (e.g. reagents) appear later in text.

complemented by a fusion strategy involving somatic fusion of lymphocytes to a murine myeloma cell line expressing the bcl-2 gene product. This product results in more robust hybridomas and usefully rescues weak hybridomas that would otherwise be lost by cell death. These procedures allow us to perform immunizations, fusions, screening and cloning within ~30 days. The corresponding time-frame for developing mAb's by conventional techniques is typically 3-9 months. We have also discovered that by fusing early before immunodominancy is established we are able to broaden the repertoire of Ab's obtained.

Drug discovery programs generally necessitate the defined and targeted specificity of mAb's; however it is not standard practice to use mAb's for drug IA's because their generally low affinity can limit assay sensitivity compared to typically higher avidity polyclonal antisera. Yet mAb's are desirable because specificity can be selected during the screening stages of production, and unlimited, consistent quantities can be produced. Moreover, mAb production may not require as rigorous purification of immunogens as would be needed for producing antisera.

Herein we describe the application of our RIMMS technique for the production of mAb's to two drug haptens (ranitidine and GI198745) and summarize applications for biological targets employed during drug therapy. Our data will demonstrate that RIMMS can have a marked impact on drug discovery and development programs by allowing mAb's to be available within ~30 days, which represents a 3- to 9-fold reduction in the time required with conventional production techniques. Earlier accounts [1, 2] give the immunological background to RIMMS and more detailed description of the production of mAb's to ranitidine, including immunizations, fusion strategy, cell culture media and conditions.

EXPERIMENTAL

Materials were from the following sources.- Standard laboratory chemicals: Aldrich or J.T. Baker. Solvents: EM Science. Drug standards and haptens: Glaxo Wellcome R&D. Carrier proteins, ALP substrate, dialysis cassettes and biotinylated materials: Sigma or Pearce. Adjuvant: RIBI Immun. Chem. Res. Inc. or (Freund's) GibcoBRL. PEG 1450: ATCC. ALP-labelled γ-chain specific secondary Ab: Southern Biotech. Associates; isotype and subclass-specific Ab's: FisherBiotech. EIA/RIA, 96- and 24-well plates and all cell culture labware: Costar. Culture media and ELISA reagents: see [1].

Synthesis of the drug-protein conjugates could not be done with ranitidine or GI198745 directly because neither possesses suitable functional groups for coupling to carrier proteins. So for ranitidine an analogue was synthesized where an S-methyl group replaced ranitidine's methylamine group. Conjugation to Bovine Thyroglobulin (BTG) or Keyhole Limpet Haemocyanin (KLH) proteins was performed according to published methods [3].

For GI198745 a hapten was synthesized with a *p*-benzoic acid moiety on the N at the 4 position of the steroid ring A, (ii) in Fig. 1 which depicts coupling to carrier proteins using a variant of the Mukaiyama reaction [4, 5]. In summary, 3.2 mg of 2-chloro-1-methylpyridinium iodide (CMPI), 7.5 mg of hapten and 6.4 µl of TEA were added to 0.5 ml of fresh, anhydrous dimethylformamide (DMF). This mixture was stirred (~2 h, ambient temp.) and then added dropwise to 100 mg BTG dissolved in 4 ml of 1:1 (v/v) DMF/0.1 M phosphate buffer pH 8, immediately giving a pale yellow precipitate. After a further ~2 h the entire reaction mixture was dialyzed against several changes of 60% (v/v) DMF (each ~0.6 L) and of 0.1 M Na_2CO_3 (each ~4 L). A conjugation with 50 mg ovalbumin was performed similarly using 7.2 mg hapten, 3.5 mg CMPI and 6.5 µl TEA. The dialyzed mixture was centrifuged and the supernatant further purified by GPC with superfine Sephadex G25 using 0.1 M Na_2CO_3 (flow 2 ml/min; 45 × 1.5 cm column). Successful conjugation was confirmed by UV spectroscopy. The presence of protein was confirmed by a positive Bradford test.

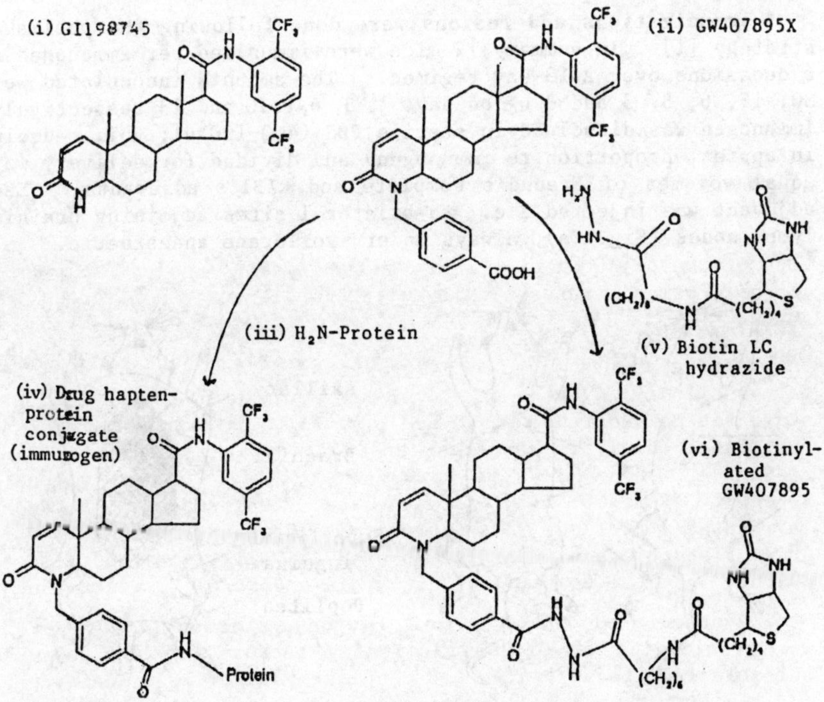

Fig. 1. Synthetic routes for the GI198745-protein and -biotin conjugates.

Synthesis of the biotinylated GI198745 hapten (Fig. 1) was done by reacting 1.1 equivalents of CMPI with 4 of TEA in 0.5 ml anhydrous DMF (~60 min, ambient temp.), then adding 3 equivalents of biotin-LC-hydrazide and 0.5 ml DMSO and stirring overnight. Solvent was removed under vacuum, residual solvent being removed by azeotroping with methanol. The residue dissolved in methanol (1 ml) was streaked onto a 20 × 20 cm Uniplate (500 µm thickness; Anatech) for separation with 20% v/v methanol in dichloromethane (DCM), letting the solvent run the entire plate length. Visualization was done by UV light: R_T's GW407895X 0.85, biotinylated product 0.63; nearer the origin, other unidentified components.

The biotinylated product band was scraped from the plate into a funnel with a sintered glass filter and eluted with 3 × 5 ml of 20% methanol in DCM. The filtrate was dried down on a rotary evaporator, then dissolved in deuterated DMSO for proton-NMR studies. The compound was then recovered by evaporating the DMSO and azeotroping with methanol. The product was then dissolved in ~5 ml methanol and stored in the dark at ~4°. NMR and negative-ion MS confirmed the identity of biotinylated GW407895X.

Immunizations and fusions were done following the published strategy [1]. In summary, 2 mice were immunized per immunogen on 6 occasions over a 13-day regimen. The amounts innoculated were 30, 15, 6, 3, 3 and 3 µg on days 1, 3, 6, 8, 10 and 13 respectively. Immunogen was dissolved in sterile PBS (400-100 µl; vol. reducing in approx. proportion to immunogen) and divided for delivery with equal volumes of Freund's complete and RIBI's adjuvants. Each adjuvant was injected s.c. to bilateral sites adjoining draining lymph nodes (Fig. 2a), always under isoflurane anaesthetic.

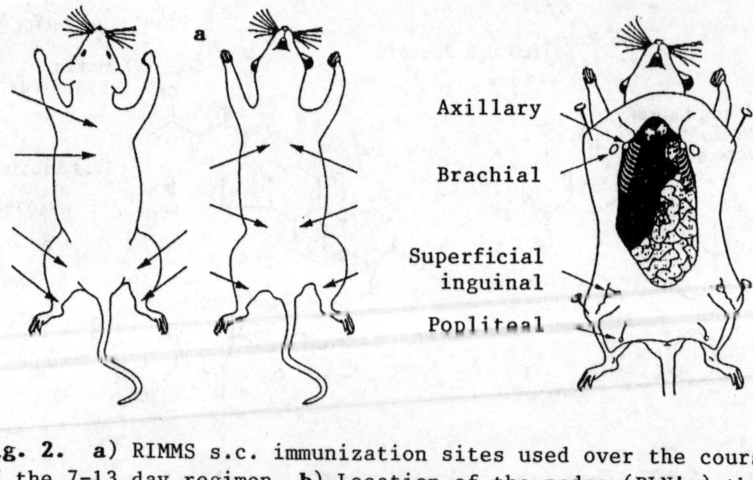

Fig. 2. **a)** RIMMS s.c. immunization sites used over the course of the 7-13 day regimen. **b)** Location of the nodes (PLN's) that are removed aseptically prior to producing a suspension of pooled PLN cells for somatic fusion with P3XBcl-2-13 cells, typically within 8-14 days following the primary immunization.

On day 14 popliteal, superficial and deep inguinal axillary and brachial lymph nodes were harvested from each animal (Fig. 2b), pooled and single-cell suspensions prepared. These peripheral lymph node (PLN) lymphocytes were fused with P3XBc1-2-13 cells at a ratio of ~2.5:1, as described [1]. The fusion suspension was distributed as 1 ml aliquots into 24-well plates. Hybridomas were 'fed' and underwent hypoxanthine, aminopterin and thymidine (HAT) selection prior to screening for Ab production ~5-7 days after fusions.

Enzyme-linked immunosorbent assay (ELISA).- A 'screening' ELISA was used to identify hybridoma populations producing Ab's to either ranitidine or GI198745. The general method involved coating a 50 ng-1 μg amount of drug-protein conjugate (KLH carrier protein for ranitidine or ovalbumin for GI198745) in each well of a 96-well plate. The plates were then blocked with goat or horse serum to prevent Ab NSB. Plates were washed, and re-washed after addition of hybridoma supernatants from the growth-positive wells. Affinity-purified ALP-labelled goat anti-mouse IgG γ-chain-specific secondary Ab was added at the manufacturer's recommended titre. Incubations for each of the above steps were performed at 37° for ~60 min or else at 4° overnight. Wells were washed and the colour development performed using Sigma 105 ALP substrate following the guide-sheet. Absorbances for each well were determined at 405 nm.

In a further method developed for GI198745 neutravidin-coated plates were used to capture the biotinylated hapten rather than a drug-protein conjugate.

The ELISA's were also used for competitive assays, except that drug standard was added to the wells before the Ab addition. Isotyping was performed using ALP-labelled isotype and subclass-specific Ab's rather than, as above, the IgG γ-specific reagent.

RESULTS AND DISCUSSION

The RIMMS technique provides to bioanalytical and research scientists an oportunity to produce and isolate affinity-matured class-switched (IgG) mAb-producing cell lines within some 30 days of the primary immunization. The stages of mAb production are: immunization and fusion, completed within 9-14 days; screening, typically during days 17-22; and cloning and mAb specificity evaluation, generally completed during days 17-30. Conventional techniques would entail 3-9 months overall. The time savings with RIMMS come mainly from the shortened immunization schedule. Besides time-savings we consider that performing fusions early with PLN-derived lymphocytes and use of P3XBc1-2-13 cells yields more robust hybridomas producing a broader repertoire of Ab's.

ELISA for GI198745

Conventional ELISA proved unsuitable as a screen for Ab's to GI198745 because in competitive assays the drug either adsorbed onto the wells of the EIA plates or bound to the coated protein.

This caused unacceptably high NSB and apparently poor displacement. To overcome this we developed an assay that immobilized a biotinylated hapten, rather than a hapten-protein conjugate, onto the 96-well plates. This immobilization was accomplished using neutravidin-coated 96-well plates (Pierce); the remainder of the method was unchanged. This reduced NSB, but unfortunately not to a level acceptable for reliable quantitative applications, whilst suitable for screening purposes. We are currently investigating other assay formats, including DELFIA and SPA, with a view to further reducing the NSB. (For DELFIA see [1] in art. #E-5n.-*Ed.*)

Polyclonal titres and screening for mAb's to either drug

All mice immunized with either ranitidine or GI198745 immunogens responded by producing anti-drug antisera, with the highest titres observed in mice immunized with conjugates prepared with BTG. Conjugates with either BSA or ovalbumin elicited lower titres. Antisera collected at euthanasia were assayed as positive controls in the ELISA's used to screen hybridoma supernatants for mAb production.

Cell culture media from each of the wells containing the post-fusion cells were screened for IgG Ab's to the drug haptens; ~8-10% of these supernatants afforded positive results. For Ab's to GI198745 the absorbance data gained in 2 assays (**BOLD** where hybridoma selected for cloning; plate refs. cited) were:-

1TB6: 0.510, 0.554; **1TD2: 3.345, 3.794**; 3TB4: 0.199, 0.487;
3TD3: 0.193, 0.294; 4TD5: 0.311, 0.551; 4TD6: 0.145, 0.478;
5TD3: 0.249, 0.621; 5TD6: 0.473, 0.514; 6TB1: 0.105, 0.538;
6TD6: 0.875, 0.929; 7TC4: 1.446, 1.430; 8TD5: 0.289, 0.560;
8TD6: 0.180, 0.480; 9TA1: 0.706, 0.847; **9TA2: 0.129, 1.053**;
9TA6: 0.850, 3.470; **9TD3: 1.844, 3.932**; 9TD6: 0.765, 1.284;
10TA4: 0.433, 0.477; 10TA5, 0.411, 0.523; 10TB2: 0.088, 0.478;
10TB3: 0.119, 0.714

The second assay was performed 3 days after the first, and the generally higher absorbance values reflect higher Ab production which was expected as cell numbers increased in the 24-well plates. These supernatants, and those containing ranitidine Ab's (obtained in a separate production program), were further tested by single concentration competitive ELISA to confirm the antigen relevance of the Ab's (Fig. 3). As expected, the ELISA absorbance decreased in the presence of drug, because Ab binding to the hapten coated onto the wells was competitively inhibited. The data in Figs. 3a and 3b indicate a broad range of displacement values within the competitive assay. These differences probably arise from Ab's having diverse affinities for the drug haptens. Some Ab's may also recognize the 'bridge' region in their corresponding immobilized conjugate, which would result in Ab's having higher affinity for the conjugate than for the unconjugated drug. Alternatively, the range of displacement values may relate to marked differences in the concentration of Ab present. However,

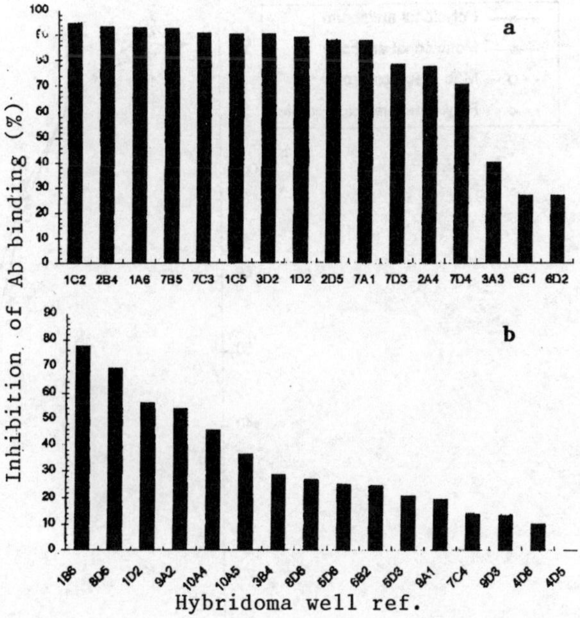

Fig. 3. Histogram showing the inhibition of ranitidine or GI198745 binding in the presence of (**a**) 10 µg of ranitidine or (**b**) 200 ng of GI198745 per well.

the latter was investigated for ranitidine; there did not appear to be any reasonable correlation between Ab concentration in the supernatants and displacement.

Cloning

The most promising hybridomas, i.e. those affording the most displacement, were selected for cloning, by limiting dilution - which, for the hybridomas secreting the ranitidine mAb's, gave 45% viability - or by flow cytometry, which gave 57% viability and was preferable because it is markedly less labour-intensive.

Calibration

Fig. 4 presents calibration graphs and precision profiles comparing the performance of the most promising ranitidine mAb with that of the polyclonal antisera used for the established RIA. Clearly the two reagents, by competitive ELISA, afford superimposable calibration graphs and precision profiles. This suggests that the affinity constant of the mAb is similar to the 1×10^9 L/mol value reported for the antiserum, and confirms that the mAb is suitable for quantitative IA applications. We are currently developing a quantitative assay for GI198745.

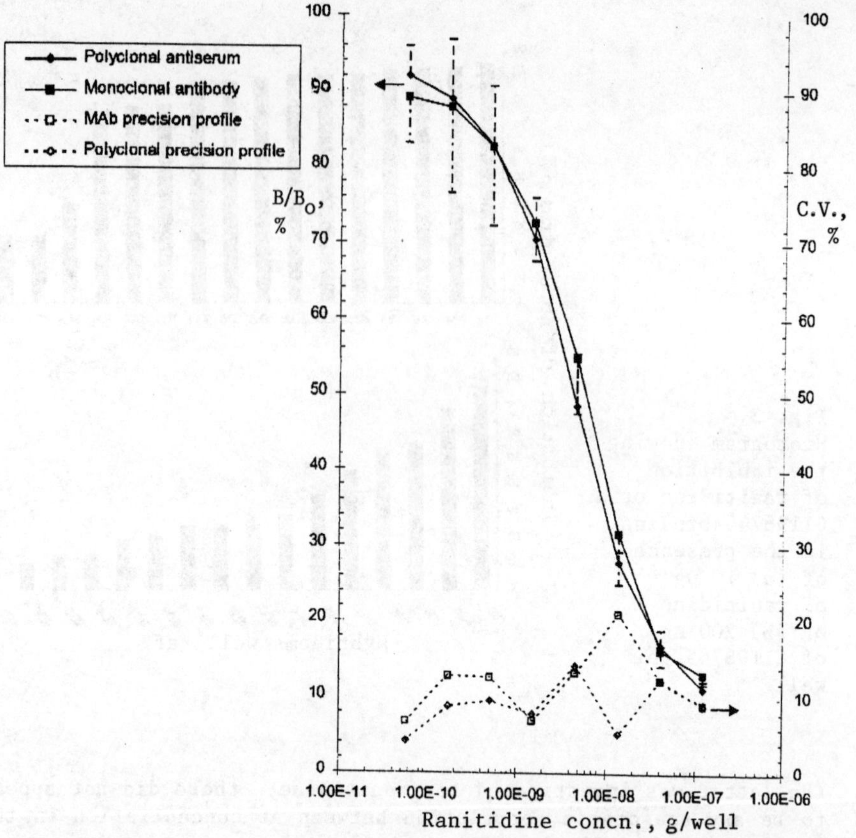

Fig. 4. Mean +S.D. calibration graphs and precision profiles (n = 8) for the optimized ELISA of ranitidine using the RIMMS murine mAb (2B4D4) and the reference sheep anti-ranitidine serum. Ab concentrations were matched to give similar absorbance readings for the zero standard (B_0) wells. *This Fig. is expected to appear also in an art. (Wring et al.) in J. Pharm. Biomed. Anal.*

Applications of RIMMS to support drug research

The RIMMS technique has been successfully employed to support several drug research programs and antigen targets. Antigen types (with no. of successful RIMMS applications, and target example) include: proteins in cell extracts (1; tyrosine phosphorylated proteins); recombinant proteins (12; RZRα, recombinant); peptide haptens (4; ROBO-1, 19 residue); *in vivo* expressed protein using DNA immunization; and drug haptens (2; ranitidine, GI198745).

For example, in a recent valuable application of RIMMS one of us was able to identify novel tyrosine-phosphorylated adhesion proteins that may play fundamental regulatory roles in human epithelial cell transformation and cancer. This was achieved by generating mAb's to an array of tyrosine-phosphorylated cellular proteins, associated with ras-transformed human breast epithelial cells. The Ab's were characterized using a novel, rapid immunocytochemical staining technique, and facilitated the identification and localization of numerous tyrosine-phosphorylated adhesion proteins in breast cancer cells [6].

CONCLUSIONS

The RIMMS technique has proved successful for the production of mAb's to a broad range of antigens, including: drug haptens, peptide haptens, recombinant proteins, cell extracts, and also for DNA-based immunizations. The development program can generally be achieved in ~30 days. which represents a marked improvement over the 3-9 months typically required using conventional mAb production techniques.

RIMMS can, we believe, have a marked impact on drug discovery and development programs by increasing both the speed of production, and the availability, of high-quality mAb's suitable for applications in research and bioanalysis.

Acknowledgements

The authors thank Mr S. Witherspoon for performing the cloning by flow cytometry.

References

1. Wring, S.A., Kilpatrick, K.E., Witherspoon, S.M., Jenner, W.N. & Serabjit-Singh, C.J. (1998) submitted to J. Pharm. Biomed. Anal.
2. Kilpatrick, K.E., Wring, S.A., Walker, D.H., Macklin, M.D., Payne, A., Su, J-L., Champion. B.R., Caterson, B. & McIntyre, G.D. (1997) Hybridoma **16,** 381-389.
3. Jenner, W.N., Martin, L.E., Willoughby, B.A. & Fellows, I. (1981) Life Sci. **28,** 1323-1329.
4. Mukaiyama, T. (1979) Angew. Chem. Lett. **18,** 707.
5. Wring, S.A., Williams, J.L., O'Neill, R.M., Ayres, D.W., Jenner, W.N., Bowers, G.D., Fellows, I., Carr, R.M., Wadsworth, A.H. & Sutherland, D.R. (1996) Anal. Chim. Acta **334,** 225-237.
6. Kinck, M.S., Kilpatrick, K.E. & Zhong, C. (1998) Hybridoma, in press.

Comments on #C-2, M.J. Hemsley: SENSITIVE/SPECIFIC CHIRAL ASSAY
#C-4, S.A. Wring: ELECTROCHEMICAL SENSORS
#C-5n, R.F. Venn: HPLC-LIF USING DEEP UV
#C-6, J.T. Stewart: POST-COLUMN PHOTO-REACTIONS
#C-7n, J.F. Stobaugh: PRE-COLUMN DERIVATIZATION
#C-8, S.A. Wring: RAPID mAb GENERATION

Chiral assay.- K-H. Lehr remarked that a good assay should need only one column step, rather than a determination of total enantiomers and then, by chiral HPLC, a ratio determination.

Lactate **sensor specificity (W. Herron** queried) is high, the dehydrogenase being inert towards glutamate, malate, etc.

Venn, replying to Stobaugh: fibre-optic delivery of the laser light isn't feasible at 229 nm, and so wasn't attempted.

Reply by **Stewart to K. Borner**: for photo-reactors there is no readily available light source for wavelengths besides 254 nm. Reply to **P. Timmerman**, who asked about the selectivity of the photo-reaction approach (compared with UV detection) in the context of Phase III clinical trial samples.- If co-administered drugs may be present in such samples, method specificity should be checked at the start. The post-column irradiation approach shows more specificity towards parent drugs than metabolites.

Replies by **Stobaugh.- To J.D.H. Cooper**, who suggested that peptide oPA derivatives might fluoresce better if excited at 230 nm.- Perhaps! but there might be matrix-related noise. **To M. Bayliss.-** In the oPA approach, Cys derivatives can serve in place of unpleasant thiols, and the toxicity of cyanide is not a problem with automated systems. **To R. Whelpton** (who mentioned that Susan Lunte had separated substance P fragments by CE with a neutral run buffer).- Post-column or CE use of oPA or NDA, with a mercaptan, is feasible in principle; but with (e.g.) substance P the α-amino labelling reaction might be rather slow for CE use.

Wring, answering W. Jenner.- Fully matured Ig Ab isotypes are obtainable even in the 8-12 day immunization period; there is no tendency to IgM formation. **H-J. Skrzipczyk** remarked that use of anti-analyte mAb doesn't preclude metabolite cross-reactivity.

Citations contributed by Senior Editor
relevant to Sect. #C, including derivatization and identification

Protein-binding (drug) assessment

Polyacrylonitrile dialysis fibres gave binding data (vacuum ultrafiltration) for various drugs incubated with plasma, as an alternative to centrifugal ultrafiltration.- Lam, H. et al. (1996) J Pharm Biomed Anal **14**, 1753-1757.

Binding to immobilized BSA was established for some indolocarbazoles with data from 2 columns, BSA (iPrOH/phos., pH 7) and C-18.- Ashton, D.S. et al. (1996) J Chr'y **B 677**, 194-198.

Frontal analysis, distinguishing free/protein-bound drug as studied with warfarin, can be incorporated on-line in a direct-injection HPLC system; it hinges on getting, with high loads, a plateau corresponding to unbound drug besides a total-drug peak. - Shibukawa, A. & Nakagawa, T. (1996) Anal Chem **68**, 447-454.

Features of packings, including affinity (cf. #A) and protein-tolerant types

For **growth-promoting drugs** (anabolic steroids and β-agonists) in bovine urine post-hydrolysis, commercial affinity packings (generally immuno) had poor specificity, weak Ab binding or poor operating instructions. Ab-coupled polymers made in-house worked well.- Crooks, S.R.H. et al. (1997) J Chr'y **B 690**, 161-172.

Hygromycin B in water or in ion-exchange SPE eluates from swine plasma or bovine serum was separable using agarose-linked lysozyme and α-lactalbumin, with post-column fluorescent derivatization.- Campbell, N.F. et al. (1997) J Chr'y **B 692**, 367-374.

An immunosensor cartridge* for neuropeptides in biopsy sonicates served as a chromatographic column (LOD 0.4 ng/ml, by UV); cell-culture mAb isolated on Protein A columns was incubated with immobilized neuropeptide-receptor complexes.- Phillips, T.M. (1996) Biomed Chr'y **10**, 331-336.

Excerpts (by Ed.) from a Forum abstract that had 5 authors - Mayes and Lowe (as in art. #A-5), S.H. Iqbal and (at Lund Univ.) K. Haupt & K. Mosbach: **MIP's for use in sensors and assays**

"...we are investigating a generic approach using competitive binding techniques that will transduce ligand binding events into measurable signals..... [Problems arise] with chemically labelled (fluorescent or EC) compounds. If the imprint is made against the unlabelled compound, then the labelled compound may be sterically excluded from many binding sites, reducing the sensitivity, and possibly also the selectivity, of the assay. [If the converse] many additional binding interactions between polymer site and label are possible, increasing the binding affinity....the parent compound becomes an inefficient displacer of the bound label and sensitivity is again lost. [We are trying] the imprinting of bulky but functionally 'featureless' derivatives of the parent compound ....to create binding sites with extra space to accommodate the chemical label, but having no functional groups which will specifically interact with the label when it is bound. This should maximize the number of binding sites being used in the assay, while minimizing the loss of sensitivity due to label interactions in the binding sites...." [Cf. ref. on p. 67, Kriz et al.]

Internal surface phenyl-boronate pre-column, for β-blockers (e.g. pindolol): with plasma/urine filtrates, effective isolation was shown.- Ohta, T. et al. (1996) J Chr'y **B 675**, 168-173.

*cf. Santandreu, M. et al. (1997) Anal Chem **69**, 2080-2085.

A protein-tolerant pre-column/switching stratagem, using C-8 silica coated with protein, served as well as a semi-permeable surface pre-column (with switching) for HPLC assay of bupivacaine in serum.- Emara, S. et al. (1996) Biomed. Chr'y **10**, 131-134.

HPLC on 'dynamically modified silica' (eluant: phos. pH 7.4/ cetyl-Me_3N^+; 220 nm) was used to study ibuprofen metabolites, including glucuronides, in human urine.- Kepp, D.R. et al. (1997) J Chr'y **B 696**, 235-241 [cf. I.D. Wilson's art., #C-1].

Pore features of C-8 silicas made from 6 commercial silicas were examined by (e.g.) N_2 ad-/de-sorption.- Bereznitski, Y. et al. (1996) J Liq Chr'y **19**, 2767-2784.[Cf. Vol. 24, this series: C.F. Poole, p. 194.] **C-18 silica stability** at alkaline pH was enhanced if a sol-gel silica support was used and if the buffers were Li^+- instead of Na^+-based; silica solubility rose with temperature.- Kirkland, J.J. (1996) J Chr'ic Sci **34**, 309-313.

Choosing HPLC packing and conditions*

A C-8 packing having 2 µm pores (TSK microspherical silica gel), tested for cyclic antidepressants in biosamples including brain and liver homogenates, gave better sensitivity (×5) than a C-8 column (Hypersil MOS) with 5 µm pores.- Tanaka, E. et al. (1997) J Chr'y **B 692**, 405-412; cf. **688**, 155-160 (carbamazepine).

An acrylate-based **co-polymer with odd temperature dependence** was tested up to 60° with water or aq. methanol as mobile phase. Retention of alcohols or PEG rose with temperature, contrary to behaviour on alkylsilica packings.- Kamiyama, F. et al. (1996) Biomed Chr'y **10**, 105-110. Steroid hormones were well resolved at 50°, but not at 5° where hydrophilicity prevailed over hydrophobicity, with a **modified aminopropylsilica** having polyisopropylacrylamide residues attached (water as eluant).- Kanazawa, H. et al. (1996) Anal Chem **68**, 100-105.

Setting eluant temperature below column temperature (dual thermostatting) was shown, for C-8, C-18 and $-NH_2$, to improve resolution (e.g. anthracene, phenanthrene) and allow faster flow, with some dependence on apparatus features.- Welsch, T. et al. (1996) J Chr'y **A 728**, 299-306.

Optimization of peptide/protein separation, including trial of raised-temp. HPLC, has been studied with various 'StableBond' packings (sterically protected; comparison with ordinary polymeric bonded-phase silica packings); their stability was good.- Richer, R.D. et al. (1995) J Pharm Biomed Anal **14**, 93-105.

Review: solvent selection.- Barwick, V.J. (1997) TRAC **16**, 293-309 [cf. Swart, P. et al., 332-342: open-tubular HPLC].

Narrow-bore systems and CE refinements

Miniaturized column systems were surveyed (1997) in J Chr'y: **A 778**, 67-75 (H. Poppe; incl. MEKC and microcapillary HPLC) and **A 779**, 1-28 (J.P.C. Vissers et al.; some bioanalytical refs.).

*Review of SFC: Erickson, B. (1997) Anal Chem **69**, 683A-686A.

Sensitivity gain by using capillary LC columns (0.32 mm), in context of automated assay of clinical samples (can be low-vol.; for PK studies), was shown with mizolastine and diltiazem using 5 µm RP packings (detection by UV - vague). Microbore (1 mm) columns were also effective. On-column focusing obviates sensitivity loss due to high injection vols. (0.1 ml).- Malavasi, B. & Ascalone, V. (1996) J High Resolu Chr'y **19**, 503-510.

Reviews on CE occupy two J. Chr'y issues (1997). On-line sample pre-concentration and micro-reaction technology features at the start of **A 744** (Tomlinson, A.J. et al.; pp. 3-15); it ends with a coated-polymer sieving approach for oligonucleotides (pp. 347-354). Surveys in **B 697** (ends at p. 254) cover peptides and other endogenous components*, drugs and metabolites, in clinical chemistry and PK contexts: pp. 3-35 (Lehmann, R. et al.; markers surveyed), 37-66 (Guzman, N.A. et al.; 'slant' as for Tomlinson, above, also detection) and 67-75 (Leveque, D. et al.; PK role).

Chiral separations (additional examples in #ABC and earlier)

REVIEWS.- Enantioselective GC, with CSP listings: Juvancz, Z. & Petersson, P. (1996) J Microcol Sep **8**, 99-114. HPLC or CE of drugs using cyclodextrins (literature on their inclusion in CE buffers is thin).- Bressolle, F. et al. (1996) J Chr'y **B 687**, 303-336.

Chiral bioanalysis with column-switching (coupled LC) has been reviewed, with various drug examples, e.g. zidovudine and metabolites.- Fried, K. & Wainer, I.W. (1997) J Chr'y **B 689**, 91-104. See **B 686**, 65-75 (1996; Ducharme, J. et al.) for a survey of HPLC for chiral drugs, covering validation and touching on biofluids; deproteinization might alter enantiomer ratios.

The 'brush-type' CSP (Sumichiral, YMC) shown to be effective for prilocaine enantiomers in serum (after SPE) acts through π-bonds, steric hindrance and dipole interactions; methanol worked better than higher alkanols as a component of the TFA-containing eluant.- Silveru, M. & Stewart, J.T. (1996) J Pharm Biomed Anal **15**, 389-392 (cf. art. #B-2).

A new CSP.- Microbore or capillary columns were used to try chiral 1,2-diaminocyclohexanes, linked to silica and then derivatized. The SFC mode (with CO_2/methanol) was also explored.- D'Acquarica, B et al. (1997) J High Resolu Chr'y **20**, 261-264.

Aprotic agents tried with a traditional CSP (Chiracel OD) as modifiers, for drug enantiomers, included THF, acetonitrile and ethyl acetate.- Kirkland, K.M. (1995) J Chr'y **A 718**, 9-26.

Urinary **ibuprofen metabolite enantiomers**, solvent-extracted, have been separated by sequential achiral-chiral HPLC (detection at 220 nm).- Tan, S.C. et al. (1997) J Chr'y **B 701**, 53-63 (cf. art. #C-1). See also 71-80 (Takahashi, H. et al.): a Chiracel OD column separated unbound **warfarin enantiomers** solvent-extracted from plasma (and 7-hydroxywarfarin from urine): cf. art. #C-2.

*See also 'endogenous' citations relating to #E at end of book.

Detection, and its achievement by analyte transformation

Fluorescent derivatization of carboxylic acids off-line can be done with a new reagent (having a piperazine moiety) which is finally removable by extraction with aqueous acid.- Wu, H-L. et al. (1997) J Chr'y **A 769**, 201-207. [For other approaches see the Index entry 'Derivatization' (acids, HPLC) in earlier vols., as far back as Vol. 7 (Horwood, 1978; ed. E. Reid) wherein acyl-action/alkylation is done in a micro-reflux device (W. Dunges).] Another new reagent, of oxazine type, served for visible diode LIF.- Rahavendran, S.V. & Karnes, H.T. (1996) J Pharm Biomed Anal **15**, 83-98.

On-line derivatization at pre-column stage, for fluorescence detection (e.g. using oPA), was done with switching and C-18 HPLC to measure amphetamines in urine.- Herráez-Hernández, R. et al. (1996) Anal Chem **68**, 734-739.

Hydroxylation positions during metabolism of nebivolol can be established by photodiode array detection after high-pH C-18 separation on urine; UV absorbance is influenced by adjacent auxochromes. NMR was rather insensitive.- Hendrickx, J. et al. (1996) J Chr'y **A 729**, 341-354.

Luminescence detection aided by lanthanides (triply charged cations) has been reviewed in relation to HPLC and CE. Specificity may benefit, but not sensitivity if the analyte is a strong luminophore, e.g. a fluoroquinolone.- Rieutord, A. et al. (1996) Analusis **24**, 349-360.

Capillary HPLC-EC (carbon-fibre electrode, amperometric) was effective as a novel approach for retinoids in bovine serum deproteinized by acetonitrile.- Hagen, J.J. et al. (1996) J Chr'y **B 677**, 225-231.

From a Forum abstract -

A.A. Hudson & M. Bayliss, Micromass UK, Altrincham, Cheshire:

MS coupled to CE or CEC (capillary electrochromatography) ...**provides** the analyst with the sensitivity and selectivity...of ESP-MS and the speed and resolution afforded by CE....minimizing sample work-up procedures. [As CE furnishes] poor resolution of neutral species, a mixed-mode separation mechanism combining RP chromatography and CE has emerged....CEC has now been coupled to ESP-MS....[allowing] on-probe capillary separation.

Detection in CE, MEKC, CEC etc. features in diverse reviews (1997; Electrophoresis **18**, 2123-2501). Thus J.F. Banks (2255-2266) considers **MS(ESP)**; **EC** advances are surveyed (P.D. Voegel & R.P. Baldwin, 2267-2278).

Section #D

APPROACHES TO BENEFIT QA OR THROUGHPUT

The suffix '**n**' as in #D-2n distinguishes a 'NOTE' from main articles. COMMENTS (starting on p. 246) comprise discussion remarks and Editor's 'annotations'.

For methods with MS-MS detection, see also Section #B.

#D-1

ANALYTICAL PITFALLS WITH DRUG STABILITY MEASUREMENTS IN EARLY ASSAY DEVELOPMENT

P. Heizmann*, W.F. Richter, F.P. Theil, E. Saulnier & P. Schenk

Department of Drug Metabolism and Pharmacokinetics, F. Hoffmann-La Roche AG, Basel, CH-4002 Switzerland

Pointers to any instability in the biofluid samples or at pre-chromatographic stages must be obtained for any drug candidate even early in its development, particularly if its structure is ominous. Studies on a β-lactam have shown variable degradation, notably pH-influenced instability in animal plasma. Apparent but artefactual instability was also encountered in the HPLC stage. Sample pH differs amongst plasma samples (likewise amongst urines) and, as shown by test storage at 37°, undergoes an adverse alkaline drift, which can be dealt with. Further investigations indicated the critical elements in the design of stability experiments. Attention must be paid to sample preparation steps and to storage in the autosampler (full, capped vials may be requisite).

In the early pre-clinical development of a new agent, valid analytical data are needed as soon as possible for the kinetic characterization of the candidate and for further decision-making. Even at the outset a minimum of information must be obtained on drug stability during sample storage (e.g. short-term stability at 37°) and during the different stages of sample preparation. The contribution of plasma instability to the plasma clearance under *in vivo* conditions can be considered. Such investigations are of particular importance where the drug's structure points to instability, as with β-lactam antibiotics which are sensitive to pH [1]. The analyte now studied was a β-lactam, selected during screening because of its pharmacological activity and *in vitro* stability found in human plasma at 37°†.

Analytical procedure.- A preliminary HPLC-UV procedure was developed. Sample preparation for plasma involved $HClO_4$ protein precipitation. After addition of water (50 µl) and 1 N $HClO_4$ (50 µl) to the plasma (100 µl) and centrifugation, 100 µl of the supernatant was analyzed. The autosampler was kept at 4°.

---

*addressee for any correspondence
†'Lability' index entries in past books are pertinent. In Vol. 23 [eds. and publisher as now], stability problems are considered by H.M. Hill & co-authors (p. 359) and, warranting consultation (the present survey being a follow-on), by P. Heizmann & co-authors (p. 351) who touched on a troublesome (pH-influenced) β-lactam. The present art. concerns a different β-lactam.

An Intersil ODS2, 5 µm, 250 × 4 mm analytical column and a 10 × 4 mm guard column of the same material were used. A linear gradient from 2% to 16% acetonitrile in 50 mM phosphate buffer pH 3 was run within 10 min. The compound eluted at ~9 min. Quantitation was based on the measured peak heights.

Procedure for stability measurements.- Spiked plasma was stored at 37° in a shaking water bath. Aliquots were taken at different times, subjected to sample preparation, and the supernatant finally stored in the autosampler until HPLC injection.

STABILITY EXPLORATION, AND PROBLEMS REVEALED

Surprisingly, initial stability investigations with the new compound in animal plasma (37°, 5 h storage period) revealed not only instability but different degrees of degradation from the same experiment carried out on different days (Fig. 1). In rat plasma, the degradation ranged from 17% (day 1) to 50% (day 2). In cynomolgus monkey plasma it was 76% (day 1) and 27% (day 2).

The raw data from these experiments were checked in more detail to find indications which possibly could explain the adverse results. The observations were:
- a change of the plasma pH from 7.40 at experimental start to 7.85 at the end (rat, day 1);
- gradual loss of sensitivity for 'control' samples (compound in mobile phase) which were spread over the analytical series; peak broadening throughout the series (cynomolgus, day 1);
- 25% loss of sensitivity (peak height) for 'time zero' samples, which were re-injected at the end of the first 50-run series (rat, day 2).

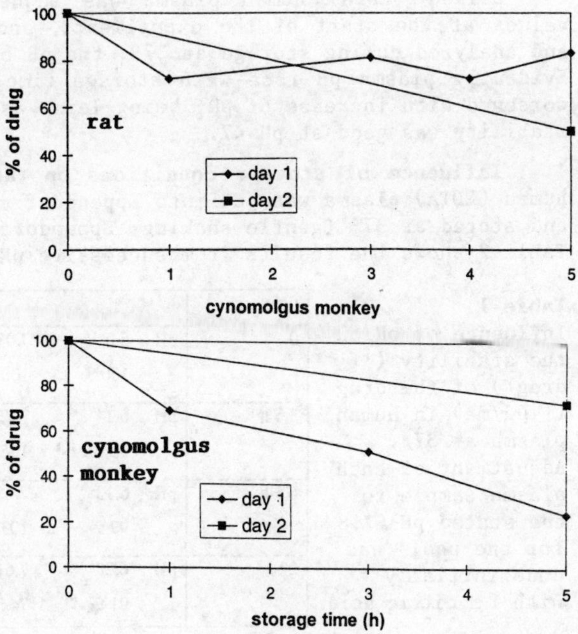

Fig. 1. Preliminary results on the stability of the drug in animal plasma (EDTA/NaF) at 37° (same experiment on different days; stability expressed as '% of drug'; 100% at start).

From these observations, we considered that the following might have contributed to the inconsistent findings:

- an instabiity of the chromatographic system (leading to permanent loss of sensitivity);
- drug instability in the final injection solution;
- pH-sensitive drug instability in plasma;
- different plasma pH's at experimental start, and possible pH increase during the experiment.

ENSUING EXPERIMENTS

Experiments were performed to verify the above assumptions, especially to obtain more information about the influence of the plasma pH on the drug stability and about the anticipated pH shift in plasma during storage at 37°.

The pH of the biofluid

Plasma.- The pH values measured in different batches of pool plasma (EDTA) were very different, the ranges being 7.1-8.5 (rat), 6.9-7.7 (rabbit) and 7.0-8.4 (human). The pH rises, above the physiological level, may relate to the storage duration, to repeated freeze/thaw cycles or to handling prior to freezing.

Urine.- Compared to the normally found narrow range of plasma pH *in vivo*, urine pH's are more variable. Urine pH is altered by food or drugs, by the clinical status, and also varies with time. The extremes are at 4.5 and 8.2 for man [2].

Influence of pH on stability of the drug in plasma (Table 1)

Spiked (EDTA) human plasma was adjusted to different pH values at the start of the experiment, and aliquots were taken and analyzed during storage at 37°, the pH being measured as well. Evidently plasma pH rose with storage time, and drug stability worsened with increase of pH, being least as the pH approached 8; stability was good at pH <7.

Influence of storage conditions on the plasma pH.- Rat or human (EDTA) plasma was put into Eppendorf vials or glass vials, and stored at 37° (gentle shaking; Eppendorf thermomixer, #5437). Table 2 shows the results from successive pH measurements.

Table 1. Influence of pH on the stability ('% drug') of the drug (1 µg/ml) in human plasma at 37°. Adjustment of each plasma sample to the stated pH (7.8 for the pool) was done initially with 1 M citric acid.

time					
0	pH	6.0	7.03	7.43	7.8
		100 %	100 %	100 %	100 %
3.5 h	pH	6.14	7.15	7.65	7.9
		96 %	106 %	98 %	79 %
6.5 h	pH	6.13	7.22	7.71	8.01
		99 %	103 %	93 %	68 %
22 h	pH	6.28	7.61	8.01	8.31
		91 %	74 %	42 %	15 %

Table 2. Change of pH of plasma (EDTA/NaF; vol. given in **left column**) when kept at 37° in 3.5 ml glass vials which, except for a), were always capped/re-capped and which in b) [full throughout] & c) were different for each time point.

		storage time (h)								
	plasma	0	1	2	2.5	3	4	5	7	8
a)	rat	7.2	7.74		7.93		8.24		8.67	
1 ml	rat	8.46	8.38		8.46		8.54		8.70	
	human	7.8	8.47		8.76		8.88		9.0	
	human	7.19	7.52		7.66		7.73		7.80	
b)	rat	7.2	7.18		7.19		7.19		7.22	
full	human	7.8	7.66		7.63		7.63		7.68	
c)	rat	7.06	7.45		7.51			7.54		7.59
1 ml	human	7.02	7.25		7.27			7.25		7.25
d)	rat	7.10							7.91	
0.5 ml	human	7.02							7.62	
e)	rat	7.10	7.12*	7.32*		7.64*		7.93*	8.24	8.40
2.5 ml	human	7.02	7.0*	7.18*		7.33*		7.64*	7.82	7.93

*times when 0.5 ml taken and vial re-capped

Evidently the plasma results showed (for both species) that:

- pH changed very quickly when storage was in uncapped vials;
- pH remained constant in completely filled and capped vials;
- pH changed in partially filled and capped vials, the change depending on the remaining air volume in the vial;
- pH shifted when sampling from the (capped) vial, full at the start, was done at subsequent times, so decreasing the plasma volume and increasing the air volume (this is the procedure often used for stability measurements, as in the preliminary experiments of the present study).

Drug stability during autosampler storage

Drug stability in the supernatant from $HClO_4$ protein precipitation was investigated by working up a large aliquot of a spiked plasma and injecting aliquots of the resulting acidic supernatant at different times after precipitation. The drug was stable under these conditions for at least 20 h.

Chromatographic instability

A larger pool of a spiked plasma was worked up. Aliquots of the resulting supernatant were filled into separate vials and then injected for HPLC, thus simulating analytical series with ~50 runs. Peak sensitivity (peak height) slowly fell within an analytical series (example: Fig. 2). Peak broadening (reduction of column efficiency) was observed in parallel. The sensitivity decrease was less in other test series. It was found later to be unconnected with the acidic pH of the large sample aliquot injected (100 µl); adjustment of its pH to 2.5 gave similar results. The effect could not be explained, but the peak height reduction could be reversed with a fresh guard column.

As a first outcome, analytical batches had only ~30 samples.

Fig. 2.
Instability of the chromatographic system, expressed as peak height reduction within an analytical series (37 injections of identical samples during a 12 h period).

For injections #4 and #36 resp.:
pk. height 145, 121;
width 0.135, 0.157;
area 1260, 1220

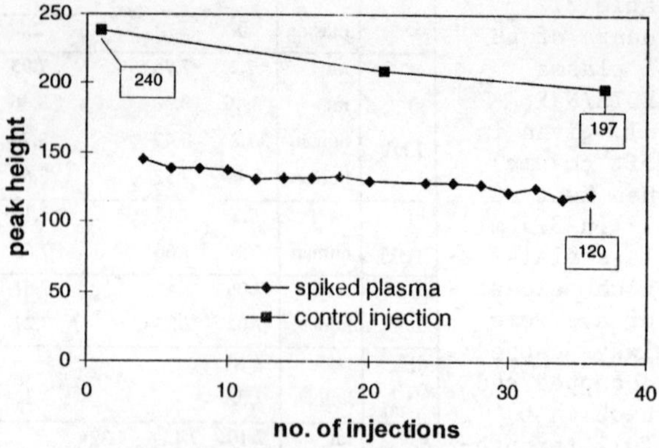

CONCLUSIONS FROM THE EXPERIMENTAL FINDINGS

Perhaps bearing on the inconsistencies in stability trials:

- a chromatographic deterioration caused a decrease in response with successive injections, simulating drug instability;
- drug stability in plasma (or other biofluids) was pH-sensitive, best at pH <7; marked degradation occurred towards pH 8;
- during plasma storage at 37° the pH underwent an alkaline drift (adverse to stability) if the vials were not full or uncapped;
- individual plasma batches differed in pH; the higher the pH at the experimental start, the higher the alkaline value attained (and the greater the risk to the analyte's stability).

Design implications for short-term stability studies (plasma, 37°):

- check plasma pH at start and, if not physiological pH, adjust;
- fill completely and cap the storage vials (plastic or glass; a separate one for each time point where pH is to be checked);
- if, from a plasma pool, aliquots are taken after different storage times, measure pH and re-adjust at each sampling point (although this procedure seems less precise and convenient).

Further studies, following this design, on stability in plasma (EDTA) showed only slight species differences, the decreases over 4 h ranging from 0% (marmoset) to 25% (rabbit; rat, 19%). Human plasma showed some decrease (15%), not fully in accord with the high stability previously found during screening.

References [In-house location of the data: PRNK journals/folders #1369, #1358]

1. Connors, K.A., Amidon, G.L. & Stella, V.J. (1986) **Chemical stability of pharmaceuticals: a handbook for pharmacists,** Wiley, Chichester.
2. Wissenschaftliche Tabellen Geigy (1977) Ciba-Geigy Ltd., Basel.

#D-2n

A Note on

ASSAY SPECIFICITY IN THE PRESENCE OF METABOLITES: A CASE REPORT ON A REQUEST FROM AUTHORITIES

P. Heizmann†, **H. Martin & D. Dell**

Dept. of Drug Metabolism
& Pharmacokinetics,
F. Hoffmann-La Roche AG,
CH-4070 Basel, Switzerland

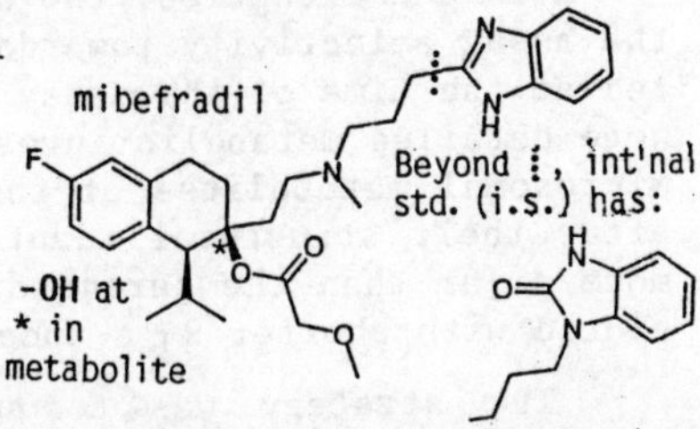

Mibefradil (Ro 40-5967) is a novel calcium-channel blocker, with certain kinetic advantages over other such drugs including good bioavailability and long plasma ½-life [1]. It is indicated for treating hypertension and stable angina pectoris.

The HPLC procedure which had first been developed for animal and clinical kinetics (1990*) was fairly straightforward, entailing liquid-liquid extraction with dichloromethane, RP-HPLC on Spherisorb C6 (125 × 4 mm) isocratically (acetonitrile/pH 3.9 phosphate buffer; 30/70, v/v), and fluorescence detection. Quantification (limit: 5 ng/ml) used internal standardization and was based on peak height measurements. [Chromatogram shown below, Fig. 1.]

The method was used to support toxicokinetic studies as well as most clinical pharmacology studies. The simultaneous determination of the major plasma metabolite, an alcohol (Ro 40-5966) arising from hydrolysis of the ester side-chain, was included at a late stage and used in a few studies only. The procedure was internally reported in 1995ª and submitted in 1996 as part of the registration document.

During the registration procedure, the authorities asked specific questions on assay validation parameters, including selectivity:

"It is claimed in your letter of ..... that mibefradil is separated from metabolites and endogenous compounds. The limited number of sample chromatograms submitted do not identify metabolite peaks. Please submit data to demonstrate that the analytical methods used are specific in the presence of endogenous compounds and significant metabolites."

Metabolism.- The drug is extensively metabolized. Hydroxylation at different sites, glucuronidation of the hydroxyl groups, O- and N-dealkylation and hydrolysis of the ester side chain were observed. Some 30 products were found after *in vitro* incubation of hepatic microsomes with the parent drug. The alcohol is the

---

†addressee for any correspondence

*H. Eggers, J. Körner & R. von Alten, in-house data file.

ªH. Eggers, A. Martin, G. Klamp, J. Körner & J. Kropf, in-house data file.

most prominent metabolite circulating in human plasma at steady state. Other (minor) metabolites are mainly secondary metabolites (derived from the alcohol) and minor amounts of primary metabolites [2].

SITUATION AND RESPONSE TO THE REQUEST

With one exception, the already available alcohol metabolite, the assay selectivity towards metabolites could not be investigated at the time of the assay development, because not till later were detailed metabolic investigations started. Three additional microsomal metabolites of the parent drug were synthesized later after their structural identification. The metabolites are all more polar than the parent drug and are therefore expected to be eluted with shorter R_T's under RP conditions.

The strategy was to answer the authority's questions as convincingly as we could with the available data (some obtained after the internal reporting of the analytical procedure) and at the same time to carry out additional investigations so as to be prepared for a possible further request.

Drug selectivity *vs.* endogenous components.- It was readily demonstrable with chromatograms from drug-free plasma that the procedure was selective in this respect.

Drug selectivity *vs.* metabolites.- The few synthesized metabolites were analyzed under the conditions of the actual assay. As expected, they were eluted with shorter R_T's than the parent drug and were separated from it. It was found later that these compounds were minor plasma metabolites.

The retention of these compounds (as well as that of the parent drug, alcohol metabolite and i.s.) was susceptible to mobile phase ionic strength, pH and composition. Variations of these parameters were later used to modify the chromatographic conditions in a few multiple-dose clinical studies, to influence selectivity.

Other metabolites had not been synthesized, and therefore could not be similarly investigated. However, our argumentation was based on the results of metabolic studies wherein metabolites were separated and identified in microsomal incubations and in steady-state plasma samples [2]. Most of the microsomal metabolites were either not present in human plasma or were minor metabolites only. The metabolites eluted at shorter R_T's than the parent drug and were well separated from it. The chromatographic conditions were similar but (unfortunately) not fully identical to those of the actual assay. However, similar retention characteristics and separation from the parent drug could be expected under the chromatographic conditions of the actual assay. This hypothesis was supported by the results of the above experiment, where the synthesized metabolites (alcohol metabolite included) and the parent drug were eluted in a similar retention order.

From these results and findings the assay for the parent drug was argued to be selective with respect to endogenous components.

Selectivity of the internal standard (i.s.) was the greater problem. It had been recognized during the analysis of clinical samples from multiple-dose studies, but not in previous analyses on samples derived from single-dose applications (cf. Fig. 1) - that the i.s. eluted in a retention region where additional peaks (possibly minor metabolites) were observed in the chromatograms (Fig. 2). some being eluted close to the i.s. The structures of these believed metabolites were then unknown; so specific investigations of assay selectivity were precluded. The decision at that time was not to investigate a different i.s., nor to modify the chromatographic conditions drastically. Precautions to minimize contributions from any such (minor) metabolites were taken:

- adding the i.s. in sufficiently high amounts;
- monitoring the i.s. peak heights in calibration samples and unknowns: a marked interference with co-eluted metabolites would be indicated by an associated increase in the i.s. peak height - which was normally not observed;
- for samples with high amounts of parent drug (and alcohol metabolites): dilution with drug-free plasma prior to sample preparation, so reducing any possible contribution from interfering metabolites to the i.s. peak height.

In a few cases the nature of the mobile phase was slightly modified (pH, ionic strength, composition) and the samples run under such improved conditions. Normally the interference at the i.s. R_T was <5%. These arguments together with representative data and figures were reported to the authorities.

ADDITIONAL SELECTIVITY INVESTIGATIONS

To gain more detailed knowledge of the retention behaviour of the minor metabolites under the actual chromatographic conditions, the parent drug and major metabolite were incubated with rat liver microsomes and the resulting metabolite mixture examined by HPLC using fluorescence and a radiomonitor. The metabolite pattern for rat microsomes resembles those with human microsomes. The resulting metabolites were sufficiently separated from the parent drug and main metabolite [P.H., M. Schleimer, R. Gasser & H.M. (1997); data on file]. Only one minor metabolite interfered with the i.s. under the selected conditions. The findings gave support to our previous arguments.

CONCLUSIONS

Analysts know how difficult (but essential) it is to find the right i.s. Here it was selected early in assay development without detailed knowledge of metabolites. Normally they are more polar and precede the parent drug under RP conditions. With hindsight, choosing an i.s. with longer R_T than the drug might have been preferable. However, detailed knowledge of what will be required in a later phase is normally unavailable at the time of assay development.

It is important that, as is normal, metabolism studies be carried out parallel to assay development. Possible metabolite-interference problems are then recognized at an early stage and so given timely attention in the assay development. Nowadays, for metabolite selectivity problems use of LC-MS might be preferable.

The above-mentioned arguments and results were submitted to the authorities, and were accepted.

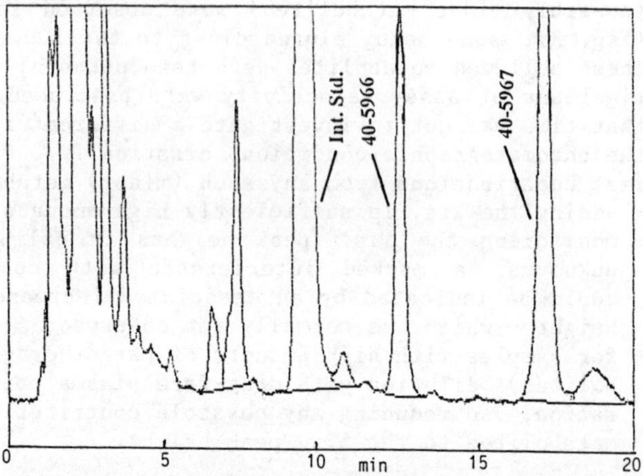

Fig. 1. Chromatogram of a patient sample after a single mibefradil dose (plasma concn. 240 ng/ml).

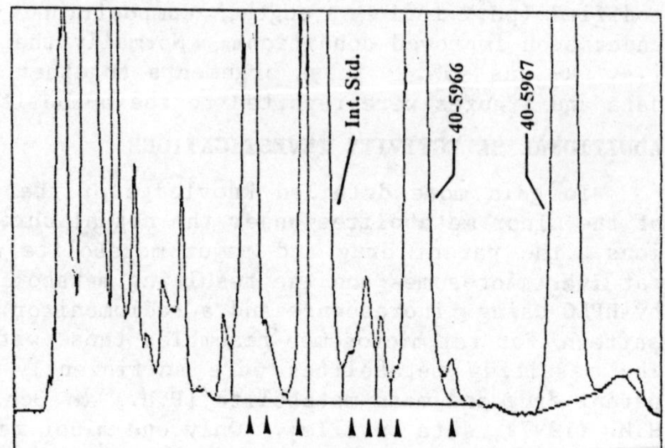

Fig. 2. As for Fig. 1 but after repeated doses (concn. 1050 ng/ml). ***Arrows:*** peaks of possible metabolites appearing after repeated dosing.

References

1. Welker, H.A., Eggers, H., Kleinbloesem, C., Erb, K., Breithaupt, K., Butzer, R. & Belz, G.G. (1989) Eur. J. Clin. Pharmacol. **36** (Suppl.), A304.
2. Wiltshire, H.R., Sutton, B.M., Heeps, G., Betty, A.M., Angus, D.W., |a Madigan, M.J. & Sharp, S.R. |b Worth, E. & Welker, H.A.| (1997) Xenobiotica **27**, |a 539-556. |b 557-571.|

#D-3

AUTOMATING '96-WELL' FORMAT SOLID PHASE EXTRACTION FOR LC-MS-MS

Stephen Pleasance† & Robert A. Biddlecombe

International Bioanalysis, Glaxo Wellcome Research & Development, Park Road, Ware, Herts. SG12 0DP

A custom-built automated system for SPE in the 96-well format and its application to high-throughput bioanalysis is described. This system comprises a Zymate XP robot, a cooled storage carousel that acts as both a warehouse for all the labware and as a refrigerated storage for the extracts, and an SPE station. The latter is a modified reagent-addition station, which incorporates a two-stage vacuum manifold. Reagents are added a row at a time and a solvent-switching-valve allows up to 9 different reagents to be used on the system. This station performs all the vacuum, conditioning, washing and elution steps, and a robotic sample processor is used simply to dilute samples, to add i.s. and to transfer samples from tubes to plates. The plates containing the extracts are returned to the cooled storage carousel to await analysis. The validity of this approach is demonstrated by analytical data obtained from a typical clinical study. The phased development of a fully automated SPE-LC-MS-MS system on-line to a triple quadrapole mass spectrometer, capable of round-the-clock operation, is also described.*

Since the development of API*, HPLC coupled with mass spectrometry (LC-MS) has rapidly replaced HPLC with UV or fluorescence detection as the technique of choice for bioanalysis in our Department. The combination of chromatographic retention time and two stages of mass selection in LC-MS-MS result in extremely high specificity. It is the combination of high sensitivity and high specificity that enables both sample preparation times and analysis times to be significantly reduced, thereby offering the potential for both fast method development and high sample throughput.

Although a number of sample preparation techniques such as protein precipitation, liquid-liquid extraction and SPE are routinely used, SPE has become one of the most widely used [1]. Its relative simplicity has led to the development of many commercial instruments for either on-line or off-line sample preparation. Unfortunately, the cycle times of these automated sample preparation systems have historically been driven by analysis times in conventional HPLC and have not kept pace with

---

†addressee for any correspondence

**Abbreviations (see also Preface)*.- API, atmospheric pressure ionization (in tandem MS context); i.s., internal standard; QC, quality control; RAS, reagent addition station; RSP, robotic sample processor.

the fast analysis times possible with LC-MS-MS. One solution has been to separate sample preparation from sample analysis and perform the preparation off-line batchwise. This has often meant that the sample preparation has become the rate-determining step, resulting in inefficient instrument utilization.

The recent development and commercial availability of SPE in the 96-well format coupled with the use of robotic sample processors has offered a solution to this problem, giving a 3-5 fold increase in sample throughput and once again making analysis the rate-limiting step. We have previously reported on the application of the Microlute® system of SPE in the 96-well format for high throughput bioanalysis using LC-MS-MS [2]. The original system uses a customized vacuum manifold located on the deck of a Packard MultiPROBE® robotic sample processor. This allows automation of the prime, load and wash steps; but the operator is required to manually place the collecting plate in the vacuum manifold prior to the elution step. In order to maximize the advantages of this approach and overcome some of the limitations, a fully automated stand-alone system has been developed†.

OFF-LINE SYSTEM

The choice of Zymark as the automation partner provided the further advantages that existing systems could be re-engineered to work with the 96-well format and there was a considerable amount of Zymate experience within the company. The system consists of a Zymate XP robot, cooled storage carousel and a custom-built SPE station (Fig. 1). This robot runs System V software (version 2.5) and has a customized Visual Basic® interface for ease of use. The user simply chooses to create a new method or run an existing method from the front screen.

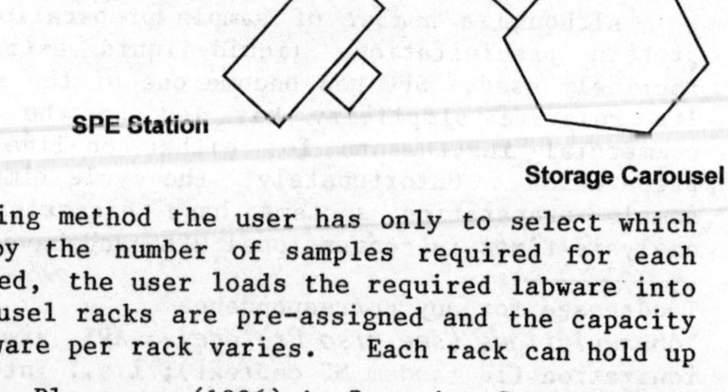

Fig. 1. The Zymark 96-well system (schematic).

To run an existing method the user has only to select which method(s) followed by the number of samples required for each method. When prompted, the user loads the required labware into the carousel. Carousel racks are pre-assigned and the capacity for each type of labware per rack varies. Each rack can hold up

†R.A. Biddlecombe & S. Pleasance (1996) in *Proc. Int. Symp. on Laboratory Automation and Robotics* - Boston, MA; now in press (pp. 445–454).

to 22 standard microtitre plates, 10 Microlute I (without the column extenders) or Microlute II or Empore® plates, 8 deep-well or square-well collection plates of 5 Microlute I (with column extenders). The cooled storage carousel acts not only as a warehouse for all the labware but also as a repository for the collection plates containing extracts.

To create a new method, the user must select the type of SPE block (e.g. Microlute II or Empore) and of collection plate (e.g. deep or square well), and specifies the volume and reagent used for up to three prime and two wash steps and one elute step. The vacuum time and a *bleed factor* are also entered for each vacuum step. The bleed factor varies the strength of the vacuum by means of an extra valve that can be opened to air. This is used to control the flow rates and ensure that the phase does not dry out before the sample load step. The vacuum can be optimized by using separate programs called *calibrate vacuum* and *calibrate bleed factor*. The latter subroutine allows the user to add reagent to the block and gradually decrease the air bleed value until the desired flow rate is achieved. The *calibrate vacuum* program allows the operator to optimize the duration of application of the vacuum. There are two additional screens, one for assigning reagents to the system and one for purging the reagent lines.

The SPE station is a modified reagent addition station (RAS) which incorporates a 2-stage vacuum manifold (Fig. 2). Reagents can be rapidly added row at a time, and a switching valve allows up to 9 different reagents to be used on the system, enabling different methods to be run in a sequence. The variable height control on the RAS allows Microlute blocks, with or without the column extenders, and Empore blocks to be used, the dispensing height depending on the labware type. The different height of the drain-pipes on the Empore plates requires the Zymate to place an adapter on the SPE station prior to running these blocks.

The variable height control on the RAS is also advantageous in that it can be used to push down on top of the blocks to ensure that the vacuum works consistently. The variable height setting on the platform allows both deep-well and standard microtitre collection plates to be used. Although the SPE station performs all of the vacuum, conditioning, washing and elution steps, the system utilizes a MultiPROBE to dilute samples, add i.s. and transfer samples from tubes to plates. The MultiPROBE configuration for transfer from tubes is shown in Fig. 3.

The robot sequence begins with the SPE block being ejected from the storage carousel. All types of SPE block except the Microlute I with column extenders are stored upside-down in the storage carousel so that they can slide easily. The Microlute I blocks with the column extenders have a special adapter that they sit on so as to enable them to slide. The block is transferred to the SPE station where it is primed, and then to the MultiPROBE for the sample transfer step. The Zymark controller and the MultiPROBE controller are networked so as to enable the transfer procedure to be initiated from the Zymark controller. When the MultiPROBE signals that it has finished, the Zymate collects the

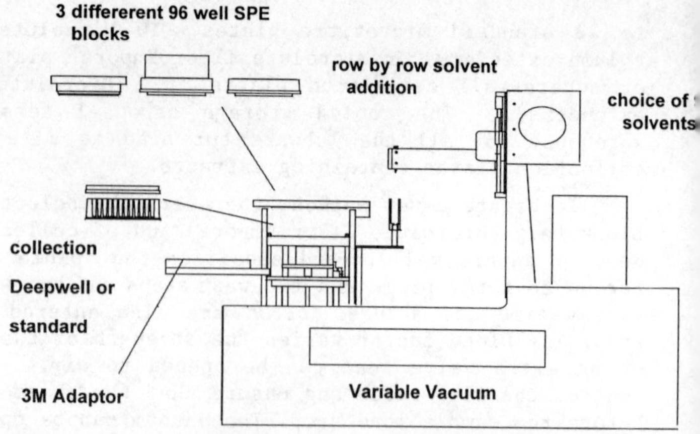

Fig. 2. The SPE station (schematic).

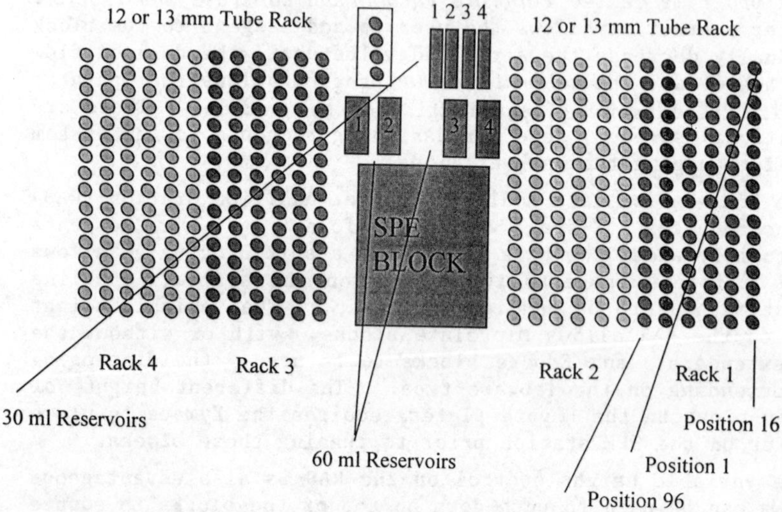

Fig. 3. MultiPROBE configuration on the Zymark 96-well system.

loaded block and returns it to the SPE station, which then applies a load vacuum, carries out the wash step(s) and the elution step. The plates containing the extracts are returned to the cooled storage carousel until manually transferred to an autosampler.

This system effectively halves the time to extract 96 samples compared to an off-line MultiPROBE, taking ~30 min/block depending on the number of steps. This system was originally designed to perform a minimum of 4 consecutive runs (384 samples) from sample tubes and to be capable of handling all the current available labware (Microlute and Empore blocks). We have so far validated methods on all the currently available formats and are supporting clinical studies for a number of compounds.

An illustrative set of analyses

The performance of the Zymark 96-well system is highlighted by looking at the data from a repeat-dose phase I clinical study for a compound being developed for rheumatoid arthritis. There were 3 dosing occasions each generating 288 samples (36 for each of 8 patients), a total of 864 samples. Rapid turn-round of the data was required so that the PK analysis could be done before the next dose occasion. The SPE method was straightforward and required no concentration step. The sample preparation was fully automated on the Zymark 96-well SPE system. The Microlute block, containing 15 mg Varian C-2 sorbent, was primed with 0.1 ml of acetonitrile followed by 0.1 ml water. The samples were diluted 1:1 with the i.s. solution and loaded onto the block. This was washed with 0.1 ml water and the analyte was then eluted into a deep-well microtitre plate with 0.1 ml of mobile phase, acetonitrile/0.1% formic acid (50:50). The plate was then covered with foil, to prevent evaporation, and placed on the autosampler.

Fig. 4 shows LC-MS-MS peak patterns (conditions in legend), the R_T for the analyte being 1.9 min. The transitions monitored were m/z 395 → 86 for the parent and 399 → 86 for the i.s. The data were processed using MacQuan® Software (PE Sciex). Peak area ratios of analyte to i.s. were used to construct a calibration curve for interpolation of sample concentrations. The regression model was linear with $1/x^2$ weighting. With 0.1 ml sample volume the validated range for the method was 1-2000 ng/ml. The SRM chromatograms in Fig. 4 for the 1 ng/ml calibration standard show the method is both sensitive and specific for the analyte. The performance of the method is illustrated by data for the calibration standards and QC's (Table 1), obtained from 13 assays during support to the phase I study. Precision and accuracy (% bias) for both sets of data are each <5% and well within our acceptance criteria of 15%.

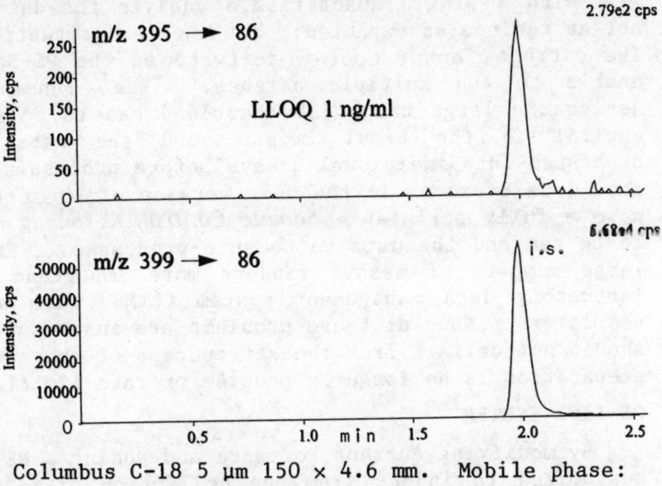

Fig. 4. SRM chromatograms for the 1 ng/ml standard.
Column (at 40°): Columbus C-18 5 µm 150 × 4.6 mm. Mobile phase: see text (1 ml/min; 1:10 post-column split). Detection: PE-Sciex API-300 with TurboIonspray. Injection vol. 50 µl.

Table 1. Precision and accuracy data from a phase I studv.

	Back Calculated Calibration Standard Concentrations								Quality Control Data		
Nominal	1.08	3.23	10.75	32.25	107.50	322.50	1075.00	2150.00	3.17	846.40	1692.80
Mean	1.09	3.13	10.29	31.86	108.49	328.71	1100.70	2182.59	3.27	862.57	1701.41
SD	0.05	0.14	0.28	0.89	2.63	11.09	28.76	51.08	0.13	36.75	40.65
%CV	4.67	4.44	2.76	2.79	2.42	3.37	2.61	2.34	4.12	4.26	2.39
%Bias	1.34	-2.88	-4.24	-1.20	0.92	1.92	2.39	1.52	2.95	1.91	0.51
N									25.	26.	25

The sample preparation time is ~25 min per block, and with an analysis time of 2.5 min/sample each block took 4 h to run. We were able to give next-day turn-round for the data on all three dosing occasions. The system has so far been very reliable; we have had one minor hardware problem and one problem with the software, resulting in only one day's down-time.

Lessons from use of the robotic system

Use of the robot has highlighted other drawbacks in work practices in our laboratory. While the choice of serum rather than plasma obviates having to centrifuge the samples prior to analysis, the sample format itself (e.g. tubes) is not ideally suited to bioanalysis with high throughput. The logistics of removing the caps, racking the samples and then replacing the caps on 400 tubes remains a laborious task for the analyst. A great advantage would be to receive the samples either directly in microtitre plates, or in tubes or strips of tubes (8 or 12) that can be plugged into a microtitre plate format. As an interim fix we are considering using a reagent sample processor (RSP) to transfer samples to microtitre plates at the sample reception point, generating duplicate plates in case repeat analyses are needed. Such a sample format would have a significant impact on the number of consecutive batches that could be run.

With a single quantifiable analyte the data processing was not as tedious as expected; but the real situation is not ideal. The current *Sample Control* software on the PE-Sciex API-3000 is unable to run multiple batches. When running 4 consecutive assays one large batch (384 samples) has to be set up in sample control *via* the Excel import, and the data retrospectively separated into individual assays before processing. This problem should be overcome in the next version of *MassChrom,* which will have a fully scriptable *Sample Control,* allowing multiple batches to be run and the data to be auto-processed. The processing of large numbers of assays renders more important the need for a laboratory data management system (LIMS) with automated report templates. None of these problems are insurmountable, and they should not detract from the key success of the robot, viz. sample preparation is no longer a problem or rate-limiting.

ON-LINE SYSTEM

By modifying current software and adding a PE 200 autosampler customized to inject from one collection plate with easy robot access, on-line LC-MS-MS is now possible (Fig. 5). The system

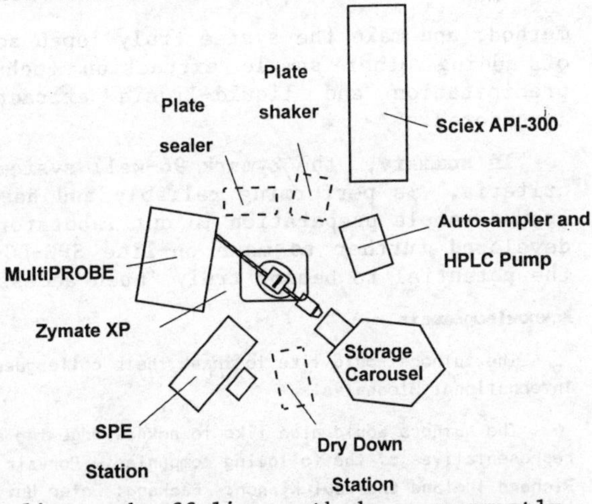

Fig. 5. Proposed on-line system (schematic).

can run samples for on-line and off-line methods concurrently. There is a manual over-ride function to enable off-line plates to be removed and manually transferred to other LC-MS-MS systems. The software polls the autosampler to check when the current plate is finished, replaces this in the carousel and retrieves the next plate for the autosampler if it has been processed. If a batch of samples is to be processed on-line, the system carries out the first extraction, places the collection plate into the autosampler and initiates the injection program. Any subsequent blocks are then processed and the extracts are stored in the carousel until they can be analyzed. The autosampler control and API-365 link still has to be fully developed to allow greater flexibility.

The biggest advantage that on-line extraction and analysis offers is more efficient utilization of instrumentation. Without adopting shift work or overtime it would provide two additional days of use of the mass spectrometer. To maximize this potential several other peripherals will have to be to be added to the system. A robot-compatible dry-down station is essential so that methods requiring a concentration step can also be run on-line. If the extracts are to be reconstituted on the system, a plate sealer and plate shaker would also be required. The proposed system is shown schematically in Fig. 5.

With samples in tubes, as now, the system is capable of running four consecutive assays. This could be expanded to possibly 25 assays if the samples in microtitre plates could be stored in the carousel and retrieved just prior to extraction. The current system also needs to be made more flexible, allowing analysts to add work to the existing work-list at any time and to schedule priority samples. Generic SPE methods using some of the new polymeric phases such as Oasis® have been developed in our laboratory. Coupling these SPE systems with generic conditions for HPLC, also under development, would afford the flexibility to run multiple methods on-line, not just large batches of the same method, and make the system truly 'open access'. The possibility of adding other sample extraction techniques such as protein precipitation and liquid-liquid extraction is already being considered.

In summary, the Zymark 96-well system has met all the design criteria, is performing reliably and has reduced the bottleneck around sample preparation in our laboratory. The system is being developed further to make on-line SPE-LC-MS-MS possible and has the potential to become truly 'open access'.

Acknowledgements

The authors would like to thank their colleagues in the Department of International Bioanalysis.

The authors would also like to acknowledge the assistance given by the representatives of the following companies.- Porvair Sciences: Tony Castleman, Richard Ireland and Paul Kinnon; Packard: Peter Hutchings, Steve Jobe and Michael Kelly; Zymark UK: Terry Roche, Alan Shingler and Paul Savage; PE-Sciex: John Robson and Frank Vandeburgh; Waters: Dorothy Phillips, Ray Crowley and Ed. Bouvier.

References

1. Berrueta, L.A., Gallo, B. & Vicente, F. (1995) Chromatographia **40**, 474-483.
2. Allanson, J.P., Biddlecombe, R.A., Jones, A.E. & Pleasance, S. (1996) Rapid Comm. Mass Spectrom. **10**, 811-816.

Trademarks

#Easyprep, #MultiPROBE: Packard Inst. Co., Meriden, CT, USA;
#Easylab: Zymark Corpn., Hopkinton, MA, USA;
#Empore: 3M, St. Paul, MN, USA;
#MacQuan: PE-Sciex, Ontario, Canada;
#Microlute: Porvair Sciences, Shepperton, Middx., UK;
#Oasis: Waters Corpn., MA, USA;
#Visual Basic: Microsoft, Redmond, WA, USA.

#D-4

AUTOMATED SAMPLE PREPARATION FOR GC-MS AND LC-MS-MS AND AUTOMATED REPORTING

J. Schmid†, D. Schütze, T. Sauter & H. Eschey

Boehringer Ingelheim Pharma KG, Dept. of Pharmacokinetics & Drug Metabolism, 88397 Biberach, Germany

Clinical studies have lately escalated in volume and the development-process time scale has shrunk. GLP and GCP requirements have forced a good documentation of such studies. This pressure demands automated bioanalytical methods, introducing LIMS and automated reporting. For human studies on salbutamol (albuterol; β_2-adrenergic) an HPLC-MS-MS method was developed using 1 ml plasma, with separation of the sulfate conjugate and parent compound after sample preparation by a Zymark Rapidtrace®: LLOQ* 15 pg/ml; precision within 9%; recovery ~80% and carry-over effects <0.3%. The Rapidtrace is a very useful instrument for method development and reliable in routine work using >100 µl.*

Data-handling and reporting time becomes critical when sample throughput is increasing. The LIMS (KINLIMS®) with a calibration program (UNICHROM®) and an Oracle database for full documentation under GLP are used, whereby extracting and writing the report can be by push-button. The report comprises the pre- and main-study calibration samples and QC's, and the unknowns. Moreover, all study deviations are documented in the audit file.¶

Salbutamol, 2-(tert-butylamino)-1-(4-hydroxy-3-hydroxymethylphenyl)-ethanol, can be determined in plasma very sensitively using GC-MS with chemical ionization [1, 2]; but this requires extensive sample clean-up and conversion to the TMS derivative. More recently, a column-switched HPLC-MS-MS method for its assay in human plasma was reported [3], the LLOQ being 0.2 ng/ml using 0.5 ml plasma.

Based on this method we developed an assay utilizing SPE with a Zymark Rapidtrace for sample preparation off-line. This allowed the clean-up of larger volumes of plasma and increased the sensitivity of the method to the extent required for the analysis of plasma samples from clinical trials.

When sample throughput is increasing, time for data handling becomes critical. We use a LIMS (KINLIMS®▫) with a UNICHROM®▫

†addressee for any correspondence

*Abbreviations.-
GLP, Good Laboratory Practice;
GCP, Good Clinical Practice;
LLOQ, lower limit of quantitation;
QC, quality control (sample).

▫Source of this Laboratory Information Management System and UNICHROM: Datasign AG, Clarastr. 12, CH 4058 Basel.

¶'Audit file' and 'GLP journal' are synonymous (cf. Oracle).

calibration program and an Oracle database for full documentation under GLP [4].

The present article describes the automated sample clean-up for salbutamol and the data handling.

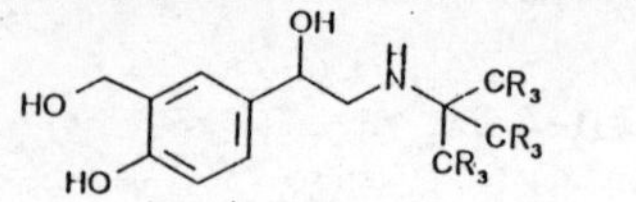

R = H or, in $^2H_9$-salbutamol (internal std., i.s.), R= D - synthesized in-house by the Radiochemistry group

METHODS

Systems used

HPLC.- Autosampler #A200 S: CTC Analytics, Zwingen, Switzerland; #1090 ternary gradient pump with integrated autosampler and diode array detector: Hewlett-Packard (Waldbronn, Germany); guard and analytical columns (Bischoff, Leonberg, Germany): stainless steel (20 × 2.1 and 60 × 2.1 mm) packed with 5 µm Purospher RP18 (Merck, Darmstadt, Germany); post-column splitter (split ~1:4): Ziemer, Mannheim - specially made T-piece and fused silica (50 µm i.d.).

Injection vol.: 90 µl. Eluants: **A**, 10 mM ammonium formate pH 4.0; **B**, acetonitrile. Gradient from 0 to 30% **B** over 5 min, then a step to 90% at 5.01 min, isocratic to 7 min, and a 1-min finish. Flow rate: 300 µl/min; temp. 40°. Salbutamol R_T ~3.2 min.

Detection, integration and data storage.- VG Quattro II mass spectrometer: Micromass, Altrincham, UK, with MassLynx software, version 2.22, equipped with an electrospray probe. Source temp. 160°; nebulizer gas flow: 18 L/h; drying gas 390 L/h. MRM ion transitions: salbutamol, 249 to 149 amu; $^2H_9$-salbutamol, 249 to 149 amu; inter-channel delay: 0.01 sec.

Sample procedure Microliter centrifuge: Hettich, Tuttlingen, Germany. Human plasma (EDTA; 1 ml) with 5 ng i.s. in 0.1 ml water added was vortexed and put into the Rapidtrace® (Zymark, Idstein, Germany) for the following steps.- To C18 BondElut cartridges (200 mg, 3 ml; Varian, Harbor City, USA), conditioned with 2 ml methanol and 4 ml water, the sample was applied slowly (1 ml/min) and the cartridges were washed with water. Elution was performed by 2 ml 0.2% (v/v) aqueous ammonia/methanol (1:1; 5 ml/min).

The transfer lines and cannula were carefully washed with water and eluting solvent. After each step 2 ml air was pushed through at doubled flow rate. Sample throughput was 10 or, with 5 Rapidtrace modules, 50 in 2 h. The liquid phase was evaporated with vortexing, reconstituted in 100 µl buffer and, with filtration, transferred into the autosampler vials.

RESULTS (Chromatograms: Figs. 1-3)

Specificity.- Virtually no interfering peaks were observed in blank plasma from humans (cf. Fig. 1). The main metabolite of salbutamol, its 4-hydroxy-sulfate [5], was separated by HPLC (cf. Fig. 3). Very small signals seen in some plasma blanks are most probably due to a small carry-over effect and minor impurities in the deuterated i.s.; the resulting peak areas were, however, only ~¼ of those found for the smallest calibration concentration.

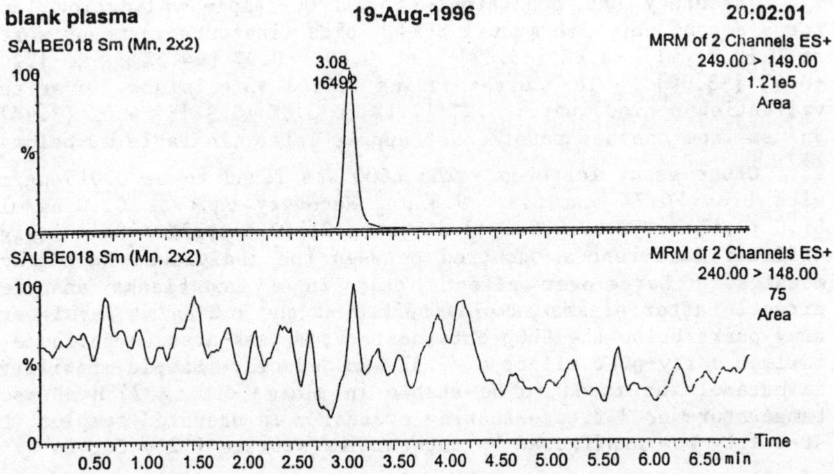

Fig. 1. Chromatogram of a human plasma sample spiked only with 5 ng i.s./ml (upper trace; N/A to lower trace). Note scale difference (automatic computer-adjustment to the highest peak).

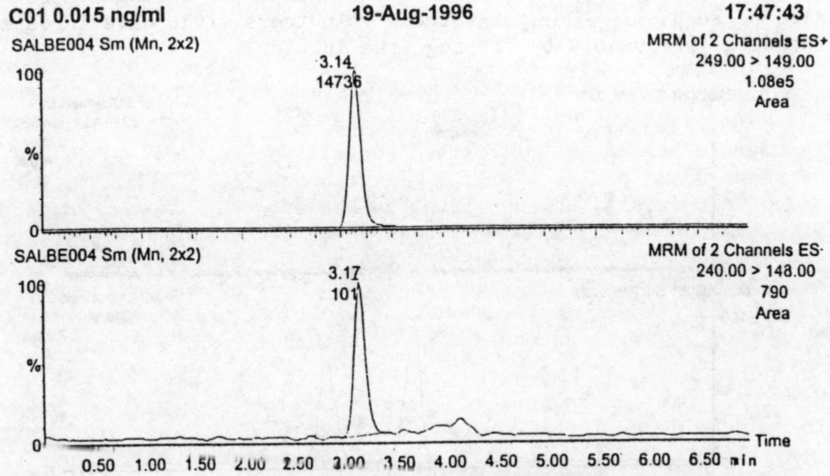

Fig. 2. Plasma sample spiked with 5 ng i.s./ml (the upper ion trace) and 15 pg/ml salbutamol (lower trace).

This accords with the mean intercept found for the calibration curves which was <$^1/_3$-rd of the lowest calibrant concentration.

Linearity.- Linear calibration curves (weighted 1/y) spanned the range 0.015-4.0 ng salbutamol/ml for undiluted samples and had a mean coefficient of determination (during validation) of 0.99940.

Accuracy and precision.- For a QC-sample validation [and for a second set with manual SPE], bias (inaccuracy) means were: at 0.10 ng/ml: -2.8% [-5.0%]; at 0.40: -0.9% [-4.5%]; at 1.20: +0.4% [-3.0%]. The corresponding C.V.'s (precisions, over the validation period) were: 5.9% [7.1%]; 3.3% [5.9%]; 4.0% [3.9%]. Values from another robotic set appear later (in Table 1 , below).

Other assay features.- The LLOQ was found to be 0.015 ng/ml with bias -0.7% and C.V. 9.8%. **Recovery** was: at 0.10 ng/ml, 81.6 +7.4%; at 0.40, 79.1 ±3.8%; at 1.2, 79.1 ±2.7%. Only insignificant differences occurred between the individual Rapidtrace modules. **Carry-over effect**: only in a few blanks analyzed directly after plasma samples spiked at the 1.2 ng/ml level were tiny peaks below the LLOQ obtained. From peak-area comparisons a maximum carry-over effect of 0.3% was deduced. **Sample stability:** salbutamol was found to be stable in plasma during 24 h at room temperature or 3 freeze-thawing cycles. In prepared samples its stability was unaffected by autosampler storage (12°) for 24 h.

Observations during the study.- With plasma from volunteers who had taken salbutamol, a second peak was always found in the chromatogram. This peak was attributable to the main metabolite of salbutamol, its sulfate conjugate [5], fragmented in the cone (Fig. 3). However, in some pre-dose samples the metabolite could also be seen, suggesting that these volunteers might have ingested the drug (presumably by 'trying' the inhaler).

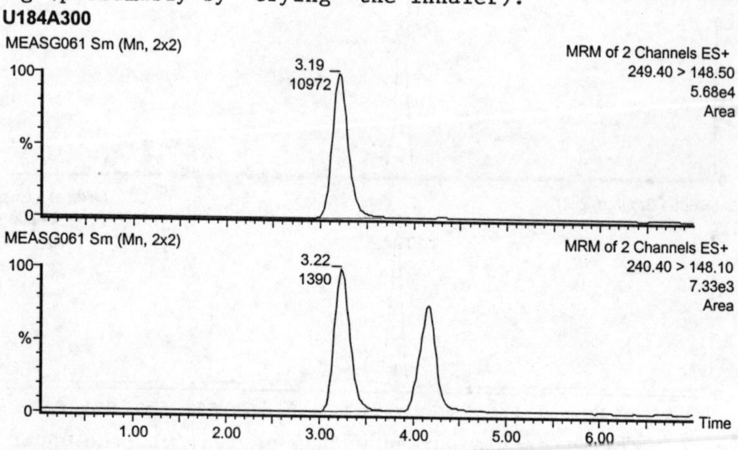

Fig. 3. Plasma from a patient treated with salbutamol. Upper ion trace: the i.s. peak. Lower ion trace: peak at 3.22 min for salbutamol and at 4.2 min for the sulfate metabolite (undergoes collisional part-fragmentation, in cone region, to salbutamol, whose ion trace therefore shows a second peak).

Data handling

HPLC-MS-MS with its high throughput has brought back to mind that analyzing a pharmacokinetic study comprises not only the purely analytical tasks but also others, now summarized.-

Activities performed during a whole study, and % relative time:

Planning and sample administration: #study protocol; #preparing labels and vials; #generating a LIMS project; #control of arrived samples. **Preparing QC & calibration samples**: #calculations; #collecting blank plasma; #analytical control of blanks; #spiking samples.	25-50%
Re-validation	
Measuring the study samples: #sample preparation; #analysis; #re-analysis.	25-50%
LIMS activities: #data collation; #calculation; transfer to GLP unit; #report.	25-35%

One point of outstanding importance is the data handling. We use a study-oriented LIMS, KINLIMS®, which was designed and developed by Uwe Timm from Hoffmann-La Roche, Basel [4]. The notable main features are: #it has a database (Oracle); #it is study-driven, which is very important for pharmacokinetic studies; #it performs instrument-independent calculations and data reduction with the interface UNICHROM®. Moreover, it furnishes an impressive GLP 'journal'.

Fig. 4 shows overall steps and the role of KINLIMS. Firstly a new study is defined therein. Then the samples, unknown/calibration/QC, are prepared. When the samples arrive, a daily worksheet is prepared in which the three types are put together. They are down-loaded to the instruments and analyzed. The results are sent to Unichrom - which calculates the concentrations, assures the quality of the data and sends the reduced data to KINLIMS.

From the KINLIMS our analytical report is printed automatically with acceptance limits and pre- and main-study calibration data. A calibration curve is plotted, pre- and main-study QC's are reported and finally the unknowns are listed. That part of the report is programmed in RS2. As an example the pre-study QC's associated with one study are listed in Table 1.

Overall, KINLIMS and UNICHROM are very sophisticated programs which have been in use in our Department for 4 years. The main advantages are that the system has an Oracle database, gives a good GLP journal, and generates the report directly out of the database. The disadvantage for the user is that the system is not of Windows type but a VAX type-based application which makes its use by untrained people rather difficult. System operation is very demanding; mistakes cannot easily be corrected after the study has been set 'on-going'. It depends very strongly on the performance of the network and the central computer department responsible for the Oracle system.

CONCLUDING COMMENTS

The salbutamol method is accurate, precise and sensitive (the high volume of plasma sample being advantageous). Column narrowness (2.1 rather than 4.6 mm) increases sensitivity ~2-fold.

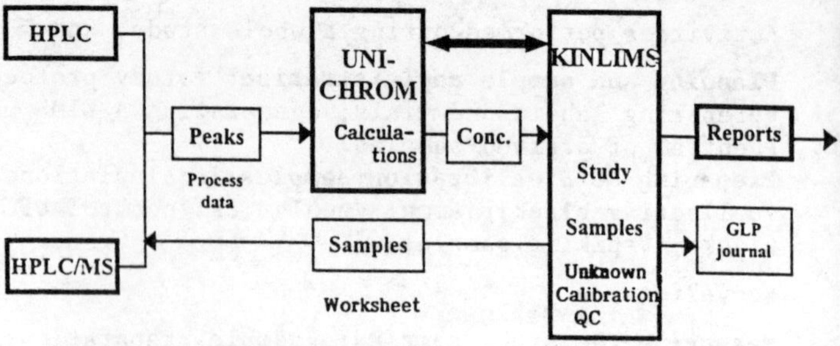

Fig. 4. Scheme of data processing in KINLIMS® and UNICHROM®. (Sources for their procurement: see footnote on p. 213.)

Table 1. Example of data processing scheme with KINLIMS and UNICHROM.

Table 5.1.1. Pre-study quality control samples

report : B757
project : P026/97SA
protocol : 243.2
analyte : SALBUTAMOL [pg/ml]
species : man
matrix : plasma Heparin
cal-range: 15-4000

AnaRep V1.6.9 Prod
Schuetze
04-APR-97
page 11/33

control no. concentration	Q01 100.0	dev.[%]	Q02 400.0	dev.[%]	Q03 1200	dev.[%]
SZ 25-FEB-1997	94.77	-5.2	369.4	-7.6	1119	-6.8
	89.98	-10.0	380.6	-4.9	1155	-3.7
SZ 26-FEB-1997	95.94	-4.1	361.5	-9.6	1160	-3.3
	93.78	-6.2	364.2	-8.9	1136	-5.4
SZ 26-FEB-1997	96.65	-3.3	388.5	-2.9	1124	-6.3
	93.47	-6.5	385.5	-3.6	1139	-5.1
mean	94.10	-5.9	375.0	-6.3	1139	-5.1
CV	2.5%		3.0%		1.4%	
N	6		6		6	

The procedure with off-line sample preparation proved to be much simpler than the reported GC-MS assay. The automated method is robust and applicable to the analysis of large sample numbers. Data handling with the KINLIMS allows reports to be printed out of the database. All events are finally documented in a GLP journal.

Acknowledgements

We thank Dr M. Stiasni for providing the radiolabelled and the deuterated salbutamol. Ths skilful technical assistance of C. Ackermann, A. Büechler and B. Schoch is gratefully acknowledged. We thank M. Birkhahn and K. Kasper for their efforts in the LIMS project.

References

1. Foerster, H.J. (1990) Fresenius J. Anal. Chem. **337**. 16-17.
2. Garcia Regueiro, J.A., Perez, B. & Casademont, G.J. (1993) J. Chromatog. **A 655**, 73-76.
3. Schmeer, K., Sauter, T. & Schmid, J. (1997) J. Chromatog. **A 777**, 67-72.
4. Timm, U. & Fuchs, K. (1996) Lab. Automation & Information Management (1996) **32**, 7-22.
5. Lin, C., Li, Y., Meglotten, J., Morton, J.B. & Symchowicz, S. (1977) Drug Metab. Disp. **5**, 234-238.

#D-5

A NEW APPROACH TO ASSAYING VERY LARGE BATCHES OF SAMPLES AND TO ACCEPTANCE CRITERIA IN BIOANALYSIS

Krzysztof Selinger

Glaxo Wellcome Inc., BioMet Division,
Five Moore Drive, Research Triangle Park, NC 27709, U.S.A.

A new 'big batch' concept for assaying a large number of samples (500-1000) in a single batch is presented. Validation of such a procedure could be done by retrospective analysis (meta-analysis) of existing data. Assay control should be verified by using QC control charts, while the acceptance criteria should be based on the inspection by variables. The concept presented is internally coherent, is based on solid statistics, and offers significant savings.

Chromatography-based bioanalysis is rooted in ideas which go back at least 15 years, and which were designed when analytical throughput was lower and the equipment used often unreliable. In conventional chromatographic methods, a typical run time of 10-12 min translated into ~5 samples/h and 100 samples/day; 100 was almost the maximum that an analyst could process manually in an 8-hour working day, using liquid-liquid extraction and possibly back-extraction into acidic or basic medium. A simple protein precipitation yielded greater output, but fewer projects lent themselves to this technique.

It is no wonder that older autosamplers were designed to inject a maximum of 100 samples, some even less. There seemed to be no need for larger capacity. The common laboratory software usually had the same limitation; in the case of larger batches of samples, merging of files was employed which rendered the data processing more complicated, clumsier and slower.

The last few years has seen a revolution in bioanalysis with the advent and maturing of LC-MS with robust and more affordable instrumentation for LC-MS, of capillary electrophoresis (CE) and of turbochromatography*. All these techniques can routinely and reliably furnish run times of 2-5 min and hence outputs of up to 500 samples/day. Similar progress has been observed in the area of automated sample preparation. New automated technologies based on 96-well plates for protein precipitation and/or SPE employ parallel processing to provide samples at a high rate, ~1/min. Coincidentally, over the same period of time the pharmaceutical industry started its metamorphosis from the noble, if somewhat

*term used by author to connote turbulent-flow chromatography with a very fast gradient (4-40 ml/min flow) - *Ed.*

old-fashioned, discovery of new medicines into the electronic-chip driven, highly efficient commercial venture we see emerging today.

What is missing in this fast-paced progress are new concepts for the method validation, assay control and acceptance criteria. In this author's opinion, today's high-throughput bioanalysis should reach into the wealth of the industrial QC literature and adopt suitable systems, based on solid statistical foundations.

METHOD VALIDATION

It is impracticable in an industrial R&D exercise to analyze several batches of 500-1000 samples to prove the point that such large batches are feasible. One may, however, do in retrospect an analytical appraisal of some particularly successful and analytically robust pharmacokinetic projects and method validations. Inspection of several projects of this nature at Glaxo Wellcome showed that the slope of the calibration curve remains virtually constant (C.V. 5-10%) over a period of days, weeks and even years. The conclusion is that in cases of well-developed and robust analytical methods, running calibration standards with every batch of 60-80 samples is excessive and contributes little to the precision and accuracy of the results. It is proposed to analyze complete studies of 500-1000 samples in one very large batch ('big batch') containing only a front and possibly back calibration curve and an appropriate number of QC's.

The validation of such an intended future approach should involve comparison of existing data generated in the traditional way (one or two calibration curves for each batch of 60-80 samples) with data calculated using only the first calibration curve from the first batch and the last calibration curve from the last one. The comparison could be done in many ways. For example, within the author's laboratory a comparison of any two sets of data obtained independently (in contexts such as method transfer or cross-validation) is done in the following manner, using generally available Microsoft Excel functionalities.-

a) A set of at least 25 samples (clinical or QC's) is subjected to two independent analyses ('reference' and 'test');
b) the test set of data is divided by the reference set, the ratio being ideally 1.00;
c) the mean ratio and 95% confidence intervals around this mean are calculated: both upper and lower confidence intervals must be bracketed by values 0.85 and 1.15 (values that merely correspond to ±15% of the nominal, which is a standard in the industry);
d) the ratio is plotted *vs.* the reference concentration, ideally giving a flat line parallel to the x-axis which means that the slope equals zero;
e) a slope and intercept are calculated for this line, and also the 95% confidence intervals around the slope: the upper and lower confidence interval must bracket zero.

ASSAY CONTROL

The quality of a project carried out in the 'big batch' mode has to be controlled very tightly. It is simply inconceivable to complete a run of several hundred samples without monitoring the quality, and perhaps have to reject the data in the end because of poor quality that endangers a costly and/or unique study. It is suggested to use any of the QC charts that are well described in many handbooks.* The simplest are based on standard deviation (S.D.) and contain a warning line at mean ±2 S.D.'s and an action line at mean ±3 S.D.'s. One QC outside the action line, or two in a row outside the warning line, should stop the run. There are also secondary rules to prevent results from drifting in one direction. It may be necessary in the future to develop software which will monitor performance of the system without a human intervention, as most of the run will be completed overnight [1].

ACCEPTANCE CRITERIA

It is further proposed to use a system involving inspection (sampling) by variables as acceptance criteria [2]. The most common acceptance criteria, termed 'fixed range' ('4-6-15'; four out of six QC's must be within 15% of the nominal) are arbitrary, confuse precision and accuracy, and are not based on statistics in any strict sense. The limitations of this approach have been discussed in the previous volume in this series [1].

Acceptance criteria in bioanalysis should be based on the principles set up in manufacturing industries. One such system is inspection by variables [3, 4], in contrast with the 'fixed range' approach which is basically an inspection by attributes - the one and only question which it answers being: "How many results are within the specification?". Inspection (sampling) by variables is a valid, internally coherent tool in establishing the acceptability of bioanalytical results [2]. As a rule of thumb the inspection-by-variables approach requires only half as many QC's as the inspection-by-attributes approach. It provides immediate answers, as it is based on internationally recognized principles (ISO 3591 & MIL-STD-414), it incorporates precision, accuracy and specification into the acceptance criteria, and it relates the number of QC's with the batch size. To render the Washington Conference Report recommendation [5] in agreement with ISO 3951, the Acceptable Quality Level of 6.50% or lower must be selected. The flexibility of the system permits accommodating larger and smaller batches of samples, and establishes statistically valid project-specific acceptance criteria.

Using the concepts described above and combining the 'big batch' approach with inspection by variables could introduce significant savings. By eliminating excess calibration standards and halving the number of QC's, the number of samples in a study

*example: 'Statistical Process Control' (1991) by G.B. Wetherill & D.W. Brown, Chapman & Hall, London.

can be reduced by ~15%, with collateral savings in analysts' effort, biological medium, reagents used and waste produced. By increasing throughput in this way, unit analytical costs can be reduced and productivity increased to meet the requirements of high-throughput screening and contribute to the shrinkage in the cost of developing new medicines.

Acknowledgement.- The author thanks Glenn Smith and Derek Parsons for their assistance in editing this manuscript.

References

1. Hill, H.M., Tennant, K. & White, S.A. (1996) in 'Biofluid Assay for Peptide-Related and Other Drugs' [Vol. 24, this series] (Reid, E., Hill, H.M. & Wilson, I.D., eds.), Royal Society of Chemistry, Cambridge, pp. 316-326.
2. Selinger, K. (1995) J. Pharm. Biomed. Anal. **13**, 1427-1436.
3. Sampling Procedures and Tables for Inspection by Variables for Percent Defective (1957) MIL-STD-414, U.S. Department of Defense, Military Standard, Government Printing Office, Washington DC.
4. Sampling Procedures and Charts for Inspection by Variables for Percent Nonconforming (1989) ISO 3951-1989, International Organization for Standardization, CH-1211 Geneva 20.
5. Shah, V.P., Midha, K., Dighe, S. & 9 other authors (1992) J. Pharm. Sci. **81**, 309-312.

APPENDIX: a drug-analyte example of the validation approach

#batches	3
Total # of samples	295
Stds and blanks	54
Quality Controls	18
Zero conc. samples	74
Non-zero conc. samples	147
Mean Ctest/Creference ratio	1.0536
95%confidence intervals of the mean	1.0362 - 1.0710
Ratio	Passed
Lower 95% CI of the slope of the regress.	-0.00000015, passed
Upper 95% CI of the slope of the regress.	0.00000656, passed

In the project presented in Appendix, three batches of approximately 100 samples each were analyzed. Each batch contained own front and back calibration curve; results obtained by reading peak height against the own calibration curve are denoted as Creference. At the end of the study the front calibration curve from batch 1 and the end calibration curve from batch 3 were combined. All non-zero clinical samples were read against this combined curve; these are denoted as Ctest.

#D-6

ACCURACY AND PRECISION: DETERMINATION AND APPLICATION[¤]

H.M. Hill†, C. Gardner & K.A. Tennant

Covance Laboratories Ltd., Dept. of Biopharmaceutical Analysis, Otley Road, Harrogate, N. Yorks. HG3 1PY, U.K.

The terminology associated with precision and accuracy is varied and rarely well defined. Moreover, the procedures used to determine these parameters tend to be poorly described in bioanalytical reports, be they for publication or submission to regulatory authorities. A wealth of definitions originating from international bodies, including ISO, ICH, USP, AOAC as well as the Consensus Washington Meeting of 1990 (which leans more to bioanalysis), has brought added confusion rather than clarification. For example,* precision *is variously defined as* repeatibility, intermediate precision, reproducibility *and forms a component of* robustness *and* ruggedness. *Similarly* accuracy *is defined as* bias *or* trueness.

Along with the diversity of terminology and definitions there are a range of procedures for calculating these parameters. In any document these should be well defined or at least appropriately referenced. While the acceptance of bioanalytical data is subject to the empirical criteria defined in Washingon guidelines,§ uses of the criteria including the 'confidence interval approach' are being proposed, each depending largely on the establishment of acceptable precision and accuracy data for the method during validation.

The terminology used to define accuracy and precision, whilst widely documented, is subject to review and refinement. The terms as such may have different meanings to different individuals. In view of the fact that these criteria, once generated, can be used to characterize the performance and acceptability of a method, it is essential that these parameters be well defined. Their impact on determining method acceptability is now discussed with respect to different acceptance criteria.

---

†addressee for any correspondence

Abbreviations.*- For **organizations see ref. [1], p. 316 (& 325 for addresses; ICH = International Conference on Harmonization; ISO, International Standards Organization). LLOQ, lower limit of quantitation; Lo/Me/Hi, low/medium/high calibrant levels; QC, quality control (sample); (R)SD, (relative) standard deviation.

¤Editor's Note.- *Dr Hill warmly responded to a plea - "Forum talk/ book article, please!" - from the Senior Editor, often faced with infelicitous phrasing - e.g. "good inaccuracy", "low precision", "imprecision was better than..." - for values deemed satisfactory. The present article, and a* ***companion article*** *[1] less focussed on accuracy/precision aspects, together furnish* ***notable guidance*** *of especial help for* ***compiling Reports and satisfying Regulators.***

§published (1992) in Pharmaceut. Res. **9**, 588-592 and J. Pharm. Sci. **81**, 309-312.

Precision (and **repeatability, robustness, ruggedness**)

The term precision has been variously defined as imprecision, intermediate precision and reproducibility, and as a component of reproducibility and ruggedness. While the term imprecision more accurately *(sic)* defines the value of this parameter, be it expressed in relative standard deviations (RSD) or coefficient of variation (C.V.)*, it nevertheless has negative connotations and as a consequence is not widely used.

Recently ICH have defined 'precision' as having 3 components - repeatability, intermediate precision and reproducibility. ISO uses similar terminology, as discussed earlier in this series [1].

Two parameters which incorporate a 'precision' element and are often used interchangeably are 'robustness' and 'ruggedness'. Both are defined in the USP but only robustness has been incorporated in the ICH guidelines.- "Robustness is a measure of the method's capacity to remain unaffected by small but deliberate variations in method parameters and provides an indication of its reliability (ruggedness) during normal use." The USP defines robustness as "degree of *reproducibility* of test results obtained by the analysis of the same samples under a variety of normal test conditions", e.g. differing laboratories, analysts, instruments, lots of reagents, elapsed assay times, assay temperatures or days. Ruggedness normally connotes the lack of influence on test results of operational and environmental variables of the analytical method.

The ruggedness parameter is widely evaluated during method validation in the pharmaceutical analysis of formulations for content and impurities as applied to product release or stability evaluation. The two most widely used techniques are multivariate analysis and, now largely replacing it, the pragmatic Plackett/Burman method [2, 3] - well exemplified in a Vol. 24 article on validation of bioanalytical methods [4]. The amount of work that is involved in carrying out such studies must be considered in the light of potential benefits.

'Ruggedness' can be considered as largely a practical reflection of the more mathematically evaluated 'robustness'. Method ruggedness may be improved without implementing the complexities of the Plackett/Burman analysis if the analyst is aware during method development of critical factors which may impinge upon the ruggedness of the assay. Thus by documenting them in the analytical methodology, a relatively fragile assay may be made more robust.

*In this book, 'C.V.' is generally used (assuming no confusion with *curriculum vitae!*) rather than 'CV' - which sometimes serves to connote 'cardiovascular' [as in art. #E-1; cf. exceptional use of 'RSD' rather than 'C.V.' therein] or 'cyclic voltammetry' [as in the MS. for art. #C-4; altered to 'CyV' therein].-*Ed.*

All approaches may be reckoned correct, provided that the terms of reference are known; each has its *pros* and *cons*. It is incumbent on the analyst to define the limits of the calculations.

Accuracy (and **bias, trueness**)

The complexities of an apparently obvious term are discussed in detail in an ISO document [5].- "The general term accuracy was at one time used to cover only the one component named trueness, but it became clear that to many persons it should imply the total displacement of a result from a reference value, due to *random* as well as *systematic* effects". Thus accuracy according to ISO is a combination of trueness and precision.

The term 'bias' has been in use for statistical matters for a very long time; but because it caused certain philosophical objections among members of some professions (e.g. medical and legal), the positive aspect has been emphasized by the invention of the term 'trueness'.

'Accuracy'-related terms as defined by ISO are as follows.-

Accuracy.- The closeness of agreement between a test result and the accepted reference value.

Trueness.- The closeness of agreement between the average value obtained from a large series of test results and an accepted reference value.

Bias.- The difference between the expectations of the test results and an accepted reference value; comprises the total systematic error in contrast with random error. There may be one or more systematic error components contributing to the bias. A large systematic difference from the accepted reference value is reflected by a larger bias value.

However, accuracy and precision are so widely regarded as separate entities that reckoning precision to be a component of accuracy remains a remote possibility in bioanalytical quarters.

CALCULATING ACCURACY AND PRECISION (For reporting, see p. 229)

Of more immediate concern than 'dictionary' definitions is the matter of how these terms are calculated. A usual procedure for Bioanalytical Validation involves the analysis of at least 6 QC's at 3 levels (Lo, Me, Hi) or 4 levels (LLOQ also) in 3 to 6 batches although there is a tendency to complete this process in 3 or 4 batches. For 3 batches and the LLOQ level, Table 1 shows data relating to precision and accuracy calculation. It gives, in relation to **intra-batch** accuracy and precision, calculated values for each batch, and overall averages, the C.V. being 16.9% (corresponding to the ±0.366) and ranging from 14.4% to 21.2% for individual batches. (Any incompleteness of the sets in one or more batches would impose a limitation.) **Inter-batch** parameters may be calculated (cf. Table 1 data) in at least two ways: either by using all the numbers and generating an overall mean and SD,

Table 1. Calculating accuracy and precision, exemplified by 3 assay batches each with 6 QC's at the LLOQ (here 2.0 ng/ml).

Batch	Values found (ng/ml) for successive QC's						Mean ±SD.
1	2.3,	1.9,	2.7,	1.8,	2.1,	2.4	2.20 ±0.335
2	1.9,	2.1,	2.3,	1.9	2.6,	1.8	2.10 ±0.303
3	2.2,	2.1	2.9,	1.5,	2.4,	2.0	2.18 ±0.462
	(first 2)		Overall, intra-batch:				2.16 ±0.366

i.e. 2.16 ±0.53 (C.V. = 16.3%) or by pre-defining any group of 6 (3 pairs; exemplified by the *first 2* of each batch) to calculate inter-batch (run) accuracy and precision, here 2.08 ±0.160 (C.V. = 7.7%). The latter is directly comparable with repeatability in an individual batch where 6 QC's are used. While the differences may be considered minimal, the implications for future use could have significant impact upon the acceptance or rejection of data. The reporting style need not be as in Table 1, as is amplified on p. 229 where a variant style is shown (nominal value = '100%').

VALIDATION ACCEPTANCE

In respect of precision and accuracy, the acceptance criteria are generally considered to be no more than ±15%, although at the LLOQ ±20% is reckoned acceptable. These criteria are definitive, albeit empirical; but the criteria used for determining their parameters vary greatly (see above). How do they compare with current observed results? Some indication can be gained from an in-house review of achieved precisions for >100 assays using HPLC-EC,HPLC-UV and HPLC-fluor as well as LC-MS-MS.

Table 2 gives data for precision at the LLOQ in validations and sample analysis studies. Evidently average precision is in general much better than ±20%, although individual methods do in some cases approach this limit. Accordingly a cut-off set at ±20%, gives an empirically acceptable threshold, both analytically and in practical application to pharmacokinetic studies.

Should QC's be used to determine precision and accuracy? Does accuracy and precision determined from each standard curve have any relevance? In general the latter values should exhibit tighter precision and accuracy than QC-derived values. They are largely of self-fulfilling nature; hence their role is of limited value although their calculation is universal and provides the reader with a degree of comfort, albeit limited in use. Under no circumstance should such derived precision and accuracy values be used to define the statistical characteristics of the method.

Should one expect precision to be the same with 'incurred' samples (samples received; *ex vivo* or 'real') as spiked samples?

Table 2. Precision values for various in-house assays.

LLOQ, ng/ml	Precision at LLOQ (C.V.) Validation	Actual studies	Analytical techniques
up to 1	±10.1%	± 11.4%	LC-MS-MS ~90% LC-EC ~10%
1-10	±8.8%	± 10.6%	LC-MS-MS ~70% LC-EC, -UV, -fluor ~70%
1-50	±9.4%	±8.2%	LC-MS-MS ~30% LC-EC, -UV, -fluor ~70%

This is rarely considered although the Canadian authorities like to see precision for at least 15% of the incurred samples evaluated in one study [6]. However this information is usually on a "nice to know" basis rather than used in the decision-making process.

Historically 'precision' has been ascertained by use of the standard curve where S.E. of the slope and intercept can be determined [6,7], and in some instances - especially in conjunction with the confidence-interval approach - can provide an alternative measure of precision. Thus, for an assay with a nominal C.V. of ~10%, 90% confidence intervals range from 20% to 30% [7].

The '4:6:20 rule', and other acceptance-criteria considerations

While in the Washington Guidelines the variability of the assay is empirically set at ±15% for accuracy and precision except for the LLOQ, the criteria for batch acceptance have been set at 20%, resulting in what has become known as the 4:6:20 rule, i.e. 4 of 6 QC's (each 6 comprising duplicate QC's set at 3 levels) should be within 20% of the nominal value, with at least one at each concentration being within these limits. The relative flaws in this empirical approach have highlighted inconsistencies with setting acceptance levels of precision and accuracy at ±15%, as discussed by various authors and summarized in Vol. 24 [1].

Other established criteria generally used by clinical biochemists might be considered as alternative acceptance criteria whose relative merits need consideration. Perhaps the most widely known criteria are those developed by Shewhart [8] and modified by various authors; indeed the current 4:6:20 rule is not unrelated to Shewhart. These variants take account of trends which may go unnoticed using the 4:6:20 rule. Thus the '1·2SD' rule where two QC's are used to determine batch acceptance or rejection has both batch-dependent and independent criteria, e.g.:

#each of 2 QC's **inside**/outside ±2 SD: **accept**/reject batch;
#one QC outside, one inside ±2 SD: reject batch;
#five consecutive QC's between 1 and 2 SD: some action needed to correct an apparent drift.

A 'fine-tuned' version of this approach [9] introduced the multi-rule Shewhart criteria, extending the inpact of 'trend' analysis.

Thus one can accept data between ±2 and ±3 SD while rejecting data only according to an extended range of criteria, e.g.:

1:3SD rule: one QC outside of ±3 SD;
2:2SD rule: 2 consecutive QC's outside the ±2 SD range;
4:1SD rule: 4 consecutive QC's exceed either +1 SD or -1 SD, i.e. exhibit consistent bias;
if one QC was outside +2 SD and the next is outside -2 SD then the assay is said to be out of control.

These somewhat complex rules, designed largely to ensure the assay remains within limits over an extended period of time, may have limited application to bioanalysis for 'small studies' (500-1000 samples) that may be completed within 4-10 batches. However, their use in extended Phase III studies where samples may be analyzed over a 2-3 year period is to be encouraged.

The confidence interval approach may prove more useful in smaller studies, Phase I/II or preclinical. Thereby an acceptable range about the mean QC outlines can be established, where the mean ±2.58 × SD is used to set the criteria. Again, however, an empirical element must be introduced in setting the interval: 99, 95 and 90% are the usual statistically 'approved' levels, but the choice must be one that all or most analysts can agree upon - or can individually justify within the context of the application of the results. Also, should settling the definitive SD be based on validation data, or retrospectively on study data, or on the SD established in a previous study? Using the confidence approach results in only the precision parameter being used; so one must add an accuracy or bias criterion, commonly ±15%. The interaction between accuracy and precision on the acceptability of QC data has been well documented by Kringle [10] and reviewed by us [1].

An approach as for quality acceptance criteria in industrial use for >50 years has been taken up by K. Selinger, and extended to the rolling batch concept (preceding art.). Here a calibration is established early in the analytical 'run' but QC's on a rolling basis are used to determine whether the assay is under control - calling for constant real-time review of QC values, best achieved by real-time total automation of data processing and acceptance scrutiny. Thereby the empirical 4:6:20, simple and transparent to use, perhaps may give way to more complex statistical procedures. In any event, determination of the basic criteria of batch acceptance, precision and accuracy remains as debate prone and varied as the workers in the field. Any or all of these observations for accuracy/precision can be used to determine acceptability of results (cf. [7] and the foregoing art. by K. Selinger).

Calculation of precision and accuracy using QC samples may be complicated (modified) by the mere expediency involved in QC and calibration standard placement (Fig. 1).

Fig. 1. Calibration (cal) and QC within-batch distribution.

Decreasing % values (sloped lines) during a run signify manifest QC drift.

A: cal curve random; QC evenly spaced
B: cal curve at start; " "
C: cal curve and QC both random
D: cal curve at start; QC random

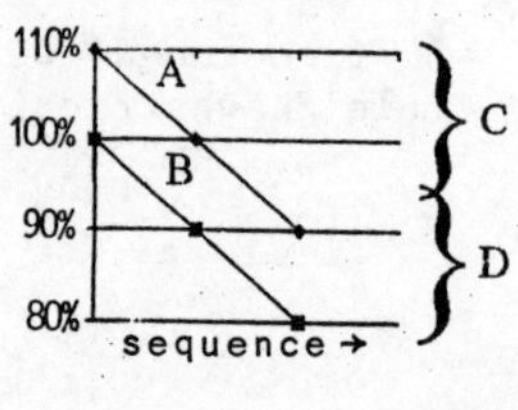

Both standards (cal's) and QC's are usually randomized (Fig. 1, C; 100% = nominal). Drift will be hard to discern/measure; likewise for D though the 'absolute values' of each QC will be lower compared with situation C. With A, any loss of signal will show only from the evenly spaced QC's, not the cal's. In B, rates of drift will appear the same, but calculated values and means will differ. Quantifying drift thus calls for evenly spaced QC's.

Terminology and calculation procedures call for consistency whatever parameter is used. Thus in bioanalysis the terms *batch* and *run* are used interchangeably to mean a collection of 'unknown' samples analyzed at the same time with one calibration curve and the appropriate number of QC's. In other disciplines not unrelated to bioanalysis the term 'batch' has varying meanings. A glossary of terms should be attached to all relevant documentation. As international submissions become increasingly common, it is essential that reviewers understand completely the terminology and the context in which it is used.

Accuracy/precision tabulations can be formatted according to the laboratory's preference provided that the headings are clear. Table 1 shows one option. In Table 3 accuracy is expressed as a total percentage, 100% representing the nominal value; the found accuracy value is the basis for precision (or 'imprecision'; %). Other authors express accuracy as % difference from nominal (i.e. bias or inaccuracy) - a minority view, yet acceptable provided that the context is defined and understood.

Table 3. Assay of a drug in human plasma: specimen format for reporting precision and accuracy of QC's.

	QC 927.0 ng/ml		Similar columns for 1 lower & 2 higher QC's
Batch	Observed concn. (ng/ml)	Accuracy (%)	
14 (e.g.)	961.8	103.8	16 other batches: similar dual rows
	967.9	104.4	
Mean (ng/ml)	967.4	-	
SD (n - 1)	32.89	-	Similar entries for other levels
Precision (%)	3.4	-	
Accuracy (%)	104.4	-	
n	33	-	

References [NOTE by Ed.: 'Validity' features strongly in 1. & in V. 20 (1990; RSC)]

1. Hill, H.M., Tennant, K. & White, S.A. (1996) in 'Biofluid Assay for Peptide-Related and Other Drugs' [V. 24, this series] (Reid, E...RSC..[as now]), pp. 316-326.
2. Box, G.E.P., Hunter, W.G. & Hunter, J.S. (1987) 'Statistics for Experimenters', Wiley, New York; notably pp. 291-434.
3. Mulholland, M. (1988) TrAC **7**, 383-389.
4. van der Stel, D.J.K. and 6 co-authors (1996) as for 1., pp. 105-117.
5. ISO 5725-1 (1994) Part I: 'Accuracy (trueness and precision) of measurement...'
6. Galliciano, K., Orrmsby, E.D. & McGilvray, I.J. (1995) in 'Biointernational, Bioavaila... ..Studies' (Blume, H.H. & Midha, K.K., eds.), medpharm, Stuttgart, pp. 397-398.
7. Leppard, J.P. (1981) in 'Trace-Organic Sample Handling [V.10] (Reid, E., ed.), Horwood, Chichester, pp. 320-335. [Is O/P. **Corrigenda** in V. 12 (1983; Plenum).]
8. Shewhart, W.A. (1931) 'Economic Control of Quality of Manufactured Products', Van Nostrand, New York.
9. Westgaard, J.O. (1982) Diagnostic Med. **61**, 741.
10. Kringle, R.O. (1994) Pharm. Res. **11**, 556-560.

#D-7

EXPERIENCES WITH HPLC-MS-MS FOR THE TRACE BIOANALYIS OF DRUGS IN PLASMA

R. Payne†, C. Smith, A.J.C. Wakefield & I.D. Wilson
Dept. of Safety of Medicines, Zeneca Pharmaceuticals, Mereside, Alderley Park, Macclesfield, Cheshire SK10 4TG, U.K.

The use of HPLC-MS-MS is now a standard method of bioanalysis in many pharmaceutical companies. Examples are presented of various problems that have arisen during the course of method development for a variety of different drug candidates of diverse structure. These examples clearly demonstrate the continued need for careful sample preparation and method development when devising HPLC-MS-MS assays for trace analysis.

There is no doubt that the recent introduction of reliable quantitative HPLC-MS-MS equipment into the drug-bioanalysis field has resulted in a dramatic change in work practices and a shift away from 'traditional' HPLC-UV and fluorescence methods. Indeed, such is the sensitivity of these instruments that in some areas they have replaced RIA as the method of choice for certain assays. However, partly because of the success of the MS-MS approach, with its potential for both high sensitivity and specificity, it has come to be assumed that such methods do not require careful elaboration. Indeed, the perception that all that is required to develop a method is some simple sample treatment such as protein precipitation, followed by an abbreviated chromatographic separation ('dilute and shoot'), is widespread.

Without doubt such a philosophy can apply under certain conditions, particularly when the intended analyte is present in the biofluid sample (e.g. plasma) at concentrations >10-50 ng/ml. However, at the ng/ml or sub-ng/ml level even MS-MS is no guarantor of specificity. As a counter to the view (questioned in the previous vol. [1]; cf. [2]) that HPLC-MS-MS provides a simple solution to trace bioanalysis, we give four examples of HPLC-MS-MS methods, developed in our laboratory, which illustrate some of the potential pitfalls that can face the bioanalyst in this area.

Equipment (Perkin-Elmer).- We used either a Sciex API-III+ or an API-300* mass spectrometer with Series 200 HPLC pumps and autosamplers. **Methods** were all for drugs in plasma and were based on RP-HPLC using a variety of commercial columns, and following diverse sample preparation procedure as detailed below. All reagents were of analytical or HPLC grade.

---

†addressee for any correspondence
**Abbreviations*.- In tandem MS (MS-MS): APcI, atmospheric pressure chemical ionization; MRM, multiple reaction monitoring. LLOQ, lower limit of quantification; MeCN, acetonitrile; MTBE, methyl t-butyl ether; TFA, trifluoroacetic acid.

Example 1: Analyte *vs.* metabolites

This example of types of difficulty that can arise in the routine use of HPLC-MS-MS for trace analysis with a difficult analyte relates to problems encountered in assaying a steroidal compound. The main reason for the development of an HPLC-MS-MS assay for this compound was that the pre-existing RIA method was time-consuming and inefficient, requiring ~5 days for analysis of 40 samples. An assay with higher throughput and sub-ng/ml sensitivity was required.

For HPLC-MS-MS, plasma samples (0.5 ml) were extracted into hexane/isopropanol, which was evaporated off (N_2 stream); the residue was redissolved in the HPLC mobile phase, methanol/water (85:15). An YMC Aquapore column was used (30 × 4.6 mm; flow rate 0.8 ml/min). The R_T was ~2 min. The method was developed on the API///+ using APcI with a heated nebulizer inlet. Quantification was performed by taking the analyte's molecular ion at 607.4 amu, performing MRM and monitoring the 589.33 amu product ion which resulted from loss of water and was the only substantial product. The assay was found to be linear up to 50 (LLOQ 0.2) ng/ml.

When the assay was applied to samples from patients, problems were observed. Firstly there was an extra peak running close to and interfering with analyte quantification (Fig. 1a). Secondly, a significant discrepancy between RIA and chromatographic results was noted for patient, but not spiked, samples. The reason for both the interfering peak and the discrepancy between methods was the presence of an interfering metabolite which cross-reacted with the antiserum used in the RIA. This metabolite was similar in R_T to the parent and also gave rise to the same ions as used to monitor the drug, causing interference in trace assays. By adjustment of the HPLC system to increase the target analyte's R_T it was separated from the interfering metabolite and a sensitive and specific method could be obtained. The resulting separation had a doubled HPLC time to obtain adequate resolution (Fig. 1b).

This example shows that, even with HPLC-MS-MS, unsuspected metabolites can cause confusion, and that there is no substitute for real samples for determining the suitability of a method. Chromatography is also clearly important, and MS-MS is not itself a guarantee of specificity.

Example 2: Interference from sample components

In the second example the analyte was an acidic Br-containing compound, such that the potential sensitivity of the method was inherently reduced by 50% because of the presence of the two Br isotopes. The pre-existing HPLC-UV method, which employed solvent extraction into MTBE and chromatography on a HiChrom HiRPB column (15 x 4.6 mm; MeCN/water/TFA, 70:30:0.2 by vol.) was used as the basis for developing an HPLC-MS-MS method. Under these conditions the HPLC run time was ~12 min. This was adjusted to provide an R_T

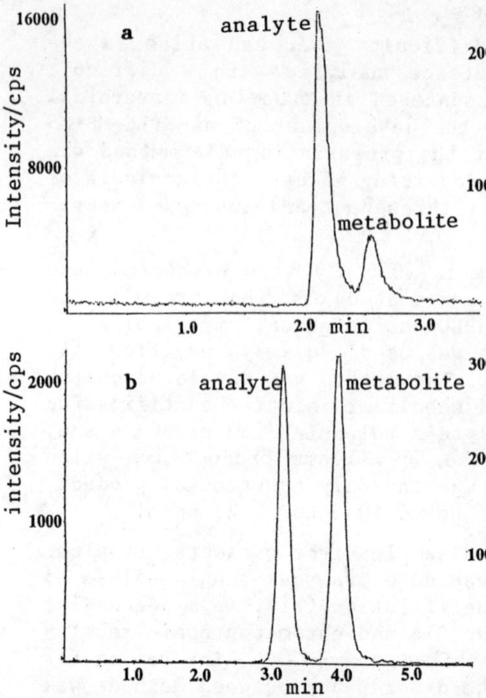

Fig. 1a. HPLC of analyte and metabolite that was beset by interference in trace analysis by HPLC-MS-MS; **b**, baseline separation of analyte and metabolite such that quantification of the former in the presence of the latter was not subject to interference.

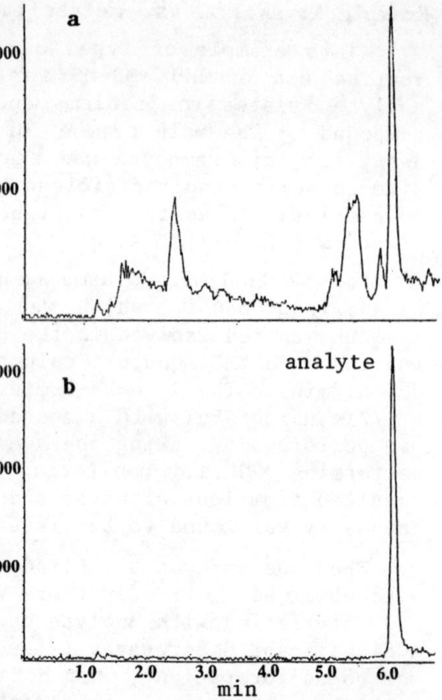

Fig. 2a. A blank extract of plasma with an interfering peak the same R as the analyte; (**b**) HPLC-MS-MS of the analyte giving R_T ~6 min.

of ~6 min for the HPLC-MS-MS assay (Fig. 2a). From preliminary work, using the fragment ion of one of the molecular ions and monitoring the product at 91 amu, the development of a method of the required sensitivity seemed readily attainable. Moreover, the linear range extended over the anticipated clinical exposure (up to ~100 ng/ml). However, when the method was applied to spiked plasma samples an interfering peak was noted running at the same R_T as the analyte (Fig. 2b). Aggravating the problem, the amount of this interference was inconstant between samples, varying in equivalent concentration from 1 to 20 ng/ml.

After extensive method development the solution to this problem was to use gradient elution, with a similar mobile phase to that employed for the isocratic method. This successfully separated the analyte from the interference but led to a run time of ~11 min per sample.

As this method shows, even in the absence of interfering metabolites it is still by no means certain that the use of MS-MS will result in the required degree of specificity.

Example 3: Problems with instrument contamination

In this case there was a pre-existing method, using HPLC-UV, which had been employed for monitoring in toxicokinetic studies; but a more sensitive method was required to support the clinical program. The sample preparation procedure developed in the HPLC-UV method was taken as the basis for the development of the HPLC-MS-MS assay, but the chromatographic system was adapted to make it compatible. Sample preparation was by SPE on mixed-mode cartridges, with HPLC on a 100 × 4.6 mm 5 µm Hypersil Elite C18 column; the mobile phase was MeCN/water/formic acid (40:60:2 by vol.; 1 ml/min). Initial investigation revealed that it should be easy to achieve the required sensitivity of ~1 ng/ml and to extend the linear range to at least 2 µg/ml, more than sufficient for the proposed clinical program. No interferences of any type were evident in plasma samples. However, it rapidly became clear that after 20-30 injections of plasma extracts onto the HPLC-MS-MS system a gradual but significant loss in the sensitivity of the instrument was occurring. This sensitivity loss was noted not only with plasma extracts but also with non-extracted standards. Clearly this was due to contamination of the instrument in some way; but the cause was not obvious.

The remedy finally developed was to send the column efflux to waste until just before the analyte was due to elute, and then switch the flow back in-line with the MS-MS for detection:

	Peak area: at start of run	at end of run
Without solvent dumping	3861	1858
After solvent dumping	1727	1773

This proved to be a simple method for preventing the detector contamination, and meant that the SPE method did not have to be re-developed with consequent loss of time.

Example 4: Positive *vs.* negative ionization

This final example of problems encountered arose in studies on a compound with metal-chelating properties. Concentrations of this analyte were expected to be quite high (>10 ng/ml), but the compound lacked a suitable chromophore for HPLC-UV. A simple protein precipitation method (2 vol. MeCN) for sample preparation prior to HPLC-MS-MS was tried. From the supernatant 100 µl was injected into a C-18 pre-column using a highly aqueous mobile phase to ensure retention. The pre-column was then back-flushed onto the C-18 analytical column using the HPLC mobile phase.

The main problem with this method resulted from the metal-chelating properties of the analyte that resulted in the detection of large, but variable, numbers of sodium and potassium adducts in Q1 (the first quadrupole) when positive ionization was used

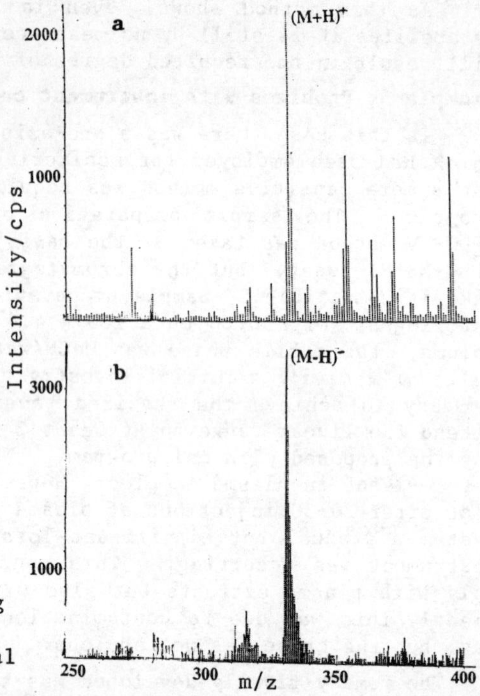

Fig. 3a. Positive-ion Q1 scan of the metal-chelating compound; **b**, negative-ion Q1 scan of the same material

(Fig. 3a). This not only reduced sensitivity but also meant that MRM could not be used since the sodium/potassium adducts would not fragment. Addition of other chelators (e.g. EDTA) or attempts to obtain a single adduct by addition of an excess of one of the adduct-forming ions were without success. Eventually it was found that the use of negative ionization in Q1, although inherently less sensitive than positive ionization, allowed the detection of the compound without the formation of adducts (Fig. 3b).

CONCLUSIONS

Attempts to assay drugs by HPLC-MS-MS with sensitivity and precision at the ng/ml or sub-ng/ml level depend critically on adequate sample preparation and/or adequate separation of the analyte from interferences. The use of MS-MS is not, of itself, guaranteed to provide specificity, only to increase it. Interferences can result from metabolites of the analyte which break down to the same ions as those being monitored for the parent, or from endogenous materials that similarly give rise to ions that are mistaken for the analyte itself. Such problems can be accentuated by short HPLC run times. Often the only way to get round these problems is to invest considerable time and effort in good sample preparation and to accept R_T's longer than desired.

References.- 1. Simmonds, R.J., Enos, T.A. & Speed, W. (1996) in Vol. 24, this series (eds. & publ. AS NOW), pp. 128-136. **2.** Mason, C. et al.: SAME, pp. 137-140.

#D-8

THE USE OF LC-MS-MS FOR MULTIPLE COMPONENT PHARMACOKINETIC STUDIES IN SUPPORT OF DRUG DISCOVERY

J.D. Gilbert[▫], D.A. McLoughlin & T.V. Olah[†]
Merck Research Laboratories, Dept. of Drug Metabolism,
West Point, PA 19486, U.S.A.

New drug candidates are being synthesized at an accelerating rate and, until recently, only some of these could be selected for PK screening. HPLC-MS-MS enables the rapid and specific quantitation of analytes even though these may not be chromatographically resolved. Our laboratories are using LC-MS-MS to allow simultaneous analysis of up to 20 individual substances in plasma. Generic extraction and HPLC conditions provide good recoveries and convenient HPLC run times. The accuracy, precision and LLOQ of these assays are carefully monitored. Studies entail either giving a single substance to individual animals and pooling plasma before analysis or co-administering mixtures of candidates to single animals. This article gives methodological details and considers the advantages and drawbacks of the PK design.*

The activities of drug metabolism scientists in the pharmaceutical industry have changed radically in the past few years. From playing, hitherto, a largely developmental role, they have rapidly evolved to fulfilling a key role in the drug discovery and selection processes, calling for their adopting and applying new technologies. Arguably the most important of these is LC-MS-MS with API*. This technique's remarkable impact on bioanalysis in the pharmaceutical industry, recently reviewed [1], is manifest in this and the preceding volume. The unparalleled resolving power of the MS-MS detector enables substances to be detected and quantitated even if not chromatographically resolved - which has proved particularly useful for drugs labelled with a stable [2] or radio [3] isotope.

Applications of quantitative LC-MS-MS have now evolved from assaying single analytes in biofluids [4, 5] to embracing 5-20 analytes simultaneously with good accuracy and precision [6, 7].

▫died late-1997 (car crash); mourned by the 'Forum fraternity'.
†addressee for any correspondence
**Abbreviations*.- In tandem MS (MS-MS): API, atmospheric pressure ionization; SRM, selected reaction monitoring. AUC, area under curve; i.s., internal standard - 'analytical' or (ref. candidate) 'biological'; LLOQ [LOQ in some Figs.], lower limit of quantitation; MeCN, acetonitrile; PK, pharmacokinetic(s); QC, quality control ('sample' usually implied); TFA trifluoroacetic acid.

Thus LC-MS-MS has been eagerly seized by drug-discovery groups as providing a method for rapidly comparing the PK properties of multiple substances after their co-administration *in vivo* [8].

Here we appraise the usefulness of this technique with close to two years' experience. We address the design of multiple-analyte PK studies, the establishment of QC criteria for analytical data and, in the co-administration context, ways to attempt to detect and compensate for interactions *in vivo*.

EXPERIMENTAL

Materials.- The drug candidates studied were synthesized in-house. Other sources: TFA and formic acid: Sigma (St. Louis, MO); methanol and MeCN, HPLC grade: Fisher (Fair Lawn, NJ); Bakerbond SPE cartridges (C-18, CN; 200 mg, 3 ml): J.T. Baker (Phillipsburg, NJ); liq. N_2 and argon (99.999%): West Point Supply (local); disposable vials and caps: Scientific Resources (Somerset, NJ).

Standards (each test substance; structurally similar i.s.).- Stock solutions, 1 mg/ml, were prepared in MeCN/water mixtures to suit individual solubilities; for basic drugs TFA was present too (0.1%). After pooling aliquots of the stock standards subsequent dilution gave a series of solutions which when added to aliquots of control animal plasma yielded concentrations generally in the range 1-1000 ng/ml. QC's were prepared by adding independently prepared solutions to further aliquots of control plasma to give high, intermediate, low and LLOQ concentrations. After addition of i.s., the test samples, standards and QC plasma were assayed together, starting with one of the following three procedures.

Sample preparation options (0.5 ml plasma) - (a) **Solvent extraction.** Phosphate buffer pH 7.5 (0.1 M, 0.5 ml) and methyl t-butyl ether (2 ml) were added. After rigorous vortex-mixing, and centrifugation, the upper organic phase was collected.
(b) **SPE** on cartridges (C-18 or CN) pre-conditioned with 1 ml methanol and 1 ml water. After washing with 1 ml water and 1 ml methanol/water (20:80 v/v), 0.5-1 ml methanol effected elution.
(c) **Protein precipitation.** After adding 1 ml MeCN, vortexing, centrifuging (2000 rpm, 10 min) and evaporating the supernatant to dryness, the residue was reconstituted in 0.1-0.2 ml of mobile phase, with ultrasonication for 10 min. Following transfer to micro-injection vials, 25 µl aliquots were analyzed by LC-MS-MS.

LC-MS-MS was performed on a Sciex (Concord, Ontario, Canada) model API III+ tandem triple quadrupole mass spectrometer having either a Sciex heated nebulizer or a Turbo Ionspray interface. The HPLC system had a Hewlett-Packard (Palo Alto, CA) 1050 quaternary pump and auto-injector equipped with a vacuum degasser. Isocratic HPLC conditions were used for all compound types. Typical RP C-8 or C-18 columns (250 × 4.6 mm) from various manufacturers were used. The mobile phase varied, depending on the complexity of the mixture, in % ratio of solvent A (90:10 MeCN/water) to B (10:90 methanol/water), both containing 0.1 % TFA, so as to achieve

5-15 min analytical run time (flow rates 0.45-1.0 ml/min). When the Turbo Ionspray interface was used, 0.1% formic acid was substituted for TFA. No splitting was performed. Typical the heated nebulizer probe was set at 500°, nebulizing gas 80 psi, auxiliary gas flow 1.2 L/min, curtain gas 1.0 L/min. For Turbo Ionspray the respective settings were 550°, 60 psi, 7.5 L/min and 2.0 L/min.

MS-MS characterization.- Intended analytes were characterized by MS and MS-MS, using flow-injection analysis, to ascertain their appropriate precursor/product ion combinations. Unit resolution of the first quadrupole mass filter (Q1) was 10% valley. In all these experiments, positive ionization was used with orifice potential set at 55 V. For background subtracted spectra Sciex's Macspec software was used. After verifying the precursor ion for each test substance, product ion spectra were collected and precursor/product ion combinations were selected for each analyte. An appropriate compromise of instrument conditions for SRM of all components in the mixture was then established. Quantitation of the SRM detection signal used Sciex's RADS and McQuan software. Dwell times were generally 150 msec and Q2 settling was normally used. The CAD gas thickness (argon) settings were typically 250×10^{13} atoms/cm^{-2} for most MS-MS analyses. For each analyte in the mixture, the analyte/i.s. peak-area ratio was used to construct a calibration curve using a linear regression weighted 1/X or $1/X^2$ (X = known plasma concentration). Plasma concentrations of a particular analyte in plasma samples were determined by interpolation from its appropriate standard curve.

Multiple assay study design

Our laboratories have used two general approaches.-

1) *Single administration with pooling*
   Advantages: #saves instrument time; #no drug interactions.
   Disadvantages: #need to pool plasma; #loss of sensitivity.
2) *Multiple administration*
   Advantages: #saves instrument time; #optimizes animal use; #no plasma pooling.
   Disadvantages: #potential drug interactions.
   Precautions: #'biological' i.s.; #lower doses.

When large numbers of animals (e.g. rats) can be dosed and bled conveniently, a single substance can be given to each animal and small blood samples (~0.2 ml) are collected serially after dosing. Plasma aliquots (50 µl) from each animal are pooled according to the time-points used in blood collection; so each pooled sample yields plasma concentration data on multiple substances simultaneously, depending on the number of experiments combined. The advantages of this approach are the avoidance of *in vivo* drug interactions and the saving of analytical run time on the LC-MS-MS equipment. The number of samples requiring preparation prior to analysis is also greatly reduced, but this is partially offset by the need to firstly pool the plasma samples - an operation which is tedious but needs skill if accuracy is to be maintained.

Quality control (QC)

The essence of multiple-component PK studies is speed. It is, however, essential that some pre-set degree of quality be maintained. Traditional assays for drugs in body fluids necessitate pre-assay validation in which accuracy, intra- and inter-day precision, LLOQ and stability of samples and stock solutions are, *inter alia,* determined. Drug-discovery activities simply cannot wait while these activities are conducted. So the concept of validation has had to be radically reconsidered in the context of PK screens supporting such discovery. The truly good analyst cannot afford to neglect the necessity of changing paradigms in the face of the scientific need for more data produced faster. Such inflexibility on the part of professional analysts will mean plainly that their role will be assumed by enthusiastic amateurs provided with superb methodology but with little or no concept of the true precepts of bioanalytical chemistry.

Recognizing the need for change/flexibility in the name of expedience, pre-study validations have been discontinued; the policy is merely to conduct assays of test samples along with a sufficient number of QC's to back up the data generated. Thus, each set of test samples is accompanied by a sufficient number of both standards and QC's to define the accuracy, precision and LLOQ for each analyte being determined. QC intra-assay acceptance criteria used in our laboratories (only for screening purposes) are as follows.

No pre-study validation .

Assay quality assessment by: #standards, n>6;
#double and single blanks;
#4 sets of QC's (n = 6): LLOQ;
low, middle, high (L, M, H).

Acceptance criteria: #5 readbacks <20% of nominal;
#blanks are clean;
#2/3 of total QC's within 25%;
#C.V. of QC's within 25%;
#2/3 of QC's at LLOQ are within 30% and
C.V. does not exceed 30%.

RESULTS

Fig. 1 shows a series of SRM chromatograms of an extract of plasma from a dog dosed simultaneously with 6 drug candidates at doses of 0.1 mg/kg, namely the reference substance - the leading candidate in the program - and 5 new drugs being compared against it for half-life in plasma. Since shorter plasma half-life was the only issue for selection of a 'better' candidate, only i.v. dosing was performed. Besides the reference substance ('biological i.s.') there is an analytical i.s. to assist quantitation.

Fig. 1 demonstrates well the remarkable power of HPLC with tandem MS detection. Peaks are 'clean' even at low to moderate ng/ml concentrations. There is no interference from endogenous

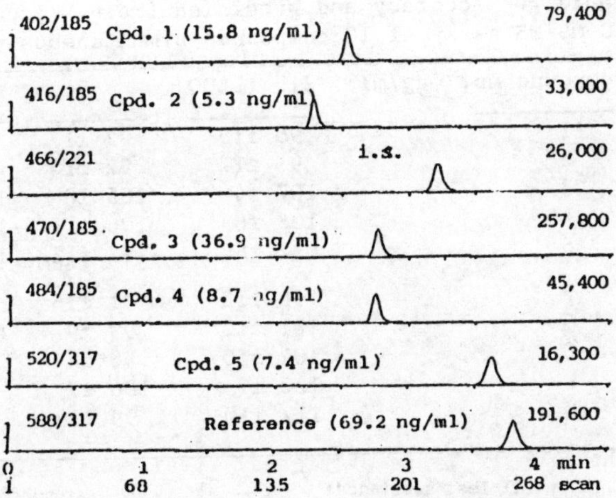

Fig. 1. SRM chromatogram of an extract of plasma from a dog simultaneously dosed i.v. with 6 drug candidates.

plasma constituents. Peaks representing substances which are not chromatographically resolved are completely distinguished by their unique parent/product ion combinations, even in cases where they share a common product ion. The mean (n = 3) plasma concentration-time profiles of dogs dosed with these 6 substances are shown in Fig. 2. Compounds L-806,245 and L-775,299 show distinctly shorter half-lives than the leading candidate (L-768,673) and accordingly were selected for further PK evaluation in single-administration studies. As stressed earlier, maintenance of good QC of these multiple-component assays is profoundly important. Table 1 shows the accuracy and precision of assays for 10 substances given simultaneously to dogs. In all cases we felt it justifiable to accept LLOQ's of 2.5 ng/ml. Fig. 3 shows mean (n = 3) profiles for dogs given these 10 substances orally. The data are meaningful only because the quality of the assays is well defined.

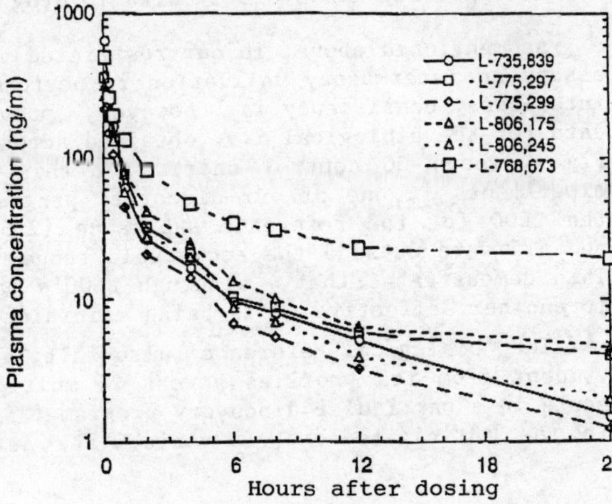

Fig. 2. Mean plasma concentration-time profiles for dogs simultaneously dosed i.v. with 6 drug candidates.

Table 1. Accuracy and precision (mean % accuracy, ±% C.V.) for LC-MS-MS assay of 10 compounds simultaneously (6 plasma QC's).

Compound no. ng/ml:	2.5 (LLOQ)	5	50	500
1	96 ±18	92 ±17	95 ±13	92 ±19
2	91 ±17	92 ±14	94 ±14	92 ±16
3	107 ±4	105 ±8	102 ±6	94 +4
4	106 ±8	106 ±11	102 ±6	93 ±9
5	92 ±8	92 ±12	102 +13	97 ±12
6	96 ±12	91 ±16	90 ±15	85 ±10
7	101 ±12	99 ±9	103 ±11	85 ±8
8	104 ±10	98 ±4	93 ±15	74 ±7
9	120 ±7	106 ±7	92 ±10	88 ±5
10	117 ±10	99 ±13	98 ±12	98 ±13

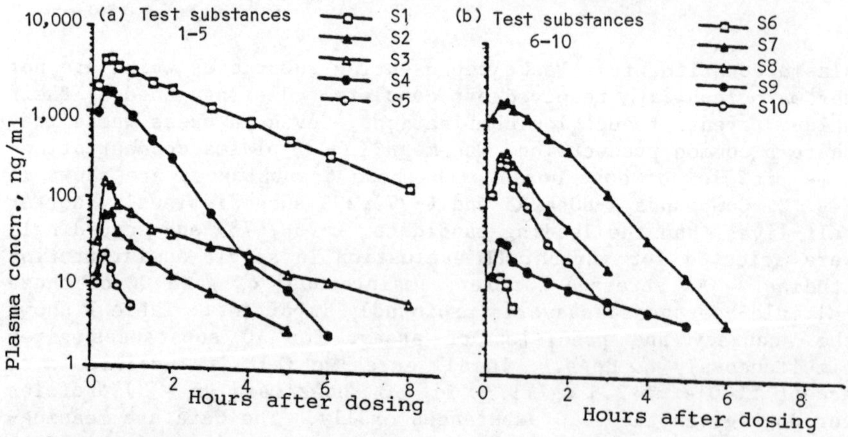

Fig. 3. Mean plasma concentration-time profiles for dogs simultaneously dosed orally with 10 drug candidates.

As mentioned above, in our restricted validation of multiple-assays no inter-assay validation is performed. Some measure of inter-assay consistency is, however, provided by analysis of QC data for the biological i.s. obtained across a series of screens. Fig. 4 shows QC control charts for this i.s. in 15 screening experiments in one of our discovery programs. The % C.V.'s at the LLOQ for the reference substance (2.5 ng/ml) in assays #14 and #15 were outside the acceptable range (30%) of our criteria. This demonstrates that, although LLOQ's may vary from one assay to another, effective QC is being maintained.

The 'biological reference' also fulfils a more important role, evident from its profiles across 26 multiple-dose (dog) experiments in a particular discovery program (Fig. 5). Although there is the anticipated *in vivo* variability across the experiments,

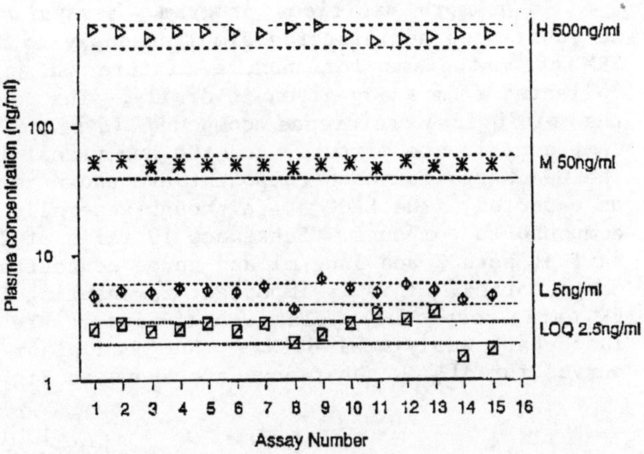

Fig. 4. Summary of mean QC data (n = 6) for 'biological' i.s. dosed in 15 co-administration studies. Dotted lines represent ±25% of the nominal QC concentration. ('LOQ' signifies LLOQ; H, high; M, medium; L, low.)

the profiles in expts. #20 and #22 are notably outside the norms. The significant perturbation of the profiles in these two (particularly in terms of plasma half-life) indicates that one or more of the 10 co-administered substances inhibits the metabolism of the biological reference. *In vitro* experiments using dog liver microsomes showed that one or more substances co-administered in expts. #20 and #22 were potent inhibitors of CyP3A4 and thus radically changed the PK profile of the reference substance.

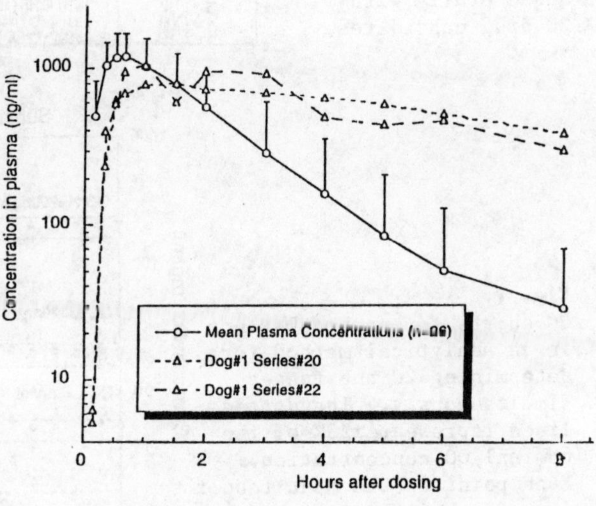

Fig. 5. Plasma concentration profiles of the biological reference compound, averaged across 26 multiple co-administration studies and shown individually for two studies (#20, #22) where drug interactions have occurred (see text).

In a more ambitious program, assays were established for 20 substances administered simultaneously to dogs. Fig. 6 shows SRM chromatograms for such a mixture in an extract of plasma collected from a dog given it orally. The peaks corresponding to the biological reference compound (synthesized to serve as a component in the mixture) and the analytical i.s. are indicated. The QC charts for each component are shown in Fig. 7. As might be expected, the LLOQ's, although generally 2 ng/ml, vary from compound to compound. Substance 19 fails at 2 ng/ml; 3, 8 and 20 fail at both 2 and 5 ng/ml and hence concentrations <10 ng/ml are not reported. This is important in ensuring that calculations of PK data, especially plasma half-lives, are not compromised by inaccurate analytical data. The mean plasma concentration-time curves for all 20 substances are shown in Fig. 8.

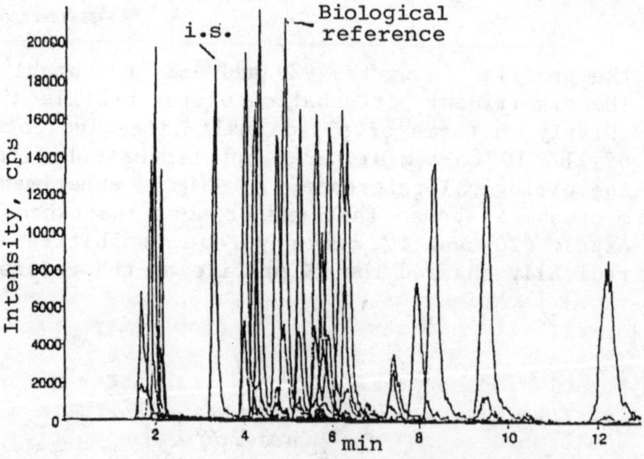

Fig. 6. SRM chromatogram for an extract of plasma from a dog simultaneously dosed orally with 20 drug candidates.

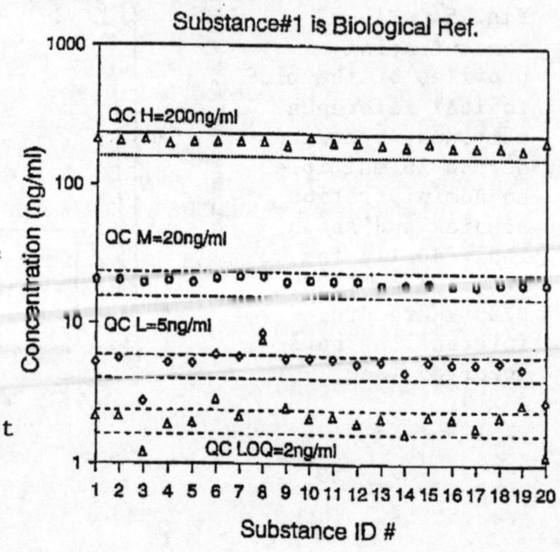

Fig. 7. QC charts for each substance in an analytical method for determining 20 substances simultaneously. The dotted lines represent ±25% of the nominal QC concentration. Each point is the mean result (n = 6) obtained for that substance.

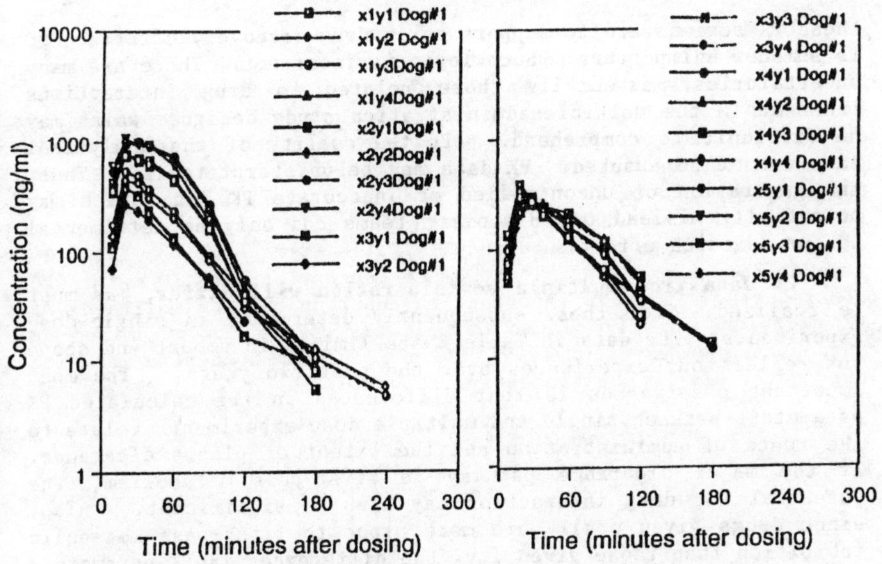

Fig. 8. Mean plasma concentration-time profiles for dogs simultaneously dosed orally with 20 drug candidates.

DISCUSSION

HPLC with API MS-MS detection is now the foremost bioanalytical technique used by pharmaceutical companies. Its remarkable specificity and sensitivity allow assays for drugs at low concentrations in biofluids to be developed in days rather than weeks. Notably important is that analytes which are not resolved chromatographically are distinguishable by their unique parent/product ion combinations - the basis of these multiple-component assays.

Despite the specificity of LC-MS-MS, attention still needs to be paid to sample preparation. Generally, the cleaner the plasma extracts, the more rugged the assays. Crude preparations by protein precipitation may cause column over-pressurization due to presence of trace amounts of protein and inorganic materials. Other endogenous ionic species can also perturb the ionization of the analytes, yielding poor sensitivity and precision especially for early-eluting substances. ESP ionization is much more susceptible to these effects than APcI, so the latter is preferred with crude extracts. For the most rugged assays, liquid-solid (SPE) or liquid-liquid extractions are preferable, and if a substance is identified from a screen as meriting further PK evaluation, a new assay will generally be established using these preferred extraction procedures.

Maintaining adequate QC is a guiding principle. To achieve this but also provide the rapid throughput so essential to the drug discovery process, we have streamlined our assay validation procedures. There seems to be a view in some companies that, since these PK screens are to support early drug-discovery efforts, QC is somehow unimportant. Such logic is incorrect. There are many uncertainties, especially those related to drug interactions inherent in the multiple-administration study design, which may be difficult to comprehend. If the quality of the analytical data cannot be assured, PK data may be uninterpretable. Thus, the generation of uncontrolled or inaccurate PK data which may potentially mislead our discovery teams can only be detrimental to our research effort.

PK data from multiple administration will differ, as must be realized, from those subsequently determined in single-dose experiments. The data in Table 2 are limited in amount and scope but reflect our experiences over the past two years. The most important observation is that differences in the calculated PK parameters between single and multiple dose experiments relate to the route of administration and the extent of plasma clearance. If the major clearance pathway is first-pass metabolism, the potential for drug interaction may also be significant. Also, since drugs given orally are more prone to first-pass metabolic inhibition than those given i.v. the differences in PK parameters calculated from oral administration will be greater than those derived from i.v. administration.

Table 2. Single and multiple component testing. Changes denoted **Gen**(erally), **Occ**(asionally). Cl_p, plasma clearance; F, bioavailability.

Program (clearance)	Admin. route	Key PK parameters	Changes from single to multiple component dosing
A (low)	i.v.	$t_{\frac{1}{2}}$	Gen: $t_{\frac{1}{2}}$ 20-30% rise
B (moderate)	p.o./ i.v.	%F, $t_{\frac{1}{2}}$, Cl_p	Gen: t 20-50% rise; Cl_p fall Occ: %F 2- to 3-fold rise
C (moderate)	p.o.	C_{max}, AUC, $t_{\frac{1}{2}}$	Gen: C_{max} & AUC 20-60% rise; $t_{\frac{1}{2}}$ 30% rise Exceptionally AUC & $t_{\frac{1}{2}}$ rise 5-fold
D (moderate/high)	p.o./ i.v.	%F, $t_{\frac{1}{2}}$	Gen: %F & $t_{\frac{1}{2}}$ 2-fold rise Occ: %F 8- to 15-fold rise
E (high)	p.o.	C_{max}, AUC, $t_{\frac{1}{2}}$	Gen: AUC & C_{max} 10- to 15-fold rise; $t_{\frac{1}{2}}$ 2- to 3-fold rise

Although based on limited data, the entries in Table 2 might provide some guidance to other users. Of particular note is that, for example, with a compound having moderate-high clearance, the AUC's and $t_{\frac{1}{2}}$ values will be reduced by perhaps a factor of 2 when one progresses from oral multiple-dose to single-dose experiments. For high-clearance compounds given orally the change is dramatic, with 10- to 15-fold decreases in AUC and C_{max}. In contrast, only minor increases in AUC are noted for low-clearance drugs given i.v. Data with high-clearance compounds might appear difficult to interpret. However, it has been found that a substance which is superior to the biological reference in PK properties will, in a multiple-analyte screen, generally also show superiority in single-dose experiments.

In general, we have found these multiple component assays to be very useful in the preliminary identification of candidates. There are many pitfalls and the PK results must be interpreted with caution; but by making sensible interpretations we can use multiple-component PK screening to predict the usefulness of single-dose PK studies. Several drug candidates recently accepted for development were initially earmarked because of their PK properties in multiple-administration studies.

References

1. Gilbert, J.D., Olah, T.V. & McLoughlin, D.A. (1996) in 'Biochemical and Biotechnological Applications of ESI-MS' (Snyder, A.P., ed.), ACS Symposia Series, **619**, 330-35.
2. Barrish, A., Olah, T.V., Gatto, G.J., Michel, M.R., Dobrinska, M.R. & Gilbert, J.D. (1996) Rapid Comm. Mass Spectrom. **10**, 1033-1037.
3. Gilbert, J.D., Greber, T.F., Ellis, J.D., Barrish, A., Olah, T.V., Fernandez-Metzler, C., Yuan, A.S. & Burke, C.J. (1995) J. Pharm. Biomed. Anal. **13**, 937-950.
4. Gilbert, J.D., Hand, E.L., Yuyan, A.S., Olah, T.V. & Covey, T.R. (1992) Biol. Mass. Spectrom. **21**, 63-68.
5. McLoughlin, D.A., Olah, T.V., Ellis, J.D., Gilbert, J.D. & Halpin, R.A. (1996) J. Chromatog. **A 726**, 115-124.
6. Olah, T.V., McLoughlin, D.A. & Gilbert, J.D. (1997) Rapid Comm. Mass Spectrom. **11**, 17-23.
7. McLoughlin, D.A., Olah, T.V. & Gilbert, J.D. (1997) J. Pharm. Biomed. Anal. **15**, 1893-1901.
8. Halm, K.A., Adkeson, K.K., Berman, J., Shaffer, J.E., Tong, Q.W., Lee, F.W. & Unger, S.E. (1996) in Proc. 44th Annual Conf. on Mass Spectrometry and Allied Topics (Portland, OR), Am. Soc. Mass Spectrometry, Santa Fe, p. 392.

Comments on #D-1, P. Heizmann/D. Dell: STABILITY-CHECKING PITFALLS
#D-3, S. Pleasance: AUTOMATED 96-WELL SPE
#D-4, J. Schmid: AUTOMATION IN MS-BASED ASSAYS
#D-5, K. Selinger: MULTI-SAMPLE ASSAY APPROACH
#D-6, H.M. Hill: ACCURACY AND PRECISION
#D-7, R. Payne: PROBLEMS WITH LC-MS-MS
#D-8, J.D. Gilbert/T.J. Olah: LC-MS-MS IN DRUG DISCOVERY

S. Wood asked Heizmann why, since decreases in peak heights occurred in his stability studies, he didn't adopt peak areas for comparison. Reply: our focus was on explanations for the observed effects; but in any case peak heights give more precise quantification if there are adjoining peaks with slight overlap at the base. Added **comment by Wood**: freeze-thawing is likely to cause plasma pH to rise. **Venn** remarked that where, in the specificity studies, the amount of a metabolite was reckoned to be <5%, from fluorescence signals, there could be doubt because of enormous variations in fluorescence quantum yields. **Dell** replied that for most of their metabolites authentic specimens were available.

Pleasance, replying to A. Edgington: the proportion of our unknown samples that gets lost in the fully automated SPE system because of failed extractions is about 10 in >900. In contexts such as LC-MS-MS, autosampler **carry-over (queried by Schmid and Timmerman**) - a major problem in (e.g.) PK studies - seems to be least with the HP 1090 instrument, compared with CTC 200 (quite acceptable performance) and Gilson; but the HP autosampler is not suitable for well-plate systems.

Schmid, answering Selinger.- 'Re-validation' is needed when a method has long been in abeyance or there have been changes in, e.g., the equipment; three calibration and QC batches have to be run. The need for 6 different blanks in one day arises where the method is new: before making a pool for setting up spike and QC samples we ascertain the blanks for 6 individuals.

Remark from **Schmid to Selinger**: because of overheads, a 15% increase in test-sample throughput doesn't decrease costs by 15%. **Hill, answering Dell**: the Snedecor formula for calculating precision of duplicates in real-sample analysis is regularly used in our lab. (Covance) but isn't widely used or well known elsewhere.

Remarks by **Pleasance to Payne.**- One should now use the term 'protonated molecule', not 'molecular ion'. MS-MS transitions entailing mere loss of H_2O lack specificity. Problems can arise in method development if sample preparation and sample analysis are done by different groups. **R.F. Venn** emphasized the risk of missing metabolites with LC-MS-MS because of its specificity, and asked about the non-use of BondElut SPE (suits for RIA) in the LC-MS-MS context. Reply by **Payne**: BondElut extraction did work for LC-MS-MS but was unnecessary since a simpler method sufficed.

Gilbert's talk on multiple-analyte PK study (SEE BELOW) preceded the following *GENERAL DISCUSSION:*

Opinions sought (now collated) by **D. Dell: LC-MS *vs.* LC-MS-MS**

For LC with single MS in quantitative analysis, experience is limited. It is much used by British Biotech (various methods developed, with a Finnigan 700Q) and, as in successful method-development done for J.D. Gilbert, hy Phoenix; altogether it does seem to have a role to play in LC, filling a niche between spectrophotometric detection and MS-MS. **M.V. Doig** advocated that, where the specificity of MS-MS is not needed, the cheaper and more sensitive LC-MS be used (ion transmission is normally 10% in tandem MS); one shouldn't expect the two instruments to have the same parameters, e.g. cone and lens focusing voltages.

Remark by **R.F. Venn**: using one triple quadrupole instrument rather than 2 single quadrupoles allows shorter run times, with no detriment to throughput. **Payne** broadly agreed, at least for requisite LLOQs in the low ng/ml range (or even sub-ng/ml); but for >25 ng/ml a triple-quad instrument need not be used. Moreover it is not always true that a triple-quad instrument is free from interference. We have many examples where a compound had to be separated away from interfering peaks when using an MS-MS method.

Gilbert (cf. #D-8) replied to several questions. Answer to **J. Moore**: dogs were used for single-dose screening studies and rats for repeated doses. To **P. Timmerman**: validation of RIA with LC-MS-MS is done using, say, ~100 patients' plasma samples, some a sample mixture after multiple dosing of different patients but also using several series of samples at different collection times from the same subject. To **H. Lehr**: we didn't pre-check the short-term plasma/serum stability of the different analytes.

Other questions/remarks put to Gilbert but, due to his post-Forum death, lacking verified replies

#**J. Schmid** queried why both AUC and C_{max} increase for a high-clearance drug when going from single- to multiple-dose studies. #Metabolites could be important; notionally an active metabolite might arise from a 'moderate' compound (**P. Timmerman**). #There could be a data-treatment problem if, on a particular analysis day, the lowest QC did not meet the criteria and all values were in that range (**P. End**). #At the preclinical stage (#perhaps a risk of selecting the wrong compound! - **P. Heizmann**), presumably the reduced criteria for validation don't apply (**S. Pleasance**).

GC-MS-MS ('ion-trap') gave, for a quinazoline drug in plasma after SPE, a lower LOD than was obtainable with quadrupole and ion-trap MS.- Rossi, D.T. et al. (1997) Anal Chem **69**, 4519-4523.

[Other Editor's citations, not MS-related, appear overleaf

Citations contributed by Senior Editor relevant to Sect. #D; e.g. automation, data handling, validation

For journals, Chr'y DENOTES Chromatog.

Miniaturized SPE apt for 96-well format (cf. art. #D-3) with membranes (Empore) or only 10 mg sorbent, tried with lamivudine, fluticasone and an acidic tricyclic, allowed only 0.2 ml plasma vol. and could be automated.- Plumb, R.S. et al. (1997) J Chr'y **B 694**, 123-133.

A pre-validation assay design, applied to develop a simple method for itraconazole in serum (with its hydroxy metabolite; initial acetonitrile/KCl step) eases evaluation of linearity, repeatability and matrix effects on recovery.- Cociglio, M. et al. (1997) J Chr'y **B 698**, 225-233. (Cf. art. by van der Stel et al. in Vol. 24, pp. 105-117.)

Validation of chromatographic methods, e.g. for drugs in bio-samples, has been surveyed by R. Causon (1997): J Chr'y **B 689**, 175-180 - and see **A730**, 381-394 (1996; Wieling, J. et al.): here experimental design is also considered, with captopril in plasma as an example [cf. above-mentioned van der Stel art., Vol. 24]. A review, citing caffeine in soft drinks as an example, addresses the **standard-addition** approach for settling accuracy (allowing for matrix influences) where no certified reference material is available.- Bagur, M.G. et al. (1996) Analusis **24**, 374-380.

'Data Analysis for Hyphenated Techniques', a book by E.J. & U.P. Karjalainen (1996): Elsevier Science, 480 pp.

Example of a **reagent artefact: TCA** is prone to break down to dichloracetic acid [Fisher, J. (1997) Drug Metab Disp **25**, 1449].

Section #E

ASSAY OF BIOMARKERS

INTRODUCTION** (editorially compiled) and **Questionnaire responses

SURROGATE/BIOCHEMICAL MARKERS: AN ANALYTICAL CHALLENGE

Increasingly the need to measure increases (or decreases) in endogenous substances, including both low mol. wt. substances (e.g. hormones, metabolites, products of intermediary metabolism) and macromolecules (e.g. proteins, DNA), is becoming important at all stages of drug discovery and development. Such 'biomarker' analytes have come to be termed either biochemical/pharmacodynamic (PD) or surrogate markers, at least amongst drug investigators though the concepts extend into non-drug biomedical fields which this Introduction does not consider, namely clinical biochemistry (disease diagnosis or status assessment) and nutritional therapy. Drug-toxicity assessment (cf. p. 280) also relies on biomarkers. The spectrum of endogenous analytes with wide-ranging biomarker pertinence has been cogently surveyed by R. Lehmann & co-authors [(1997) J. Chromatog. **B 697**, 3-35].

Terminology in drug contexts is rather a 'grey area'. Two biomarker types may be distinguished (cf. the version in #E-4).- **Biochemical marker** is taken to be the applicable term where an effect on the substance(s) is related directly to the expected pharmacology of a particular compound (such as a fall in plasma cholesterol in response to a cholesterol-lowering agent). **Surrogate marker**, on the other hand, means an indirect measure (indicator) of a pharmacological effect, e.g. gene activation caused by drug interaction with a receptor affecting the gene's activity, or an effect on cell division resulting from receptor interaction (cf. inborn errors of metabolism, where no drug is involved; p. 279). Such markers are used where a biochemical marker is not easily found.

Measurement of biomarkers is important for various reasons - the construction of PK/PD relationships; an early demonstration of pharmacological effect in man (where a clinical endpoint may be difficult to achieve in a short timescale); as a demonstration of efficacy to drug-regulatory authorities. There are, however, many problems in measuring biomarkers. These include the availability of suitable standards, the need for methods unfamiliar to bioanalysts used to determining low mol. wt. drugs in biofluids, and conceivable difficulty in obtaining a calibration curve where a fall in concentration of an endogenous substance is concerned.

The present-day scene.- A novel survey by a Questionnaire to 1997 Forum participants (from pharmaceutical and also contract-research organizations) was conducted, to look for any trends or common practices that might be indicative as to how bioanalytical groups are coping with this challenge. Of the **16** respondents (some replies mailed after the Forum), some represented several participants from a single company (U.K. or overseas); in respect of the company spectrum at the Forum the response was near 50%.

Some issues that were posed.- #Whether the company monitors pharmacological activity *via* biochemical or surrogate markers: positive, **14**; not at present (nor intended in the foreseeable future): 2. #A need to sometimes determine subnormal levels: yes, **10**; no response or didn't know: 6. #Quantitative rather than merely qualitative measurement (where answered): 2:1 ratio.

As to **where** the analysis is performed (in the bioanalytical group itself/within another group in the company/at contract), there were a variety of, it is surmised, pragmatic responses in respect of such determinations, with no clear trend. Thus, **10** drug-bioanalysis groups claimed to perform some of these marker determinations themselves, 4 did not and 2 furnished no response. However, as many as 9 of the 10 groups also made use of various 'in-house' expertises including Clinical Chemistry (3 groups), Molecular Biology, Pharmacology (3 groups) and Toxicology [cf. a reply by Dr End to a questioner at the Forum, p. 277]; there were 3 negative responses, and 4 groups did not respond. In addition, contract research facilities were also used by some laboratories (5) but not by others (6) [4 non-responders; some responders were actually from a contract facility].

Validation of the method in some depth, within the group if applicable, was reported to be attempted by the majority (9) of the positive responders, although a significant minority did not subject their methods to such in-detail evaluation (5; no reply from 2 groups). The level of validation clearly varied between groups, but in 7 was in conformity with GLP (3 groups did not validate to GLP standards and 5 gave no indication). Obtaining reference materials or standards, blank matrices [cf. arts. #B-7 & #B-8n], etc., posed a problem to a number of the respondents, as foreshadowed above and pursued in #E-4 below.

The overall picture.- The increasing reliance of pharmacologists on the use of *in vivo* markers of one sort or another to give an indication of an effect of a drug, or drug candidate, is likely to mean that bioanalysts will increasingly have to face up to the need to develop or 'import' methods for non-drug analytes. As yet there seems to be no clear consensus on just what should, or needs, to be done in such instances, although an appreciation of the problems of this type of work does seem to be emerging. The Forum-based material that appears in Section #E, pioneering but far from comprehensive, gives a flavour of the variety of work that may be involved and perhaps provides some pointers to future developments in this problematic area.

#E-1

MEASUREMENT OF SOME MARKERS RELATING TO THIOLS AND NITRIC OXIDE IN BIOFLUIDS BY HPLC AND CE

Wendy King, Paul Hamilton & David Perrett†

Dept. of Medicine, St. Bartholomew's & the Royal London School of Medicine & Dentistry, West Smithfield, London EC1A 7BE, U.K.

With a view to using total homocysteine (HCys) as a CV disease marker, a method has been developed for determining its level in plasma, entailing derivatization and HPLC-fluor. Since NO could likewise have CV relevance, maybe after intracellular interaction with GSH, a CZE-UV method has been devised to assay plasma for NO adducts and also nitrite and nitrate. Failure to find S-nitrosothiols in biosamples, even after concentration by freeze-drying, is attributed to insufficient sensitivity of the present method.*

In CV studies, two compounds of strong current interest are the simple inorganic gas, nitric oxide (NO), and the naturally occurring amino acid HCys. The effects of NO on the vasculature have led to the view that it is the long-sought EDRF (endothelin-derived relaxing factor). Levels of HCys in plasma might prove to be an important biomarker of CV disease in man.

HOMOCYSTEINE STUDIES IN CARDIOVASCULAR DISEASE

Children with the genetic disease homocystinuria have long been known to be affected by adverse thrombolytic events [1]. A recent major finding has been recognition that hyperhomocystinaemia is an independent risk factor for thrombosis and vascular disease in the general population [2]. Recent studies have shown that a 4 μM increase in plasma HCys levels gives a 40% increase in cerebrovascular disease risk [3]. There is now sufficient evidence to suggest that plasma HCys levels could form a useful biomarker to be measured in those at risk of CV problems.

In man the circulating levels of HCys are only 15-25 μM. Like all thiols, HCys occurs in plasma as the free thiol, as various disulfides with other thiols including itself, and in protein-bound form [4]. Due to the low levels of HCys in plasma and the variety of other naturally occurring thiols, selective and sensitive analytical methods are required. HPLC with EC detection or with derivatization of the thiol to a fluorophore (with a variety of agents, e.g. bimanes, maleimides) is favoured for determining free thiol, to which all oxidized thiols must first be reduced [4].

†addressee for any correspondence

**Abbreviations (see also Preface)*.- CE, capillary electrophoresis (CZE if 'zone'); CL, chemiluminescence; CV, cardiovascular; cf. RSD, precision ['C.V' in other arts.; *Editor's change]*; EC, electrochemical; GSH, glutathione (usually reduced form; GSSG if oxidized); HCys, homocysteine [*factor:* 1 μM = 135 ng/ml]; LOD, limit of detection.

HPLC ASSAYS FOR HOMOCYSTEINE

The measurement of all circulating forms of HCys would have been preferred; but since HCys rapidly binds to plasma proteins following collection it was decided that only the determination of total HCys in plasma was realistic. Initial studies attempted to measure plasma HCys after reduction with sodium borohydride and derivatization with N-(1-pyrenyl)maleimide followed by RP-HPLC according to Winters and co-workers [5]. However a large number of side-reaction products were always generated and these gave a long and complex chromatogram. The derivatives were also very hydrophobic and this caused precipitation onto the RP-HPLC column in the eluent employed. Subsequently an RP-HPLC assay was developed with the following step after borohydride reduction.-

derivatization with monobromobimane to a fluorophore:

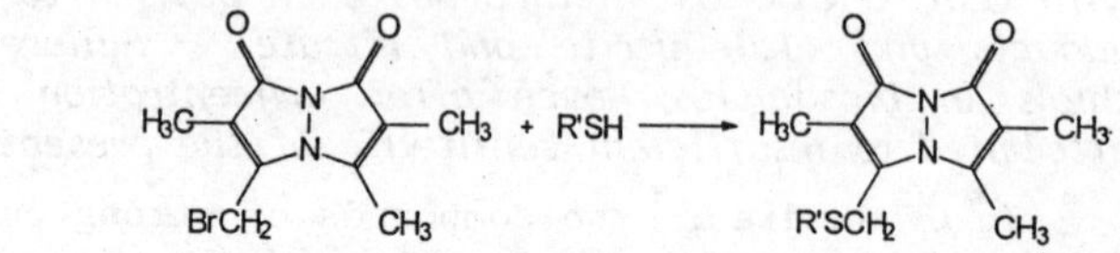

Procedure

Venous blood samples collected into EDTA tubes were centrifuged, and the plasma stored at -55°. For analysis, 200 µl was pipetted into a 1.7 ml Eppendorf tube and 70 µl 1.4 M NaOH added. After gentle vortexing, 70 µl 1 M HCl was added and the mixture centrifuged (13000**g**, 10 min). To 240 µl of the clear supernatant 70 µl 10 mM monobromobimane (Calbiochem) was added. After 30 min (room temp.) in the dark, proteins were precipitated by adding 70 µl 12% (v/v) perchloric acid and centrifuged (13000**g**, 15 min). The clear supernatant was analyzed for HCys.

The final optimized HPLC conditions were: column, 5 µm ODS-2 Spherisorb (Phase Separations, Deeside, UK), 150 × 4.6 mm; 20 µl injected; mobile phase, 25 mM sodium formate/methanol (85:15), 1 mM EDTA titrated to pH 4.5 with 6 M HCl; LKB 2015 pump; flow rate 1.5 ml/min; a Jasco 821AS fluorescence detector with xenon lamp - ex/em 385/475 nm; output was to a Midas 22+ computer integrator (Comus Instruments, Hull, UK).

Derivatization conditions for thiols are always a compromise between the reaction rate towards the reagent and the tendency of the thiol to form a disulfide. The effect of varying the reaction temperature on the yield is shown in Fig. 1. Maximum yield of fluorophore was obtained at room temperature for 30 min. The derivative so formed was stable at room temperature for at least 5 h, allowing the use of an autoinjector.

Optimum eluent composition was determined by investigating the R_T's and selectivity of the column towards bimane derivatives of HCys, GSH and Cys over the pH range 2.5-5 and methanol concentration 15-20%. Fig. 2 shows a typical separation of a standard using the final conditions. HCys eluted at ~6 min.

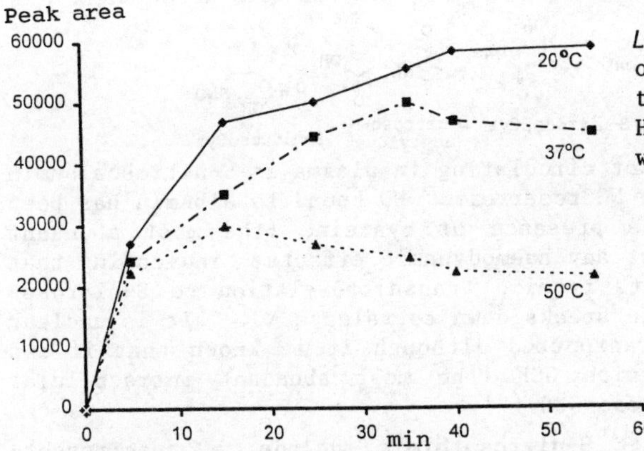

Left: **Fig. 1.** Effect of varying reaction temp. on yield when HCys is derivatized with monobromobimane.

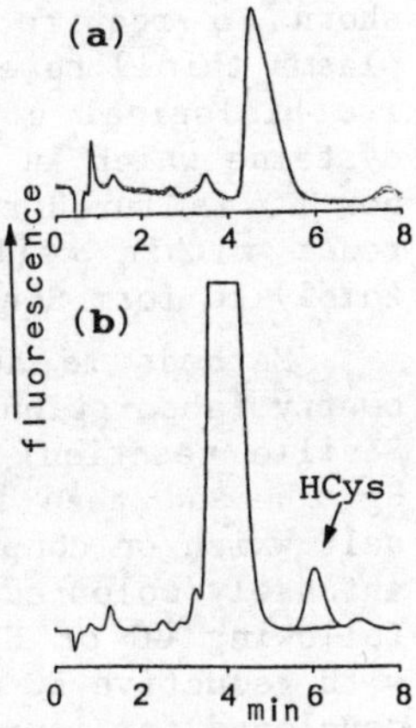

Right: **Fig. 2.** Typical HPLC runs after processing: (a) blank, (b) 10 mM HCys standard.

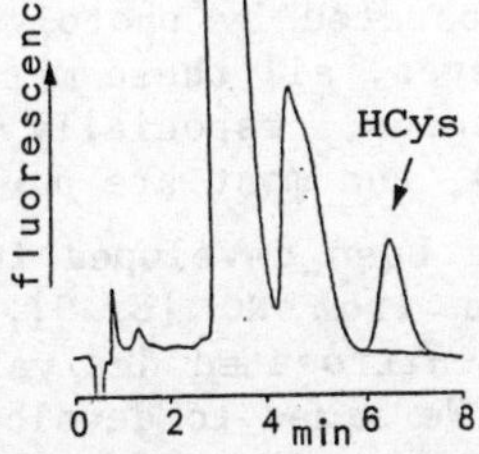

Left: **Fig. 3.** HPLC of HCys in a typical plasma analysis.

Assay performance was tested (n = 6).- With solutions of standard, linearity was found up to 40 μM HCys, and variation was 4.75% (RSD; for either 50 or 6 μM) intra-assay and 5.4% (3.3% for a spiked plasma) inter-assay. The assay LOD was 0.2 μM at S/N = 2 and the LLOQ was 0.5 μM. The assay was suitable for routine use with plasma (example in Fig. 3). Total HCys was determined in a normal population (17; mean age 30) and in a group of patients attending a thrombosis clinic at St. Bartholomew's Hospital (14). The respective μM values found were 14.6 ±4.1% and 18.4 ±8.7% (mean ±RSD). The mean HCys was higher in the patients but the difference was not significant, the groups being small.

The main problem with the assay at present is a late-running fluorophore derived from the monobromobimane, eluting at ~45 min; it restricts the number of samples that can be run in a batch before the late runner appears. Methods to overcome this, e.g. SPE or step washes, are at present being investigated.

STUDIES ON *S*-NITROSOTHIOLS USING CAPILLARY ELECTROPHORESIS

How NO transfers to its site of action (e.g. vascular smooth muscle cells) is unknown. Given the extreme lability of NO there may well be an intermediate species that stabilizes NO *in vivo*.

Candidates include endogenous thiols reacting with NO *in vitro:*

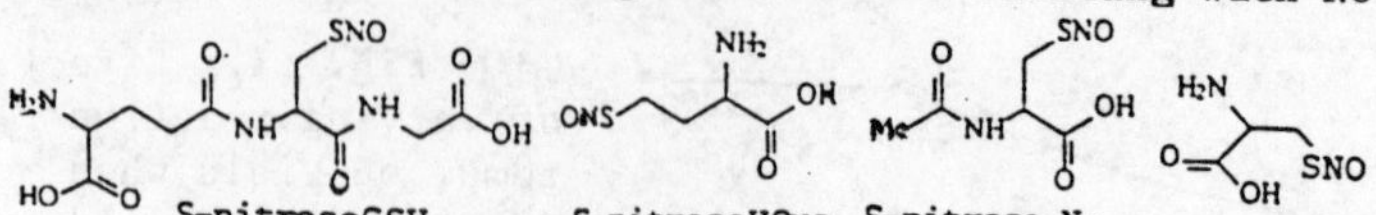

S-nitrosoGSH S-nitrosoHCys S-nitroso-N-acetylCys S-nitrosoCys

The main NO adduct circulating in plasma is S-nitrosoalbumin which could act as an NO reservoir. NO bound to albumin has been shown to require the presence of cysteine (the most abundant plasma thiol) to exert any haemodynamic effects, indicating that its biological effects require transnitrosylation to S-nitroso-cysteine which in turn breaks down to release NO. It is unclear how NO is further transported although it is known that it can react within cells with GSH (the most abundant intracellular thiol) to form S-nitroso-GSH [7].

Methods to detect **S-nitrosothiols** include: a) spectrophotometry (absorption maxima exist at ~215, 335 and 545 nm); b) the Saville reaction: NO is first displaced from S-nitrosothiols by Hg ion and then reacts with sulfanilamide to form a diazonium salt which on coupling to N-(1-naphthyl)-ethylenediamine forms an intensely coloured azo-dye absorbing at 540 nm; c) CL detection, following GC or HPLC, of NO liberated by photolysis; d) HPLC with reductive EC detection. However, all these methods have been developed for 'pure' chemical species, especially notably stable ones (e.g. S-nitroso-N-acetylCys), and most are unvalidated.*

A number of CE methods have been developed for determining the **nitrite and nitrate** derived from NO [8, 9], and CZE has been used to separate standard S-nitrosated derivatives from the corresponding disulfides [10]. We aimed to develop a CZE method for simultaneously separating S-nitrosothiols and the corresponding thiols and disulfides besides nitrite and nitrate, and to apply it to biofluids so as to establish whether S-nitrosothiols exist *in vivo*. We also planned to use the method to optimize conditions for the formation of S-nitroso-thiols *in vitro*.

Experimental (All chemicals were from Sigma)

CZE was performed on a Spectra 2000 System (TSP, Stone, UK) with rapid-scanning UV detection using a 44(37) cm × 75 µm silica capillary, at -10 kV, 25°; run time was 20 min as previously used for nitrite [9]. The final background electrolyte (BGE) comprised water containing 15 mM Tris, 2 mM tetradecyl-trimethylammonium bromide (TTAB), 2 mM EDTA and 150 mM NaCl; the pH was adjusted to 7.4 with 1 M HCl. The capillary was washed with BGE for 3 min prior to hydrodynamic injection of sample over 2 sec. Peaks were detected by scanning from 200 to 340 nm.

S-nitrosothiols were prepared by reacting 100 mM thiol with 50 mM HCl and 100 mM $NaNO_2$ at 25°. Solutions instantly turned red, indicating completion of the reaction. Each S-nitrosothiol was synthesized and run separately to ascertain migration order.

*Cf. methods for captopril etc. in #ABC (group 3) and in past vols.; survey by A. Rakhit in V. 18 (1987; ed. Reid *et al.;* Plenum).- *Ed.*

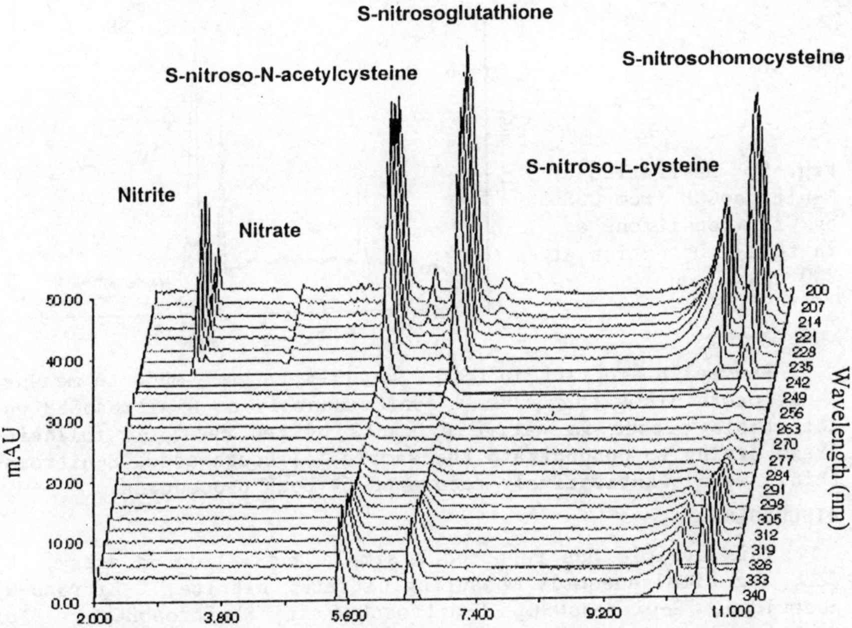

Fig. 4. CZE separation (method as in text), with spectral scanning, for a group of nitrosothiols, nitrite and nitrate.

Initial problems with separating S-nitrosoHCys and S-nitrosoCys were overcome by optimizing the buffer pH. Fig. 4 shows the final CE separation with spectral scanning.

Table 1. Validation data for a mixture of nitrosothiols (n = 10; each 7 mM; detection 215 nm; -10 kV, injection time 2 sec HD).

S-nitroso compound	Migration time RSD	Corr'd peak area RSD	LOD µM
-N-acetylCys	3.69%	2.88%	654
-GSH	2.51%	5.51%	646
-Cys	4.46%	rapid breakdown	1258
-HCys	6.94%	15.7%	425

Validation (Table 1).- The RSD's were poor for S-nitroso-Cys and -HCys due to their gross instability in solution. For the stable nitroso compounds linearity extended up to 60 mM. The same CZE conditions enabled the separation of S-nitrosothiols from the corresponding **disulfides**, their most likely breakdown products, as shown for GSH in Fig. 5.

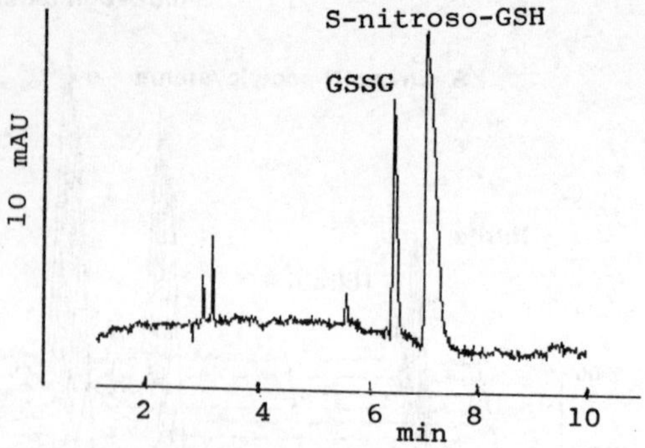

Fig. 5. Separation of S-nitrosoGSH from GSSG by CZE (conditions as in text). Detection at 200 nm.

Red cells are rich in GSH; so attempts were made to measure S-nitrosothiols within them. Unfortunately no S-nitrosoGSH was detectable either in native extracts of red cells or following freee-drying to concentrate the sample, although added S-nitrosothiols were quantitatively recovered by both procedures.

DISCUSSION

This article has reported a simple and validated assay, by CZE, for simultaneously measuring nitrate, nitrite, S-nitroso-N-acetylCys, S-nitrosoGSH, S-nitrosoCys amd S-nitrosoHCys. For quantitation of nitrite, nitrate and the S-nitrosothiols 215 nm was the optimum wavelength, but 335 nm was used for confirmatory evidence of S-nitrosothiols. This CZE method was extended to give a successful separation of each S-nitrosothiol from its thiol and the corresponding disulfide.

Using UV detection this method proved to lack sensitivity (Table 1); hence freeze-drying was performed to attempt to find S-nitrosothiols in biosamples - the main purpose of this study. Gaston and co-workers [11] measured S-nitrosothiol in broncheo-lavage fluid by CL plus the Saville reaction. No measurements were done to specifically identify the S-nitrosothiol whose concentration was reported (~500 nM). Scharfstein and co-authors [12] measured S-nitrosoCys in rabbit plasma by RP-HPLC-EC with dual Au/Hg electrodes; basal levels were claimed to be ~200 nM, ~0.5% of the plasma Cys level, but no validation of the method was reported. Other reports using UV detection are unlikely to be meaningful, for the same reasons as given here - i.e. low sensitivity of the detection method.

The fact that no S-nitrosothiols were measurable by CZE in biofluids is not surprising. Although CE is very sensitive, capable of detecting pmole quantities of S-nitrosothiols, it is not very sensitive with respect to concentration, the LOD's being in the 100 μM range. However, levels of total S-nitrosothiols have been estimated to be of the order of 100 nM. We attempted to concentrate the samples by freeze-drying to a level at which they could be detected; but our maximum degree of concentration was only ×80 for plasma and ×30 for red cells, which is clearly inadequate for detection in the light of published information.

However, pre-concentration was shown to be a viable way to get an improved concentration. In plasma the S-nitrosothiol most likely to be detected would be S-nitrosoCys, and in red cells S-nitrosoGSH, because these are the most abundant thiols in these biofluids. Other methods of concentrating samples such as SPE or liquid-liquid extraction may prove to be more efficient.

Improvement in sensitivity could be achievable through the detection mode. Mizra and co-authors [13] have recently reported MS-ESI (electrospray interface) as a method for S-nitrosothiols. Linking this to CE may provide a method by which S-nitrosothiols could be separated and measured with both high resolution and high sensitivity. Measurement using reductive EC detection could also be linked to CE.

CONCLUSIONS

Homocysteine assays are becoming increasingly important and are predicted to become a routine biomarker approach. However, such a development will mean that immunoassays will be required, and these will replace HPLC assays in future. The measurement of nitrate/nitrite by CZE is simple and easy to use, although in some circumstances sensitivity might need improvement. Several S-nitrosothiols can be readily resolved by CZE from both nitrate and nitrite as well as the corresponding disulfides and thiols. CZE is a useful means of studying their formation and instability but is rather insensitive, and a better assay is needed to determine their levels in biofluids.

Acknowledgements

Wendy King thanks the Drapers Company of the City of London for a studentship to study for a B.Med.Sci. degree at our hospital. Drs Ian James and Mark Hows are thanked for their help with parts of these projects. Dr Peter Macallan is thanked for patient samples.

References

1. Guba, S.C., Fink, L.M. & Fonseca, V. (1996) Am. J. Clin. Path. **105**, 709-722.
2. Clarke, R., Daly, L., Robinson, K. **et al.** (1991) New Engl. J. Med. **324**, 1149-1155.
3. Arnesen, E., Refsum, H., Bonaa, K.H., Ueland, P.M. **et al.** (1995) Int. J. Epidemiol. **24**, 704-709.
4. Perrett, D. & Rudge, S.R. (1985) J. Pharm. Biomed. Anal. **3**, 3-28.
5. Winters, R.A., Zukowski, J., Ercal, N., Matthews, R.H. & Spitz, D.R. (1995) Anal. Biochem. **227**, 14-21.
6. Leone, A.M., Francis, P.L. Rhodes, P. & Moncada, S. (1994) Biochem. Biophys. Res. Comm. **200**, 951-957.
9. Perrett, D., Alfazema, L., Hows, M. & Gibbons, J. (1997) Biochem. Soc. Trans. **25**, 273-278.
10. Stamler, J.S. & Loscalzo, J. (1992) Anal. Chem. **64**, 779-785.
11. Gaston, B., Reilly, J., Drazen, J.M., Fackler, J., Ramdev, P., Arnelle, D.R. **et al.** (1993) Proc. Nat. Acad. Sci. USA **90**, 10957-10961.
12. Scharfstein, J.?S. & Keaney, J.F.J. (1994) J. Clin. Invest. **94**, 1432-1439.
13. Mizra, U.A., Chait, B.T. & Lander, H.M. (1995) J. Biol. Chem. **270**, 17185-17188.

#E-2

ASSESSMENT OF CHOLESTEROL STATUS: IDENTIFICATION OF SMALL DENSE LDL

Bruce A. Griffin† & John W. Wright
Centre for Nutrition & Food Safety, School of Biological Sciences, University of Surrey, Guildford GU2 5XH, U.K.

The benefit of lowering plasma cholesterol levels in patients with CHD is now definite; but high- and low-risk subjects cannot be distinguished merely on the basis of cholesterol levels. Also, emphasis on cholesterol as the single major lipid risk factor is over-simplistic, and disregards a common phenotype, ALP, where the notable rise in cholesterol occurs in small dense LDL. ALP along with truncal obesity is an insulin-resistance syndrome feature.*

Identification of small dense LDL in relation to ALP requires preparative ultracentrifugation, too costly for routine use - as also applies to HDL-C and TG as surrogate measures of ALP. Such measurements are, however, essential for research on how ALP arises and to assess the response to drug or dietary treatment. Gradient gel electrophoresis can help identify CHD risk but is not suited for quantitating LDL subclasses; this can be achieved by density-gradient centrifugation, an approach now described.

CHOLESTEROL AND ASSESSMENT OF CORONARY RISK▫

Prospective epidemiological studies in the 1970's identified an association between raised serum cholesterol and increased mortality from CHD which established the 'cholesterol hypothesis' [1]. More recent clinical trials have successfully tested and substantiated this hypothesis by showing that death from CHD is unequivocally reduced in response to cholesterol-lowering therapy in groups with and without existing CHD [2, 3]. This considerable weight of evidence should have placed the role of serum cholesterol as a risk factor for CHD beyond dispute. However, in practice, serum cholesterol or, more specifically, levels of its principal transport lipoprotein, LDL, do not discriminate between individuals at high and low risk of CHD within populations [4]. Besides, the cost-benefit ratio of cholesterol-lowering therapy

---

†addressee for any correspondence

▫Hyperlipidaemia and the role of drugs was surveyed, without consideration of assay procedures, by K.E. Suckling in Vol. 23, this series, pp. 73-82 (1994; eds. & publisher as now).-*Ed.*

**Abbreviations (see also Preface)*.- ALP, atherogenic lipoprotein phenotype; CHD, coronary heart disease; CV, cardiovascular; LDL, low-density lipoprotein; TG, triglyceride.

in CHD, particularly in those without existing clinical signs of disease, remains highly controversial. Many individuals who should be 'at increased risk' through raised LDL levels remain healthy throughout their life (60% of the U.K. population have "undesirably high LDL cholesterol levels"). Conversely, others with lower LDL levels succumb to premature CHD. Whilst other genetic and environmental factors may underlie this apparent anomaly, the existence of structural and metabolic heterogeneity in LDL is of major aetiological significance to atherosclerosis and the differentiation of cholesterol-mediated CV risk.

IDENTIFICATION OF SMALL, DENSE LDL

LDL exists in the circulation of all healthy and diseased individuals as a number of discrete subclasses, the resolution of which has led to significant advances in our understanding of LDL-mediated CHD risk [5]. Growing interest in the role of LDL subclasses in CV disease fuelled the development of a range of biochemical techniques for the separation and quantitation of these macromolecules. Although ultracentrifugation is still recognized as the definitive procedure for the separation of lipoproteins, the practical constraints of this technique make it unsuitable for the analysis of large numbers of human samples in epidemiological and clinical studies.

Gradient gel electrophoresis

Ultracentrifugation was superseded in the 1980's by cheaper and more practical procedures such as gradient gel electrophoresis, which separates LDL subclasses on the basis of particle size [6]. In combination with densitometric scanning of protein- or lipid-stained gels, this technique was used to identify three LDL subclass patterns or phenotypes [7]: a normal pattern 'A' consisting of predominantly larger LDL, an abnormal pattern 'B' consisting of predominantly small, dense LDL, and an intermediate pattern 'I'. Small, dense LDL (pattern 'B') was independently associated with a ~3-fold increase in risk of CHD and shown to be a major component of a high-risk dyslipidaemia (a phenotype, ALP) [8]. An ALP is potentially the greatest source of cholesterol-mediated CHD risk in the healthy population at large. Gradient gel electrophoresis is thus a very effective tool for identifying increased CHD risk, although this method suffers from a number of disadvantages. The level of resolution achieved is inadequate for quantitating LDL subclasses. There are also difficulties with the size calibration of lipoprotein particles, and the procedures have now been largely superseded by rapid density-gradient centrifugation which offers a favourable alternative.

Density gradient centrifugation

Optimized density gradients consistently resolve visibly discrete lipoprotein heterogeneity which may be quantified if the separated lipoproteins are eluted through a purpose-built fraction recovery system linked to a programmable spectrophotometer. The continuous monitoring of the eluate generates lipoprotein spectra from which concentration values for individual LDL subclasses may be derived [9]. The separation of discrete subclasses may be achieved either directly from fresh plasma or serum or indirectly from the purified lipoprotein by the protocols given below. The direct method is faster and achieves higher resolutions, though buoyant TG-rich lipoproteins can interfere with the fractionation of LDL subclasses. This problem can be overcome by the low-speed separation of TG-rich material from plasma prior to the density-gradient separation. The indirect method requires one or more additional steps of sequential ultracentrifugation to purify the lipoprotein of interest. Fig. 1 illustrates the salt density gradient which was optimized for the quantitative separation of LDL subclasses, the run conditions and density distribution of separated LDl subclasses.

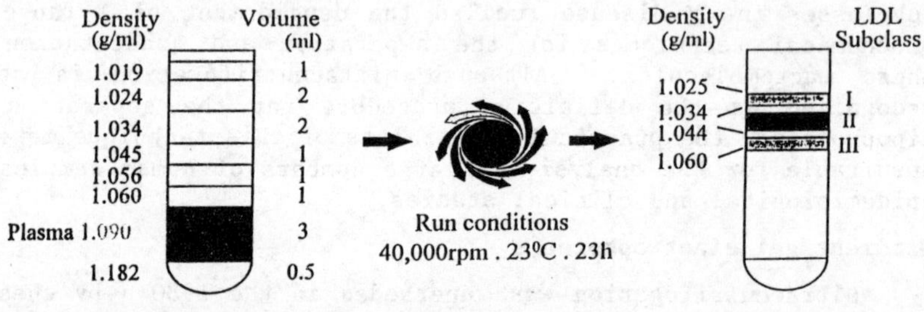

Fig. 1. Six-step, discontinuous salt gradient for separating human plasma LDL subclasses by density gradient centrifugation.

Gradient preparation.- Blood is collected by venepuncture into EDTA (1 g/ml), and plasma is harvested by low-speed centrifugation. For the direct separation of LDL subclasses, plasma is density-adjusted to 1.090 g/ml by adding solid salt (NaBr; 0.38 g required for 3 ml plasma). The plasma is then carefully overlayered with salt solutions of defined density and volume (Fig. 1) using a multi-channelled peristaltic pump, the pump tubing being fitted with glass capillaries. This facilitates the simultaneous preparation of 6 gradients. It is essential that all salt densities be checked to the third decimal place with an electronic density meter (PAAR Scientific Ltd., UK).

The discontinuous gradients are centrifuged as prescribed (swing-out rotor, Beckman SW-40). To allow for longer run-down time when using a 'frictionless' ultracentrifuge (Beckman Optima range), total spin time may need adjustment, empirical or by re-calculating run time from the accumulated angular velocity ($\omega^2 t$). Each gradient's density distribution is determined by running a blank salt solution of appropriate density in place of plasma; after collection into fractions of known volume, the salt density is determined. The discontinuous layers of increasing salt density form a curvilinear gradient on centrifugation. The lipoproteins do not reach equilibrium on these gradients.

Recovery and spectrum of LDL subclasses (Figs. 2 & 3).- After centrifugation the tube containing the separated lipoproteins is placed in a fraction recovery device linked to an infusion pump and spectrophotometer equipped with a 50 µl microflow cell and appropriate software for data capture and graphic manipulation (Fig. 2). The tube is pierced at its base by a needle assembly and the gradient upwardly displaced by the introduction of a dense, hydrophobic displacement fluid (Maxidens, Nycomed Ltd.). The eluate is continuously monitored at 280 nm to generate an LDL subclass spectrum (Fig. 3). The lipoproteins can be collected either into a number of fractions or into pools which correspond to the lipoprotein of interest. The areas beneath individual lipoprotein subclasses are integrated and expressed as relative percentage areas. Measures of lipoprotein mass (total lipid + protein, or specific apolipoprotein) on the collected pools are proportionated on the basis of the relative % areas, to yield plasma lipoprotein subclass concentrations. The lipoprotein spectra are stored on a local disk drive and simultaneously relayed to PC and network file server using Beckman 'Data Capture/Network' software.

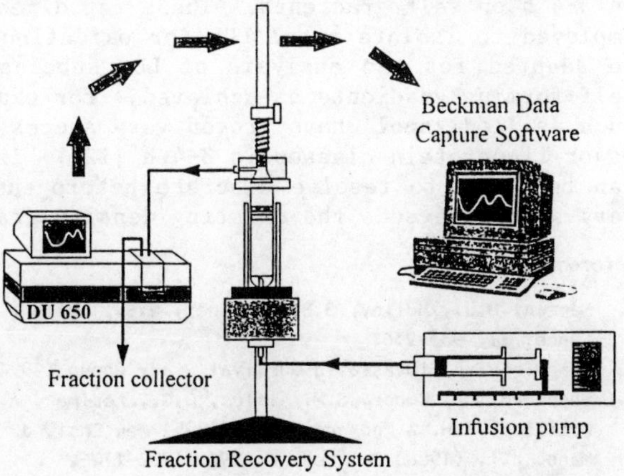

Fig. 2. Fraction recovery system and spectrophotometric detection of LDL subclasses.

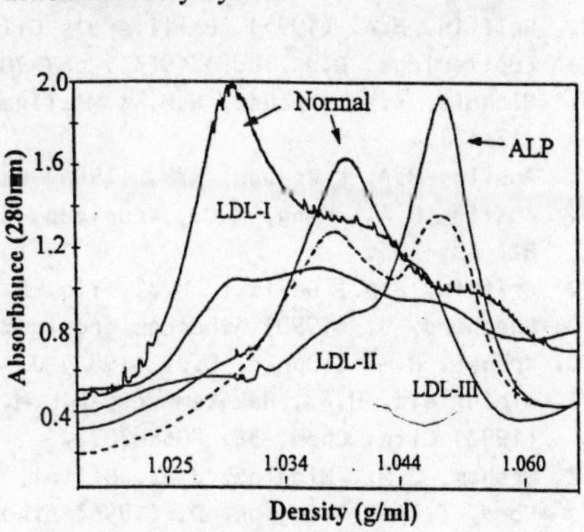

Fig. 3. Plasma LDL subclass profiles from normal and dyslipidaemic (ALP) subjects.
I: largest and least dense; predominates in subjects with very low TG levels.
II: intermediate size/density; main cholesterol transporter; predominates in healthy males.
III: small, dense; predominates in subjects with raised TG as in ALP.

Comments

The gradient described here identifies a reproducible pattern of discrete heterogeneity in LDL subclasses which corresponds to those previously defined by Krauss [10]. Although relatively time-consuming and labour-intensive, density gradient centrifugation is unique in having the capacity to resolve differences in the distribution of LDL subclasses between different individuals (between-sample heterogeneity), in the same individual at different times (temporal heterogeneity) and within one sample (single-sample heterogeneity). The method also allows the quantitative determination of plasma lipoprotein subclasses and permits their physiological isolation for compositional analysis, metabolic and cell-culture studies. Once established, the fraction recovery and data capture systems are easily adapted for the analysis of other lipoprotein species or biological molecules of interest. For many this procedure now represents the method of choice for the analysis of LDL subclasses whilst having the limitations of prolonged run-time and restricted sample capacity. Ultracentrifuges now achieve very high **g** forces and prepare lipoproteins in 3-4 h on salt gradients. These rapid techniques are routinely employed to isolate total LDL for oxidation studies [11] and may be adapted for the analysis of LDL subclasses. Alternatively, self-forming gradients as achieved, for example, with materials such as 'Iodixanol' have proved very successful for isolating the major lipoprotein classes in 3-4 h [12]. If the latter technique can be shown to resolve discrete heterogeneity in LDL it should very soon supersede the existing density-gradient procedure.

References

1. Martin, M.J., Hulley, S.B., Browner, W.S., Kuller, L.H. & Wentworth, D. (1986) Lancet **ii**, 933-936.
2. Scandinavian Simvastatin Survival Study Group (1994) Lancet **349**, 1383-1389.
3. Shepherd, J., Cobbe, S.M., Isles, C.G., Lorimer, A.R., MacFarlane, P.W., McKillop, J.H. & Packard, C.J. (1995) New Engl. J. Med. **333**, 1301-1307.
4. Mann, J.I. (1986) Br. Med. J. **296**, 1702-1706.
5. Griffin, B.A. (1995) 'Bailliere's Clinical Endocrinology & Metabolism' (Betteridge, D.J., ed.), **9**(4), 687-703.
6. Nichols, A.V., Krauss, R.M. & Musliner, T.A. (1986) Meths. Enzymol. **128**, 417-431.
7. Austin, M.A. & Krauss, R.M. (1986) Lancet **ii**, 592-595.
8. Austin, M.A., King, M.C., Vranizan, K.M. & Krauss. R.M. (1990) Circulation **82**, 495-506.
9. Griffin, B.A., Caslake, M.J., Yip, B., Tait, G.W., Packard, C.J. & Shepherd, J. (1990) Atherosclerosis **83**, 59-67.
10. Krauss. R.M. & Burke, D.J. (1982) J. Lipid Res. **23**, 97-104.
11. Kleinfield, H.A., Hak-Lemmers, H.L.M., Stalenhoef, A.F.H. & Demacker, P.N.M. (1992) Clin. Chem. **38**, 2066-2072.
12. Graham, J.M., Higgins, J.A., Gillot, T., Taylor, T., Wilkinson, J., Ford, T. & Billington, D. (1996) Atherosclerosis **124**, 125-135.

#E-3

BIOMARKERS FOR BONE DISEASE STATUS

B.J. Starkey*, D.T. Hardy & J. Brown

Clinical Laboratory, Royal Surrey County Hospital,
Guildford GU2 5XX, U.K.

Bone resorption exceeds bone formation in diseases such as osteoporosis, where bone mineral density scanning remains the diagnostic method of choice although it is too insensitive to detect density changes resulting from increased disease activity or from response to 1-2 years of therapy. Biochemical components that could serve as urinary markers for bone resorption include collagen breakdown products which are released into the circulation when bone is eroded. Urinary pyridinolines from the cross-linking region and the associated telopeptides, C- or N-terminal, are now being evaluated in view of debate as to which marker is the most useful for monitoring purposes.

The Osteomark assay for the N-telopeptide of Type 1 collagen (NTx) had an inter-batch precision better than 11% even at low analyte concentrations, and adequate linearity. Correlation with hydroxyproline was better (r = 0.86) than with free deoxypyridinoline (Dpd), as was diagnostic significance amongst 22 patient samples. Consideration is given to the use, for bone-disease diagnosis and status-monitoring, of biochemical markers in conjunction with bone scanning; it could be advantageous that their urinary excretion reflects bone resorption throughout the body.

Bone is a metabolically active tissue which is continuously being broken down and resynthesized by processes collectively known as bone re-modeling. Except at times of active growth or healing of fractures, bone synthesis and resorption are usually kept in balance, so that there is no net gain or loss of bone. Bone disease is associated with synthesis/resorption imbalance.

Metabolic bone disease includes many disorders ranging from renal osteodystrophy and hyperparathyroidism to Paget's disease. From an economic viewpoint, osteoporosis is much the most important and has been the focus for study of biochemical markers of bone metabolism. The W.H.O. defines osteoporosis as a disease characterized by an excess of bone resorption over bone formation. The imbalance in the bone re-modeling process results in bone loss and in changes in bone structure which reduce its strength and increase the prevalence of bone fractures.

*Addressee for any correspondence

Osteoporosis affects an estimated 200 million people worldwide - ~25 million in the U.S.A., where it is responsible for ~1.5 million fractures and ~50,000 deaths p.a. and develops in 30-50% of all post-menopausal women and ~12% of men. Figures may well double in the next 25 years as life expectancy increases. In the U.K. it accounts for ~60,000 hip and ~40,000 vertebral fractures and 15,000 deaths p.a. The so-called 'silent killer' disease occurs to some degree in >25% of women over 45; it is estimated to cost the National Health Service >£750M p.a. [1].

By definition, bone mineral density scanning is currently the method of choice for diagnosis; but biochemical parameters can be used to detect and assess the severity of bone disease and are of potential use in monitoring the course of disease.

Biochemical markers of bone metabolism

To reflect the imbalance in bone re-modeling associated with bone disease the ideal biochemical marker should reflect either synthesis or resorption. It should be bone-specific and respond quantitatively to the bone metabolism changes brought about not only by disease but also by treatment, to facilitate therapeutic monitoring. The reference range should be well established with minimal overlap between concentrations present in health and disease; finally the assay should be simple, accurate, precise, interference-free, inexpensive and easily automated.

Markers of bone formation (osteoblastic activity) include:
#enzymes: bone alkaline phophatase (Ostase, AlkPhaseB; reflects activity of osteoclasts, the main bone-synthesis mediators);
#cellular products (from osteoblasts): osteocalcin (e.g. Novocalcin), bone sialoprotein;
#collagen synthesis products - propeptides of Type 1 procollagen (fragments released into the circulation when procollagen is converted to collagen during bone formation):
C-terminal (P1CP, prolagen C), N-terminal (P1NP).

It is paradoxical that in osteoporosis and other diseases bone synthesis often actually increases and the markers usually show a rise. This increase in synthesis, however, is far outweighed by the increase in resorption; so the markers for resorption are likely to be of greater use. Several biochemical markers for resorption are currently being assessed, for diagnosis and, perhaps more importantly, for monitoring responses to treatment with bisphosphonates, oestrogen replacement (HRT), calcitonin or calcium and vitamin D supplements.

Markers for bone resorption (reflecting the osteoclastic activity responsible for it).- The major ones include:
#tartrate-resistant acid phosphatase;
#breakdown products of collagen (comprises 90% of the organic matrix): hydroxyproline, hydroxylysine glycosides; pyridinoline + deoxypyridinoline (Dpd), TOTAL; free Dpd (Pyrilinks D); telopeptides, C-terminal (Crosslaps) and N- (NTx, Osteomark).

Of the substances listed, hydroxyproline and hydroxylysine are of mainly historic interest since they are not bone-specific and their excretion is affected by dietary intake. The most specific bone-resorption markers are those associated with the cross-linkage region of the collagen triple helix, and include pyridinoline and Dpd in the free and bound form, and the above cross-linked telopeptides. Until recently the pyridinolines were measurable only by HPLC with fluorescent detection after SPE [2].

The advent of ELISA methods, initially for the total pyridinolines and more recently for free Dpd and the telopeptides, has greatly simplified the procedures; thus the NTx protocol is:

#add 25 µl calibrant, control or urine, to duplicate NTx-coated wells, then 200 µl Ab/enzyme conjugate;
#incubate for 90 min, wash (×5) and drain;
#add enzyme substrate/chromogen and incubate, 15 min (**blue**);
#add 100 µl Acid Stop solution; read at 450 nm (**yellow**);
#plot calibration curve and calculate NTx (nM) for the sample (second-voided urine).

Capacity: 40 duplicate samples in 2.5 h.
Precision: between-assay C.V. 5-12%.

The current availability of several of these assays means we are faced with the dilemma as to which is the best marker to use.

Background clinical studies.- Several groups have attempted to answer the choice-of-marker question. Garnero & co-authors [3] measured 4 potential bone-resorption markers in 46 healthy pre-menopausal women, and 85 post-menopausal osteoporotic women who were assessed for response to therapy with alendronate, a new anti-resorptive aminobisphosphonate. The post-menopausal and response changes (r = bone mineral density correlation) were:

NTx (EIA):	increase: +171%;	drug response, -71%;	r = -0.6;
F-Pyr (EIA):	+83%;	0%;	-0.1;
D-Pyr (HPLC):	+63%	-38%	not recorded;
ICTP (RIA):	-10%	-10%	-0.3

They concluded that NTx was the most sensitive marker for bone loss (mean rise 171% in the osteoporotic post-menopause group) and also for reflecting/predicting therapeutic response (mean decrease 71% within 3 months); only NTx correlated significantly with density. Other authors studying resorption markers in osteoporosis have made similar findings and shown that NTx is also useful in monitoring response to hormonal therapy (HRT) and vitamin D supplementation [4].

COMPARATIVE STUDY OF THREE BONE RESORPTION MARKERS

In a pilot study, we have investigated the analytical performance of two commercial ELISA assays: Pyrilinks D - free Dpd (Metra Biosystems, U.S.A.), and Osteomark NTx (Ostex, Seattle). For comparison, hydroxyproline was measured by colorimetry [5].

Inter-batch precision (C.V.) was 7.2-13.8% for hydroxyproline, 4.3-11.3% for NTx, and 3.9-10.1% for Pyrilinks D. The r value (correlation) was 0.76 for NTx *vs.* Pyrilinks D, 0.86 for NTx *vs.* hydroxyproline, and 0.72 for Pyrilinks D *vs.* hydroxyproline. So the two ELISA kits showed similar precision, better than with hydroxyproline assay, and in accord with other workers we found NTx to correlate reasonably with hydroxyproline but less well with Dpd - as further shown in Fig. 1.

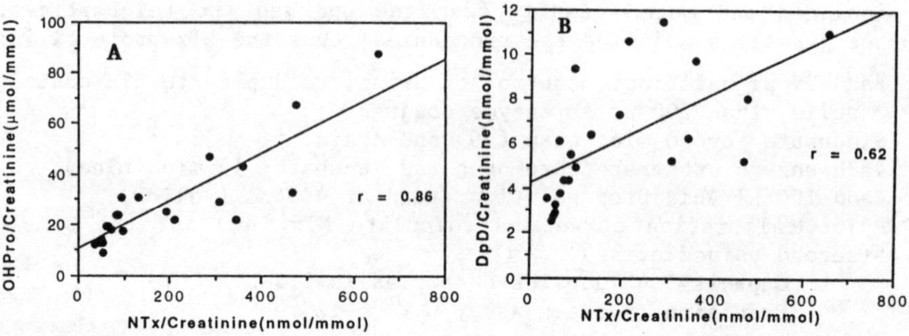

Fig. 1. Comparison of NTx with (A) hydroxyproline, (B) Dpd.

The clinical utility of the three assays was also compared, using samples from 22 patients with known or suspected metabolic bone disease. All three, run in 16, gave results of the same clinical significance, viz. normal in 6, abnormal in 10; likewise for NTx *vs.* hydroxyproline only, in 20: normal, 6; abnormal, 14. From the clinical details available, NTx appeared, but not with certainty, to be the best predictor of patient status.

ADVANTAGES AND USES OF BIOCHEMICAL BONE MARKERS

The main advantages of these markers would appear to be that they are cheaper than bone density scans and give an integrated view of the whole skeleton rather than a single or multiple site assessment (e.g. hip or spine from a scan). Moreover resorption markers convey a representation of current bone kinetics, giving an estimate of how much resorption is occurring at a particular time; a bone scan indicates only current bone structure. From the literature, resorption markers appear to be more sensitive than bone scanning, being able to detect changes in resorption resulting from therapy within 6-8 weeks compared with 1-2 years for scanning. Probably markers and bone scanning are best used in conjunction, rather than being viewed as alternatives. The potential uses of markers in this regard would be to identify fast bone losers and those requiring treatment, to monitor efficiency of therapy and to serve as a pointer to patient compliance or non-compliance, possibly improving compliance since when patients realise that they are being monitored frequently they are more likely to take the therapy as directed.

Strategies for patient management

We envisage using one or several biochemical markers in conjunction with bone mineral-density scanning so to improve patient management. Patients identified as osteoporotic by scan will have baseline bone marker values measured. After commencement of treatment, serial bone marker assays will be performed, at 3 months and every 6 months thereafter. Decisions on therapy will be taken on the basis of the marker values: if no reduction in resorption can be demonstrated, therapy will be dose-adjusted or changed. A repeat scan after 18-24 months will confirm whether or not therapeutic intervention has been successful and will demonstrate whether the information given by the bone marker measurements is correct.

Conclusions

There are a number of candidate markers for bone metabolism. They have potential uses in the diagnosis and assessment of disease severity and in the monitoring of therapy. More studies are required to ascertain which are the best markers and the most effective way to exploit them in conjunction with other clinical tools (e.g. bone mineral density scanning) to improve the diagnosis and treatment of patients with metabolic bone disease.

References

1. Barlow, D.H. Advisory Group on Osteoporosis (1994) Report from Department of Health. Department of Health, London.
2. James, I.T., Perrett, D. & Thompson, P.W. (1990) **J. Chromatog. B 525,** 43-57.
3. Garnero, P., Shih, W.J., Gineyts, E., Karff, D.B. & Delmas, P.D. (1994) **J. Clin. Endocrin. Metab. 79,** 1693-1700.
4. Price, C.P. & Thompson, P.W. (1995) **Ann. Clin. Biochem. 32,** 244-260.
5. Nobbs, B.T., Walker, A.W. & Davies, T.J. (1995) **Clin. Chim. Acta 64,** 219-221.

#E-4

PHARMACODYNAMIC MARKERS: EVALUATION, VALIDATION AND APPLICATION

H.M. Hill†, T.A.G. Noctor, T. Thompson & J.E.C. Burnett
Covance Laboratories Ltd., Dept. of Biopharmaceutical Analysis, Otley Road, Harrogate, N. Yorks. HG3 1PY, U.K.

Biomarkers are variously designated as surrogate or pharmacodynamic markers or indicators. In the present context, they are endogenous biochemicals which can serve to indicate therapeutic efficacy, or substitute for measuring the therapeutic agent where technical reasons preclude measuring plasma levels. Indeed, plasma levels themselves act as surrogate markers; but added value is obtainable by correlating the levels with a surrogate PD marker.*

Since small changes in the concentration of biochemical markers may be significant therapeutically, it is appropriate to aim to validate these analytical methodologies to the same level as that required for PK indicators, as defined in the Washington Consensus document ([1]; cf. art. #D-6). Unlike PK studies, dosing the NCE may result in reduced concentrations of the target endogenous analyte; so validating the method below endogenous levels can pose technical problems - largely because blank matrices for spiking over the proposed calibration range are difficult or impossible to obtain. Alternative pseudo-blank matrices must be used and validated.

Many endogenous analyte assays are available in kit format; but it is usually necessary to modify the application when used as markers of therapeutic efficacy. The methodology should be revalidated in the context of its application. Issues such as this are illustrated by examples. (Introduction to biomarkers: pp. 249-250.- Ed.)

Surrogate markers can range from endogenous analytes, such as cholesterol (art. #E-2), through physiological indicators, such as forced expired air volume (FEV), to physical endpoints such as reduction in tumour mass and the ultimate marker, death. Usually the true biological endpoint can only be evaluated after prolonged clinical trials. However, by measuring relevant biomarkers in Phase II trials an early indication of efficacy can be obtained prior to initiation of expensive clinical trials.

---

†addressee for any correspondence

**Abbreviations.*- ACE, angiotensin converting enzyme; HBV, hepatitis B virus; i.s., internal standard; LLOQ, lower limit of quantitation; MVA, mevalonic acid; NCE, new chemical entity; PD, pharmacodynamic; PK, pharmacokinetic; REA, radio-enzymic assay; SD, standard deviation.

Comparisons of PK profiles and their corresponding PD marker /response are frequently referred to as the PK/PD interaction. The integration of this concept into the drug development process was the subject of a major conference in 1991†. There the impact of PD data on PK/PD models was reviewed (Verotta & Sheiner, in [2]). In many cases a direct relationship between PK and PD is expected - but in some instances this is not the case, and a hysteresis plot results. Fig. 1 is a simplified model of the interaction between midazolam levels and specific EEG response. Such simplistic plots serve to illustrate the value of PK/PD data; e.g. there may be an indirect relationship between plasma levels and the PD response due to, say, the presence of a metabolite or an indirect receptor response.

This article reviews the problems associated with generating PD data of a similar quality to PK data - which is essential if meaningful interpretations are to be made. It is indeed possible, should it be necessary, to validate biomarker assays to the same criteria as those for PK assays, where accuracy and precision should be less than or equal to ±15% (Washington Consensus Guidelines; art. #D-6). Or is it the case that any sensible value is acceptable, as long as the method has been thoroughly characterized and evaluated? This is now discussed with respect to assays for HBV, ACE activity, cortisol and mevalonic acid.

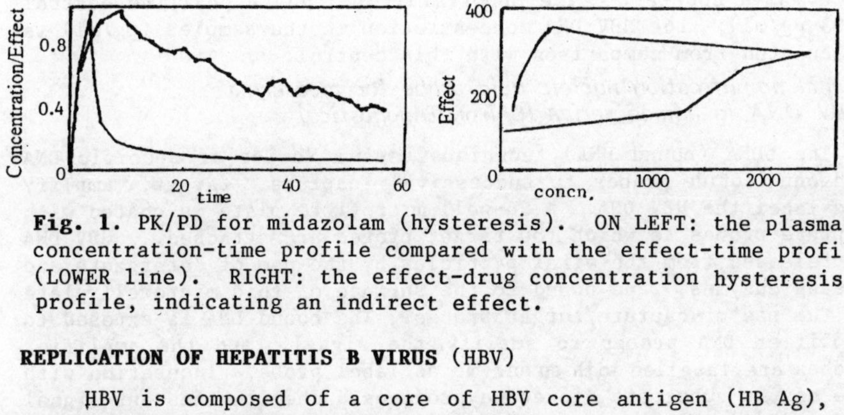

Fig. 1. PK/PD for midazolam (hysteresis). ON LEFT: the plasma concentration-time profile compared with the effect-time profile (LOWER line). RIGHT: the effect-drug concentration hysteresis profile, indicating an indirect effect.

REPLICATION OF HEPATITIS B VIRUS (HBV)

HBV is composed of a core of HBV core antigen (HB Ag), a double stranded DNA molecule of 3200 bases and DNA polymerase, surrounded by a surface antigen designated Hbs Ag. The presence of antigens in plasma does not necessarily correlate with active disease. The use of DNA probes specific to HBV has therefore been developed. The two major technologies are the Abbott Hepatitis B DNA assay and the Chiron B DNA signal amplification assay. Kapke & co-authors [3] compared earlier forms of these assays and found both to be adequate for quantifying HBV DNA in serum, although the Abbott assay had a slightly higher inter-assay C.V. and was less sensitive than the Chiron assay.

†organized by Amer. Assoc. of Pharmaceutical Scientists, the FDA and the Amer. Soc. for Clinical Pharmacology & Therapeutics

Outline of the two assays

A radiolabelling molecular hybridization assay for detecting and quantifying HBV DNA in human serum (Abbott Laboratories)

The DNA can be exposed enzymically by incubation with a proteinase and solubilized with NaOH. Detection and quantitation depend upon the general principle of molecular hybridization using a labelled probe. For the Abbott assay the probe is an ^{125}I-labelled section of DNA which will bind with complementary nucleic acid sequences with a high degree of specificity. Once this hybridization has occurred the labelled hybrid DNA molecules are separated from the unhybridized molecules by size exclusion chromatography, and the ^{125}I counted in a gamma counter.

Method.- Samples and controls (100 µl) are diluted with NaCl-containing buffer (100 µl) and incubated at room temperature with proteinase solution (10 µl) to release the viral DNA strands into solution. These are incubated with NaOH solution (20 µl) to solubilize them. ^{125}I-labelled nucleic acid probe (70 µl), prepared immediately before use by heating in a boiling water bath for 5 min, is added to each tube, and the mixture incubated overnight at 65° to allow for hybridization of the probe to the HBV DNA derived from the sample. The ^{125}I-labelled hybridized molecules are separated on Sepharose columns, collected into marked plastic tubes, and the label counted for 10 min in a gamma counter. The Abbott kit includes a positive control (103 pg/ml); the HBV DNA concentration in the samples (pg/ml) is calculated from comparison with this control.

Signal amplification nucleic acid probe for measuring HBV DNA in human serum (Chiron Diagnostics)

The bDNA (bound DNA) technique uses a series of specific DNA oligonucleotide probes to successively capture, target, amplify and label the HBV DNA. A 96-well microtitre plate is coated with capture probes to which the target probes are attached. HBV DNA is released from the viral particles by the use of detergents and lysing enzymes, and bound to the surface of this microwell plate by the use of capture/target probes. The bound DNA is exposed to amplifier DNA probes to amplify the signal, and the amplifier probes are labelled with an enzyme as label probe. Incubation with the enzyme substrate dioxetane produces a chemiluminescent signal in proportion to the amount of HBV DNA bound to the wells, and this signal is measured using a luminometer.

Method.- Samples (10 µl) are incubated at 63° with lysis reagent (10 µl) and then with target probes/denaturing reagent (10 µl). Following the addition of the neutralizing reagent (10 µl) the samples are incubated for 16-20 h at 63°; then HBV amplifier reagent (40 µl) is added and the samples incubated at 53° for 15 min. Dioxetane substrate solution (30 µl) is added, and the samples incubated at 25° in a plate luminometer for 25 min. The luminescence of the samples is measured, and the test results interpolated from the standard curve using the data management software (DMS) supplied by Chiron.

Appraisal of the two assays, and validation aspects

The Abbott assay uses a one-concentration calibrator (in duplicate) at 103 pg/ml with three negative controls, while the Chiron assay contains a set of four calibration standards and positive and negative controls. The calibration function spans the range 2.5 pg/ml to 22.5 ng/ml and uses a 4 PL fit although point-to-point may be more appropriate. Inter-assay precision (C.V.) of the assay based on incurred samples (unknowns) was 17% at 83 pg/ml and 18% at 10 pg/ml. This compares with 4% for the Chiron assay at 10 pg/ml and 20% at 2.5 ng/ml using QC samples; the same samples assayed by the Abbott procedure yielded values of 10 pg/ml and 70 pg/ml respectively.

At lower concentrations the two assays exhibit excellent correlation; but there is an almost 30-fold difference at concentrations only an order of magnitude higher. Such differences may be irrelevant when the assay is used merely to determine the status of the disease in broad terms. However, if it is being attempted to make a quantitative correlation, the conclusions could be extremely suspect. Conversely, if the assay is used to determine the presence or absence of HBV DNA - such that any detectable amounts may indicate presence of the disease - then assay sensitivity becomes paramount. Hence it is essential that this can be reliably validated, using the IUPAC definition that sensitivity = mean blank signal ±10 SD's of the background. The Abbott assay's LLOQ is 3 pg/ml, which compares favourably with the lowest calibration of 2.5 pg/ml for the Chiron assay.

A major requisite in validating any assay is analyte stability under varying storage conditions. Thus although at room temperature the 'DNA' degrades by only 5% in 24 h as measured by either assay, subjecting it to the stress of freeze-thaw cycles results, for 1, 2 or 3, in a response reduction of 12, 18 and 30% respectively. Thus transporting at ambient temperature would appear to be legitimate, so reducing transport costs, but not -20° storage or repeated freeze-thawing which can have a significant effect.

In order to evaluate these parameters one must 'validate' or at least characterize assay performance. Many kit-based assays are largely research-orientated and can be expensive; $60-120 is not unusual for a single assay. Thus a full validation as in the Washington Guidelines can cost >$20,000 for materials alone; also there may be optimization and/or further modification costs. So the validation criteria must be as focussed and relevant to the assay application as possible; validating a large calibration range where only a limited span may be of interest seems unwarranted.

While the two assays superficially perform equally well, the Abbott assay, largely as a consequence of its radiolabel, has a limited shelf life with shrinking performance.

ANGIOTENSIN CONVERTING ENZYME (ACE)

ACE inhibitors (CV-active dipeptides, e.g. captopril) have long been measured by their ability to inhibit ACE derived from rabbit lung. Two closely related assays are described, each with a synthetic substrate - hippuryl-L-his-L-leu or its ^{14}C version [4]. The product of ACE activity is measured either by HPLC-UV or by measuring the ^{14}C product; incubation conditions are similar. It is vital that initial kinetic rates pertain; hence optimizing the rabbit-lung extract is pivotal and should be re-assessed on an ongoing basis. Once the reaction is stopped (HPLC: methanol; REA: sulfuric acid); an aliquot of the methanol supernatant is loaded onto the column, or for REA the ^{14}C-product is extracted into pentyl acetate and subjected to scintillation counting.

The assays are similar in precision and accuracy, 4-16% over the concentration range for the REA and up to 10% for the HPLC assay. However, the REA is in general more sensitive, depending on substrate ^{14}C content, though obviously less specific than the HPLC assay. Yet one can still validate such an assay in the drug development process provided that one can assess the effect of any co-administered drugs and their metabolites. While the REA surpasses the HPLC assay in throughput when both are run manually, automated methodologies make their productivities comparable.

As the range of the enzyme hydrolysis component of the assay is limited, many samples have to be diluted into range; hence the inclusion of dilute QC's as a quality control requisite for each batch is essential. Both methodologies, run essentially as described [4] (in-house enzyme preparation), have been successfully validated according to the Washington Guidelines, and used to quantitate several thousand plasma samples from human subjects. Operating costs are relatively small (cf. the ACE assays).

CORTISOL

An assay for cortisol and related steroids was required to quantify an NCE's effects on cortisol metabolism and excretion. While a wide range of colorimetric immunoassays is available, the possibility of interference was considered to be important. Hence a rapid LC-MS-MS assay was developed. It involved mixing 200 µl of urine with i.s. and deionized water, and then centrifuging and injecting a 50 µl aliquot onto a Hypersil C-18 BDS column using a mobile phase of acetonitrile and ammonium acetate (35:65), at 1 ml/min flow rate with 20% split to the mass spectrometer. The detector was a Sciex API 300/365 (tandem) operating in Turbo IonSpray mode with a run time of 2 min (Fig. 2).

A major difficulty in this case and similar cases where the endogenous analyte is likely to fall below endogenous 'normal' levels is the obtaining of a suitable blank matrix. Problems in producing 'artificial matrices' devoid of the analyte of interest were considered earlier in this series [5]. In general, attempts to remove the analyte also change the nature of the matrix. Hence a 'matrix' devoid of analyte - buffer or water - was chosen. But it is essential to show that the artificial matrix behaves in a similar manner to real/incurred samples. This is achieved by spiking a series of different urines and of water with the same

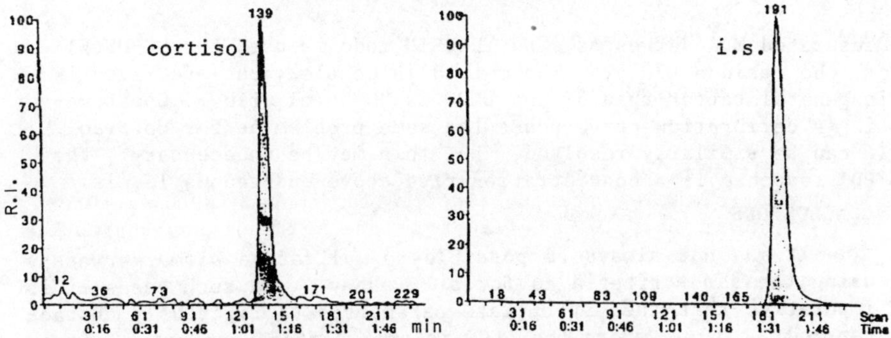

Fig. 2. Cortisol and i.s. (m/e 139, 191) quantitated by LC-MS-MS (see text).

RIGHT: **Fig. 3**. Spiking with increasing amounts of cortisol: curve with water, and (1-6) curves with 'control' ('blank') urines from 6 different individuals. The response at zero concentration is proportional to the endogenous level of cortisol in each of the 6 urine samples.

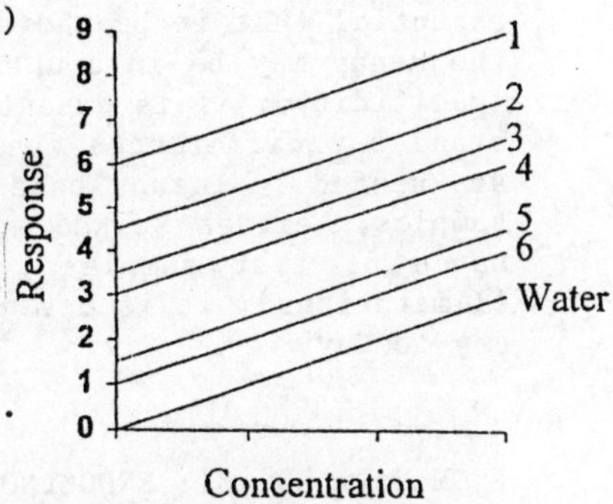

concentrations of cortisol to produce identical calibration curves each with a different 'offset', except for the water curve which goes through the origin. If it is valid to use water, the matrices from different sources should have the same or similar slopes, i.e. they should exhibit parallelism (Fig. 3). On this basis the artificial matrix can be used effectively to 'validate' or (maybe a more appropriate term) characterize the assay.

MEVALONIC ACID (MVA)

Many of the statins classified as cholesterol-lowering agents act by inhibiting 3-hydroxy-3-methylglutaryl CoA reductase, as well explained by K.E. Suckling earlier in this series [6]. In theory this might result in accumulation of a major intermediate in cholesterol synthesis, MVA. Thus the 24-h urinary excretion of MVA may correlate with the effectiveness of the statin. Amongst the many techniques for measuring MVA (critically reviewed: [7]), interestingly one of the earliest used an REA redolent of the ACE assay. Fortunately with the advent of MS detectors it has become possible to use chromatographic separation.

MVA is a very water-soluble compound, as for many endogenous small molecules. Conversion to the lactone at pH 3.5 increases its lipophilicity. Addition of 8 ml of ethyl acetate/acetone (2:1) to 400 µl urine elegantly transfers it from the aqueous to the organic phase. This is freed from water by azeotropic evaporation (N_2, 50°) and reduced to 0.5-0.8 ml. Derivatization is done with **N**-methyl-**N**-t-butyl-dimethylsilyl-trifluoroacetamide (MBDSTFA) using imidazole as a catalyst to ensure completeness.

Deuterated MVA serves as i.s. The SIM mode is used for GC-MS(EI) on the residue (70 eV). The method is complex; but precision is in general better than 5% and LLOQ is 20 ng/ml urine. Constructing a calibration curve poses the same problem as for cortisol. It can be similarly resolved, but this may be unnecessary; the 'PD' response is a concentration rise above endogenous levels.

CONCLUSIONS

It may not always be possible to validate a biomarker assay using the same criteria as for a PK assay. In such cases it is imperative that the appropriate parameters be rigorously characterized. Thus, where the LLOQ is an important parameter, it is essential that it be thoroughly validated. Where the accuracy of the assay may be in doubt (perhaps as a consequence of a lack of specificity), it is essential that some attempt be made to understand any differences even though relative data may be all that are needed. In any case it is important to include relevant QC samples, either spiked and/or from an incurred pool, with each batch of test samples. Whatever the assay's characteristics (limitations), it is essential that it performs reproducibly from day to day.

Editor's insertion:

FROM Vol. 23 [8]: ENDOGENOUS OCCURRENCE OF A THERAPEUTIC AGENT in plasma - 3-guanidinopropionic acid (GPA)

> The endogenous GPA level in plasma was reckoned as a first approximation to be 2 ng/ml, calculated by extrapolating the calibration line to the (negative) x-axis intercept, using standards prepared from a plasma pool (7 donors). A screen of control plasmas (12 donors) produced a range of baseline values distributed around the intercept found for the pooled plasma.

References

1. Shah, V.P., Midha, K.K. & 10 co-authors (1992) Pharmaceut. Res. **9**, 588-592. These 'Washington Guidelines' also appeared in J. Pharm. Sci. **81**, 309-312.
2. Yacobi, A. et al., eds. (1993) 'Integration of Pharmacokinetics, Pharmacodynamics and Toxicokinetics in Rational Drugs Development', Plenum, New York; includes art. by D. Verotta & L.B. Sheiner, pp. 159-177. (Proc. of 1991 Conf., 'The Integration of Pharmacokinetic, Pharmacodynamic and Toxicokinetic Principles in Rational Drug Design'.)
3. Kapke, G.F., Sheffler, S., Hunt, D. & Frederick, C. (1994) Clin. Chem. **40**, 1045
4. Cushman, D.W., Cheung, H.S., Sabo, E.F. & Ondetti, M.A. (1977) Biochemistry **16**, 5484-5491.
5. Burnett, J.E.C., Thompson, I. & Hill, H.M. (1996) in 'Biofluid Assay for Peptide-Related and Other Drugs' [Vol. 24, this series] (Reid, E., Hill, H.M. & Wilson, I.D., eds.), Royal Society of Chemistry, Cambridge, pp. 327-335.
6. Suckling, K.E. (1994) in 'Biofluid and Tissue Analysis for Drugs including Hypolipidaemics' [Vol. 23, this series) - as for 5.; pp. 73-82.
7. Lindenthal, B. & von Bergmann, K. (1994) Biol. Mass Spectrom. **23**, 445-450.
8. Parton, A.H., Brown, N.K., Enos, T.A. & Speed, W. (1994) as for 6.; pp. 247-250.

#E-5n

A Note on

IMMUNO-PCR: PCR-AMPLIFIED IMMUNOASSAY FOR ULTRA-SENSITIVE DETECTION OF THERAPEUTIC DRUGS AND SURROGATE MARKERS IN BIOLOGICAL FLUIDS

Peter End†, Richard Schindler & Daniel Gygax

Preclinical Development, Novartis Inc., 4002 Basel, Switzerland

We have developed an ultra-sensitive system for antigens such as proteins and low-mol. wt. drugs by utilizing the enormous amplification power of the PCR* during the signal-amplification step in IA's, especially those of the sandwich type. As a first model system, assays for interleukins 3 and 6, respectively, were designed and compared with the very well established and highly sensitive ELISA (DELFIA [1]) assay already developed at Novartis. Depending on the detection system applied for the PCR products and experimental conditions, the PCR-amplified assay is not only ~1000 times more sensitive, but also allows to monitor concentrations down to a few fg/ml of sample and over a wider dynamic range of concentrations. Next to the striking increase in sensitivity, a major advantage of this detection system over standard IA is the possibility to run an internal standard (competitive) for every sample during the signal amplification step. Signal amplification by sensitive enzymes such as alkaline phosphatase is a major source of error and low precision in standard IA's. In addition, the current project includes a comprehensive study on the detection and quantification of the DNA fragments (PCR products). The most appropriate method for this purpose in terms of reproducibility and automatization was found to be a dedicated HPLC method, developed in our laboratory.

Our immuno-PCR method was developed to monitor small changes in the endogenous cytokine blood levels (IL-2, IL-3, etc.) so as to monitor putative surrogate markers for the immuno-suppressive activity of drugs such as cyclosporin, rapamycins and derivatives. Nevertheless, there seemed to be a variety of potential applications to detect trace amounts of drugs (e.g. pharmacokinetics, drug abuse), rare metabolites (e.g. for ADME studies), by-products (e.g. from chemical processes) or contaminants (e.g. waste water and food analysis).

Endogenous plasma levels of IL-3 in humans were analyzed as a first application of the immuno-PCR assay system. Seemingly IL-3 and IL-5 might serve as surrogate markers for the efficacy of cyclosporin A on early and late reactions and associated development of bronchial hyperresponsiveness following allergen challenge.

†addressee for any correspondence *Abbreviations are at foot of next p.*

Reference

1. Gygax, D. & 6 co-authors (1996) in 'Biofluid Assay for Peptide-related and Other Drugs' [Vol. 24, this series] (eds./publ. AS NOW - E. Reid.../Roy. Soc. Chem.), pp. 11-19.

METHOD NOTES *compiled by Editor from unpublished account sent by Dr End*

Fig. 1 shows principles of the approach. Fig. 3 illustrates an HPLC separation, preferable to CE (Fig. 2) or other techniques such as gel electrophoresis.

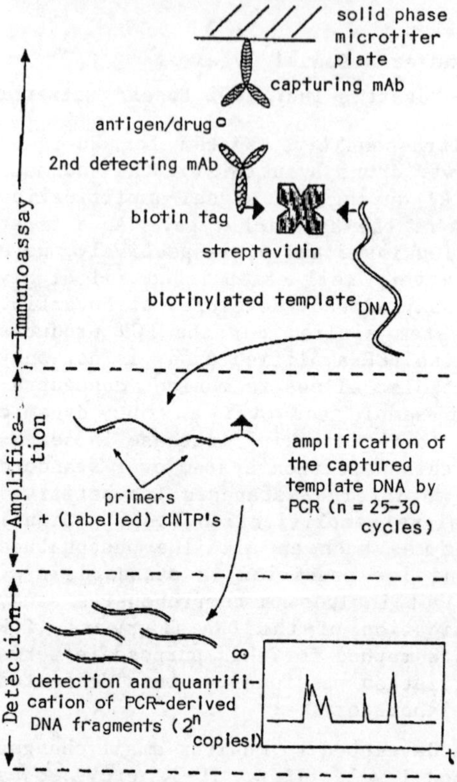

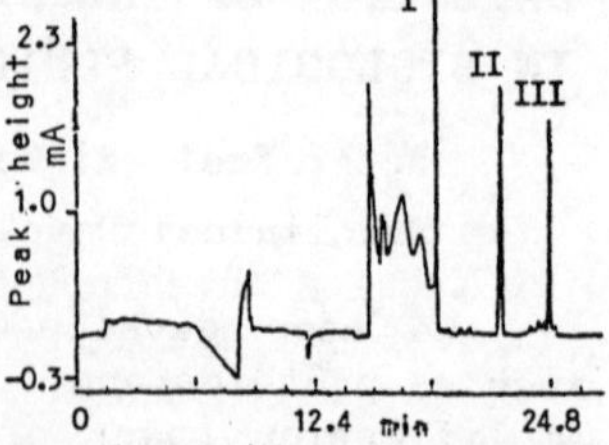

Fig. 2. CE of a PCR (competitive) mix. Buffers & capillaries from a BioRad PCR kit. See Fig. 3 legend for I-III.

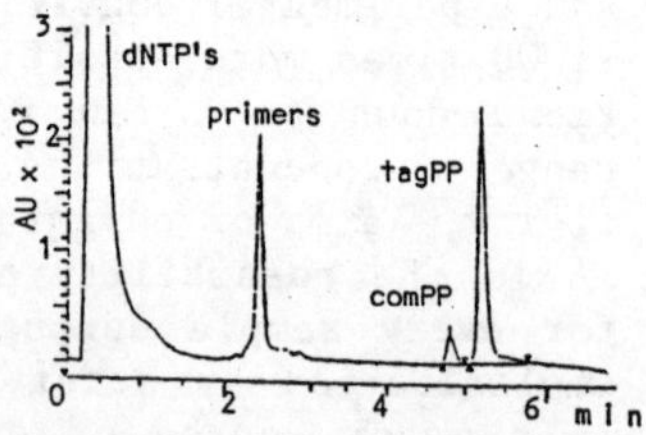

Fig. 3. HPLC of PCR mix. Products from competitor-DNA/biotin-labelled tag-DNA [compPP/tagPP; resp. II & (amplification intl. std.) III in Fig. 2 (I = nucleotides & primers)].

Fig. 1. Principles, for (e.g.) IL-3 (or IL-6, salm calcitonin; sandwich-type assay [competition or direct-coating assay may suit for drugs]). Basic IL-3 level found to be 2 pg/ml; C.V.'s ~10%.

Some details (especially for IL-3).- Ab's: see [1]. Taq-DNA polymerases: several suppliers (Promega for Klenow fragment of I, relevant to biotinylation). **Primers** (4, 1 pair 'nesting' with the other, for amplification): synthesized in-house; used to generate labelled tag- & competitor-DNA (separable) from 'cTNx' constructs based on characteristic DNA stretches. Plate-mode assay (better than beads); tagDNA (level critical), Ab-captured, transferred to tube or thermostable plate. Washing/amplification steps critical.

*_Abbreviations_ in text [SEE Fig. 3 legend for others]: IA, immunoassay; mAb, monoclonal antibody; PCR, polymerase chain reaction.

Comments on #E-1, D. Perrett: THIOL- and NO-related ANALYTES
#E-2, B.A. Griffin: CHOLESTEROL STATUS
#E-4, H.M. Hill: PHARMACODYNAMIC MARKERS
#E-5n, P. End: IMMUNO-PCR, e.g. for IL-3

Perrett, answering M. Bayliss concerning the nitrosothiols - decay and NO fate: the final products are the thiol (then the disulfide) and, one would guess, nitrate/nitrite (plasma, urine). Remark by **R.F. Venn**: our contribution (art. #B-3) bears on the late-eluting interferant problem in your bromobinane studies.

Griffin, answering Wilson.- For dietary lowering of TG/LDL, oily fish is advantageous; fibrate-type drugs also modify the LDL patterns. **B. Starkey remarked** that fasting for TG assay is non-physiological, and wondered whether, especially for small high-density LDL, post-prandial assay would be better, and at what time. **Reply**: the choice of 'standard meal' is under study; for TG assay 8 h post-prandial might be optimal, although there may still be raised TG levels in some 'at-risk' individuals. **Reply to R. Briggs**, who asked about plaque formation in relation to LDL observations and oxidative processes: there is much published work on the role of antioxidants and dietary lipids, but no clear picture has so far emerged.

Remark by **K. Borner to Hill.**- QA schemes in the clinical chemistry area are pertinent to validating biomarker assays.

Remark by **H-J. Skrzipczyk to End.**- Sensitivity comparable to immuno-PCR assay might be obtainable by direct non-isotopic IA, e.g. fluorescence or CL. **Reply to D. Dell**: it is indeed in our Drug Metabolism Dept. that surrogate marker assays are conducted; it developed them, and has the expertise.

For journals, Chr'y DENOTES Chromatog.

Citations contributed by Senior Editor relevant to Sect. #E, including problems with 'blanks'

Possible pertinence indicated thus: [ ]

Nitrite [NO-synthase marker]; cf. art. #E-1

Taking account of rapid conversion to nitrate by red cells, blood is alkalinized when drawn, but adjusted to pH 7 for centrifugation and ultrafiltration prior to anion-exchange HPLC; EC and (nitrate) 220 nm detection.- Preik Steinhoff, H.& Kelm, M. (1996) J Chr'y **B 685**, 348-353. Another anion-exchange method (for plasma or urine) entails a post-column colour reaction; nitrate is reduced using a Cu-plated cadmium column.- Muscara, M.N. & de Nucci, G. (1996) J Chr'y **B 686**, 157-164. For nitrite/nitrate in biosamples (acetonitrile supernatants), **CE** has also been used.- Friedberg, M.A. et al. (1997) J Chr'y **A 781**, 491-496.

Homocysteine [paediatrics; cerebrovascular status - cf. #E-1]

Assay in plasma.- (1) Use an amino-acid analyzer, with temp. gradient. (2) Reduce -S-S- (dithiothreitol); derivatize; GC-MS. (3) Reduce (Bu_3phosphine); TCA pptn.; derivatize (for fluor; exclude light); C-18 col.: MeCN/phos. pH 1.75; fluor. J Chr'y (1997): (1) Candito, M. et al., **B 692**, 213-216; (2) Sass, J.O. & Endres, W., **A 776**, 342-347. (3) Clin Chem (1996/7): Dudman, N.P.B. et al. **42**, 2028-2032; Vilaseca, M.A. et al. **43**, 690-694.

Cystathionine and related compounds isolated from urine by SPE were measured by LC-MS(API-CI).- Yu, S. et al. (1997) J Chr'y **B 698**, 301-307. Cf. **A 778**, 347-353 (Nozal, M.J. et al.): GSH and CysH (rabbit eye tissues), also **acetylcysteine** - for which (and other thiols) a pre-LC or pre-CE derivatization procedure has been described: Cavrini, V. et al. (1996) Chr'ia **42**, 515-520.

Cholesterol, especially LDL [cardiovascular disease]; cf. #E-2

LDL and HDL cholesterol can be measured directly on plasma by high-performance gel permeation chromatography (GPC; confusingly termed 'HPGC' by the authors) with colour generation in a post-column enzymic reactor; good correlation but not agreement was found *vs.* values by conventional methods ('Friedwald formula'; ultracentrifugation; selective precipitation).- Krause, B.R. et al. (1996) Clin Chem **42**, 1996-2001. [Lipoprotein kit sources include bioMerieux and (with LDL immunoseparation) Sigma.- *Ed.*]

Urinary pyridinolines [bone resorption]; cf. art. #E-3

(1) Dpd (deoxy) by ELISA kit; also (derivatize; HPLC) hydroxyproline. (2) Dpd and pyridinoline itself by hydrol.; clean-up on microgranular cellulose in BuOH/acetic acid medium; HPLC (48°) on 'SEPARON HEMA-B10', acet./sulfate pH 3.35; fluor. (1) Leino, A. et al. (1996) Clin Chem **42**, 2037-2039. (2) Spacek, P. et al. (1997) J Lq Chr'y **20**, 1921-1930.

Eicosanoids, etc. [inflammation/respiratory status]

Consult Index entries (Arachidonic, Eicosanoids) in past vols. (ed. Reid et al.): #22 (diagram: p. 225); #19 & #21

Urinary **LT-E$_4$**: SPE before and after HPLC-UV (C-18); finally EIA by a kit.- Mita, H. et al. (1997) J Chr'y **B 692**, 461-466.

Thromboxane/prostacyclin metabolites, PG's and isoprostanes in urine: SPE; 3-stage derivatization with clean-up on silica; GC-MS(nCI).- Wuebert, J. et al. (1997) Anal Chem **69**, 2143-2146.

In studies with tepoxalin (inhibitor of oxygenases), **TxB$_2$** was measured in serum by RIA, and **LT-B$_4$ formation** by RIA on MeOH-deproteinized plasma from incubates (+ Ca^{2+} ionophore) of whole blood.- Despré, M. et al. (1996) Int J Clin Pharmacol **16**, 1-8.

Dipalmitoyl **phosphatidylcholine** measurement in amniotic fluid entails enzyme treatment (PL-C), solvent extraction, HPLC (Ag^+ ligand for cation-exchange) and detection by light-scattering.- Alvarez, J.G. et al. (1997) J Chr'y **B 690**, 338-342.

For **matrix metalloproteinase-9** (inflammation marker) an ELISA kit and information are available from Biotrak (for IL's also).

Cytokines [inflammation/cartilage status; cf. art. #E-5n]

Using kits (Perspective Diagnostics, Cambridge, MA, USA), serum from osteoarthritis patients was assayed for IL-1, IGF-1 and TNFα - Bellometti, S. et al. (1997) Int J Clin Pharm Res **17**, 149-153 [also 127-132 (Matsukawa, Y. et al.): serum **pepsinogen I & II** determined by kit in peptic ulcer cases, to evaluate effect of lansoprazole]. In general, RIA is used to assay cytokines; TNFα and IL-6 are promising markers - Bienvenu, J. et al. (1989) in 'Biologie Prospective' (7th Colloq.; ed. Galteau, M-M. et al.), John Libbey, Montrouge & London, pp. 287-293.

IL-12 assay on biosamples (e.g. peripheral blood lymphocytes) using anti-IL-12 can be done by RIA, ELISA, or elaborated versions, e.g. based on IFN-γ production; greater sensitivity is attainable by subtle enzyme-linked immunospot 'bioassay' methods ('ELISPOT'), not based on lymphocyte proliferation, which were developed in connection with a study of rIL-12 expression using cDNA sequences to construct a vector.- Cui, Y. & Chang, L-J. (1998) Prep Biochem **28**, 23-36.

'Advanced Glycosylation End-products' [chronic diabetes/uraemia]

Pentosidine (an 'AGE' with arg/lys/ribose moieties and cross-links; accumulates in tissues) is measurable in urine or serum (hydrol.), after derivatization with HFBA, by use of a TSK-GEL column and then, after switching, an ODS-TSK-GEL column, with UV detection; the aq. MeCN eluants contain HFBA.- Takahashi, M. et al. (1996) Clin Chem **42**, 1439-1444.

Mevalonic acid, etc. [metabolic disorders]

Diagnostic guidance is obtainable from urinalysis either by CE direct (with diode array detection) or, more usefully, by solvent extraction, diazomethane derivatization and GC-MS. Glyceric acid, methylmalonic acid, and mevalonic acid/lactone have been thus determined.- Jellum, E. et al. (1997) J Chr'y **B 689**, 155-164.

Ref. to neonatal **screening for metabolic diseases* by MS-MS**: Johnson, A.W., Mills, K. & Clayton, P.T. (1996) Biochem Soc Trans **24**, 934-938. In a Forum talk, with these authors as co-presenters, **Mira V. Doig** advocated direct tandem-MS (without HPLC) which could, if automated, allow routine screening for various disorders. This facile approach, already being explored *re* fatty acid metabolism (acylcarnitine profile), liver disease (bile acid conjugates) and primary amino acid defects, might have some applicability in drug-development laboratories.- *Ed.*

[Reply to a Forum query on reagents: Aldrich furnished n-BuOH (>99.4%), also acetyl chloride (>98.5%) which must be 'fresh' or freed from absorbed water by redistillation.]

*see top of next p.

For a markers survey (cancer, infections, etc.; including peptides) see Lehmann, R. et al. (1997) J Chr'y **B 697**, 3-35.

Questionable markers [cancer, or transplant rejection]

Oncopterin, a supposed cancer marker, falsely appears to occur in urine from patients given methotrexate, due to a metabolite occupying the same C-18 peak position.- Hibiya, M. et al. (1997) J Chr'y **B 691**, 223-227. For heart-transplant rejection during cyclosporine treatment, serum **neopterin** (by RIA) is a poor index.- Cosson, C. et al. (1989) in Galteau et al. (eds.) - a Libbey book cited above (Cytokines para.), pp. 877-880.

Markers for toxicity

Dihydropyrimidine dehydrogenase (DPD) deficiency, as may be encountered in cancer patients given 5-fluorouracil, can be shown by radioassays on lymphocyte cytosol incubated with ^{14}C-5-FU, the ^{14}C-dihydro-5-FU product being separated by HPLC for the final counting.- Johnson, M.R. et al. (1997) J Chr'y **B 696**, 183-191.

The sphinganine/sphingosine ratio in human or animal urine, altered with exposure to fumonisin (a mycotoxin), is measurable by HPLC-fluor after clean-up steps.- Solfrizzo, M. (1997) J Chr'y **B 692**, 87-93.

Function assessments (especially hepatic) in patients entail alertness to drug toxicology. From a summary by **G.P. Mould**: #oral hypoglycaemics may overact on blood glucose; #aminoglycosides may compromise renal function; #GGT rises more than other hepatic enzymes with toxic doses of (e.g.) paracetamol or anticonvulsant drug combinations.- *Ed.*

Some methods for peptides, particularly endogenous
(see also #IIa in Analyte Index & some General Index entries)

Bioactive peptides featured in Vol. 24 of this series, with assay refs. (pp. 60-62). An approach reported for bradykinin and met-enkephalin in rat brain entails deproteinization by PCA, on-line derivatization, and HPLC-LIF (C-8 column).- Kanazawa, H. et al. (1997) J Chr'y **A 763**, 23-29. An ala-leu-enkephalin, given in ^{3}H form to rats, has been measured in plasma along with metablites by use of a synchronized accumulating **radioisotope detecttor** following TFA deproteinization, RP-HPLC-UV and scintillant addition.- Hasegawa, H. et al. (1997) J Chr'y **B 689**, 287-294.- See also **697**, 149-162 (Chen, C. et al.): [D-Pen^{2,5}]enkephalin in the residue from SPE of MeCN-treated rat serum was reacted with a rhodamine isothiocyanate in DMSO and analyzed by CE-LIF. RP-HPLC assay of opioid **neuropeptides** in bio-samples has been reviewed, with emphasis on detection by RIA, RRA, MS and MS-MS.- Desiderio, D.M. (1997) Anal. Chim. Acta **352**, 85-105.

Refs. elsewhere to assay of endogenous constituents include:
Various incl. peptides: J Chr'y **B 697** - cited on p. 193.
Retinoids: #ABC, within Class 7; also Wyss, R. & Bu ıeli, F. (1997) J Chr'y **B 700**, 31-47.

==============

Analyte Index

structured as in past Vols. (#22 has Cumulative Index; see p. 292)

Almost all entries refer to bioanalysis of drugs in 'real' samples; **'ch'** if chiral investigation. For [1] [2] SEE OVERLEAF; likewise for **Key to the chemical classification,** based on some analytically relevant features, e.g. an acid group (actual or, as with esters/ lactones, potential). Any category re-think (p. 292) is denoted °.

Entries in the page-no. range 114-140 refer to the Assay Compendium (#**ABC**; 'Cm'), based on **therapeutic classes** (p. 113) which may also appear in the General Index - wherein **endogenous analytes** also appear (few listed in the present Index).

Each category ends with putative drugs (e.g. 'CGP 48664') for which a name may emerge (try successive editions of the Merck Index or Martindale).

ASSIGNMENT 'CATECHISM' based on chemical categories

If **metabolites** studied, [1] ([2] if a conjugate) precedes the parent-molecule entry. *Merck Index* helps for formulae not in book text.

'Acidic' assignment (to #Ia, #IIa or #IIIa) applies where the pKa is <6, so excluding phenols; conjugate acidity is ignored. 'Acidity' embraces lactones and esters (hydrolysis being facile and likely to occur *in vivo*) except where the 'potential' acid group is reckoned (sometimes arbitrarily) not to reside in the molecule's *main* moiety.

Cyclic N, ineligible for an amine sub-category (#IIb" or #IIb"'), qualifies for #IIIy/#IIIz (and *may* be basic/UV-absorbing) *only* if *not* 'imide'/'amide' [-CO-N-C(O)-] and hence #Iz. [-CO-NHR groups (e.g. in peptides; cf. esters) resist chemical hydrolysis, as do sulfonamide groups.] Any non-amine N-heterocyclic compound that also has, say, hetero-S (not hetero-O) is put in #IIIz, not the vast #IIIy group.

The Scheme was devised in the GC heyday, when amenability to derivatization and to ECD or NPD was pertinent. It remains relevant to liquid-phase operations in respects such as pH choice for extraction and amenability to fluorescent derivatization (facile for primary amines). Bioanalysts faced with solvent-extracting an amphoteric compound might get ideas from examples in #IIa or #IIIa. In general, by-analogy method development may be aided by the Scheme, along with General Index entries such as 'Basic', 'Hydrophilic' or 'Lactones'.

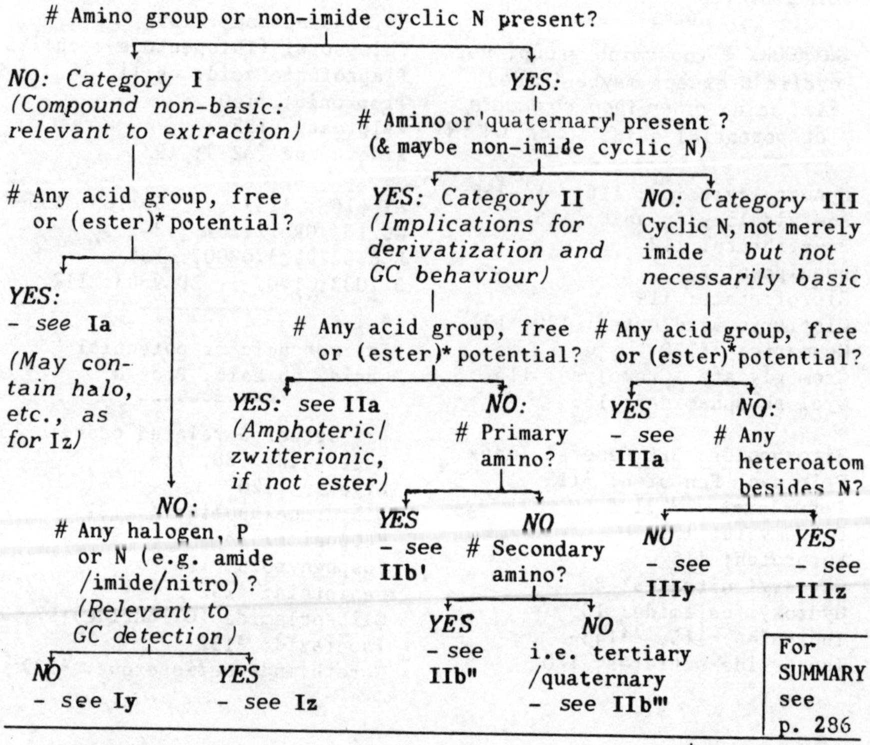

For SUMMARY see p. 286

*or lactone

Brief KEY [full version & chemical classes: pp. 281-282]: metabolite(s) also: [1] ([2] if conjugate); ch, chiral study; °, re-categorized (see p.)

#IIIz: heteroatom *(O disregarded)* besides N; otherwise as for #IIIy (*no* amino)

SUMMARY OF CATEGORIES
(Full version and 'logic': p. 282)

	I	II	III
Amino?	no	✓	no
Non-imide hetero-N?	no	maybe	✓
Acid or potential acid (not conjugate)?	✓ = Ia	✓ = IIa	✓ = IIIa
– no! (and not an ester)	Halo, P or N? – no: Iy – ✓ = Iz	Primary amino? ✓ = IIb' If no: 2^y = IIb'' 3^y or 4^y = IIb'''	Hetero atom besides N? (not O) – no = IIIy, ✓ = IIIz

Only **parent compound** listed even if metabolites ([1] [2]) studied. ch, chiral; °, p. 292. Endogenous compounds mostly in General Index.

General Index

This Index deals with features studied and with points of technique, indexed as in previous volumes so as to facilitate back-searching - likewise for the Analyte Index (AnIx). Where an item is in the 'Compendium' (Cm; pp. 113-140), on the actual Cm page the item may be asterisked, *. Entries such as 'Assay approaches' or 'Sample preparation' focus on method comparisons.

In a page entry such as '17-' the '-' connotes covered in depth /look at next page(s) too.

= = = = = = = = = =

[ctd.

AnIx denotes Analyte Index;
Cm denotes Compendium, pp. 113-140

AnIx denotes Analyte Index;
Cm denotes Compendium, pp. 113-140

AnIx denotes Analyte Index;
Cm denotes Compendium, pp. 113-14

================

Correction to Vol. 24 text, p. 220: the 2-part Fig. lacks **A, B**

Re-categorizations, marked ° in **present Analyte Index**:

Ia, Spirinolactone: previously Iy
Iz, Cyclosporin: IIb' until V.24
IIa, Topotecan: previously IIb"'
IIb', Acyclovir: previously Iz
IIIy, Mitomycin C: previously IIIa
Ia, Cyclophosphamide: previously IIIz
IIIz, Ifosfamide: previously Iz or IIb" (arguable!) and mis-spelt Ifosamide

For some other compounds, ***Vol. 24 Analyte Index*** *gives corrected categorization or spelling [alerting ▫ on left of name]; also a names clarification* ***re*** Carbamazepine *(p. 344). The corrections affect the Cumulative Index, in Vol. 22 [NOT 23; error on spine! Vol. 12 (1983; Plenum) also has a Cumulative Index of analytes].*